ADVANCES IN CHEMICAL ENGINEERING

Volume II

Advances in
CHEMICAL ENGINEERING

Edited by

THOMAS B. DREW
Department of Chemical Engineering, Columbia University
New York, N. Y.

JOHN W. HOOPES, JR.
Atlas Powder Company
Wilmington, Delaware

VOLUME II

1958
ACADEMIC PRESS INC. · PUBLISHERS · NEW YORK

TP
145
.D7
1958
v.2

Copyright © 1958, by
ACADEMIC PRESS INC.
111 FIFTH AVENUE
NEW YORK 3, N. Y.

ALL RIGHTS RESERVED

NO PART OF THIS BOOK MAY BE REPRODUCED IN ANY FORM,
BY PHOTOSTAT, MICROFILM, OR ANY OTHER MEANS,
WITHOUT WRITTEN PERMISSION FROM THE PUBLISHERS.

Library of Congress Catalog Card Number: 56-6600

PRINTED IN THE UNITED STATES OF AMERICA

CONTRIBUTORS TO VOLUME II

ERNEST F. JOHNSON, *Department of Chemical Engineering, Princeton University, Princeton, N. J.*

BERNARD MANOWITZ, *Brookhaven National Laboratory, Upton, L.I., N. Y.*

GEORGE SOFER, *Nuclear Development Corp. of America, White Plains, N. Y.*

THEODORE VERMEULEN, *Department of Chemical Engineering, and Radiation Laboratory, University of California, Berkeley, California*

SHERMAN S. WEIDENBAUM, *Corning Glass Works, Corning, N. Y.*

HAROLD C. WEINGARTNER, *Arthur D. Little, Inc., Cambridge, Mass.*

J. W. WESTWATER, *Department of Chemistry and Chemical Engineering, University of Illinois, Urbana, Illinois*

PREFACE

The editors, in this second volume of *Advances in Chemical Engineering*, continue the policies inaugurated for Volume I: to provide the chemical engineer with critical summaries of recent work—some that bring standard topics up to date, some that gather and examine the results of new or newly utilized techniques, some that survey the needs and possibilities of newly important segments of his industry. As in Volume I, each author has endeavored to provide a critical analysis rather than an annotated bibliography and to address not so much the narrow specialist in his own subject as those currently active in broader areas. For the strength of the fast-changing chemical industry depends upon its engineers' facility in bringing the findings of all areas quickly to bear upon their own. Most of the chapters of this volume were well advanced, if not completed, before the reaction to Volume I was known. Therefore, in any case, no major change in policy could have been made in Volume II. Now that the returns are in, we judge our plan to have been well considered and we shall follow it in future volumes.

With great pleasure the present editors announce that as their work on Volume III begins, they will be joined by Dr. Theodore Vermeulen, Professor of Chemical Engineering in the University of California at Berkeley. Thus, in the future, there will be three of us, to any one of whom may be directed suggestions of topics appropriate for review and of potential authors.

New York, New York and THOMAS BRADFORD DREW
Wilmington, Delaware JOHN WALKER HOOPES, JR.
January, 1958

CONTENTS

Contributors to Volume II v

Preface vii

Boiling of Liquids
J. W. Westwater

I. Transition Boiling 1
II. Film Boiling 8
III. Boiling of Subcooled Liquids 21
IV. Bumping during Boiling 27
    Nomenclature 29
    References 30

Automatic Process Control
Ernest F. Johnson

I. Introduction 34
II. Nature of Control and Control Problems 36
III. Treatment of Control Problems 41
IV. Present Status and New Directions 76
    Nomenclature 78
    References 79

Treatment and Disposal of Wastes in Nuclear Chemical Technology
Bernard Manowitz

I. Introduction 82
II. Waste Disposal as a Consideration in Site Selection 87
III. Waste Treatment and Disposal Practices 90
IV. Recovery of Fission Products from Radiochemical Wastes . . 108
V. Other Waste Problems 113
VI. Future Problems 114
    References 115

High Vacuum Technology
George A. Sofer and Harold C. Weingartner

I. Introduction 117
II. Historical Development of High Vacuum Technology 119
III. Chemical Engineering and High Vacuum Technology 124
IV. Vacuum Pumps and Gages 136
V. Conclusion 145
    References 145

Separation by Adsorption Methods
Theodore Vermeulen

  I. General Survey 148
 II. Physical Factors Affecting Separation Performance 153
III. Binary Fixed-Bed Separations 167
 IV. Chromatographic Separations 194
     Nomenclature 203
     References 205

Mixing of Solids
Sherman S. Weidenbaum

   I. Introduction 211
  II. Related Process Steps 211
 III. State of Mixedness of Batch of Solids 212
  IV. Theoretical Frequency Distributions 259
   V. Rate Equations 274
  VI. Equipment 287
 VII. Overall Concluding Comments 320
      Nomenclature 321
      References 321

      Errata to Volume I 325
      Author Index 327
      Subject Index 333

BOILING OF LIQUIDS

J. W. Westwater

Department of Chemistry and Chemical Engineering
University of Illinois, Urbana, Illinois

I. Transition Boiling	1
A. Description from Photographic Studies	2
B. Experimental Values	3
II. Film Boiling	8
A. Description from Photographic Studies	8
B. Theoretical Treatment	10
C. Experimental Values	13
1. Type of Liquid	13
a. Water	13
b. Other Common Liquids	14
c. Liquid Metals	14
2. Type of Hot Solid and Surface Texture	16
3. Geometric Arrangement	16
4. Pressure	18
5. Surface Tension	18
6. Agitation	20
7. Impurities	21
III. Boiling of Subcooled Liquids	21
A. Description from Photographic Studies	21
B. Data and Correlations	22
IV. Bumping during Boiling	27
A. Torpidity Theory	28
B. Nucleation Theory	28
C. Measurement of Bumping	29
Acknowledgment	29
Nomenclature	29
References	30

This chapter concludes the review on "Boiling of Liquids," the first part of which, published in Volume I of *Advances in Chemical Engineering*, 1956, included: I. Introduction; II. Nucleate Boiling and the Critical Temperature Difference.

I. Transition Boiling

Rarely has the transition type of boiling received attention. It cannot be studied by researchers using electric heaters because of the inherent instability of electric heaters in this region. As recently as ten years ago one writer, familiar with tests with electric heaters, argued that the transition region does not exist.

To study the transition region, one must employ a hot fluid or a condensing vapor as the source of heat. Drew and Mueller (D3) obtained transition boiling data for six liquids boiling on a steam-heated copper

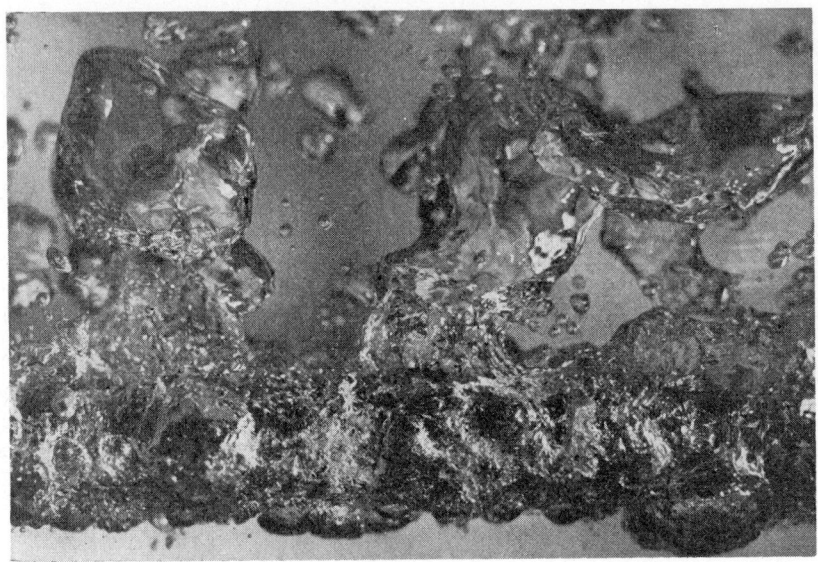

Fig. 1. Transition boiling of methanol. Point C on the boiling curve, Fig. 15. Overall $\Delta T = 112°F$. Heat flux = 69,000 B.t.u./hr. sq. ft. (W2).

tube. The heat transfer coefficients and ΔT values measured were overall values. McAdams (M2) reports individual values for h and ΔT in the transition region for two liquids, but he does not describe in detail the method of determining the wall temperature. Some studies of the transition region are now under way at the University of Illinois. Certain of the observations are included below.

A. Description from Photographic Studies

The only photographs of transition boiling are those of Westwater and Santangelo (W1, W2, W3). The photographs prove that transition boiling is entirely different from both nucleate boiling and film boiling. Figures 1 and 2, for methanol on a ⅜-in. copper tube, show that no active points exist on the metal surface. In fact none of the photographs obtained so far shows a clear-cut contact between the liquid and solid. The photographic evidence demonstrates that the tube is completely blanketed with vapor. However the vapor blanket is not a smooth, stable film. It is irregular and in violent motion.

The high-speed motion pictures show that vapor is formed by explosive bursts which occur at random locations on the tube. Apparently the vapor film becomes very thin at some location, the liquid moves in

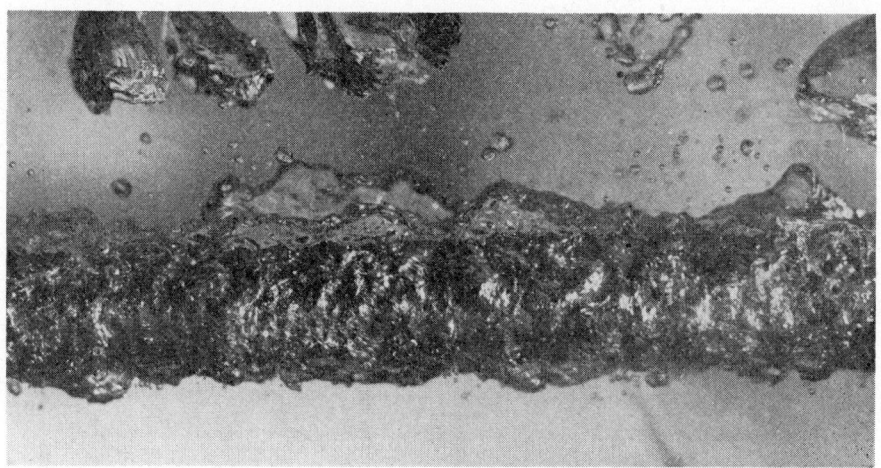

Fig. 2. Transition boiling of methanol. Point D on the boiling curve. Overall $\Delta T = 124°F$. Heat flux = 27,200 B.t.u./hr. sq. ft. (W2).

toward the tube, and then a miniature blast occurs. The blast creates a slug of vapor which forces the liquid back. The vapor slug eventually ruptures and leaves the tube. Two explosions are evident in Fig. 1. One is in profile below the tube at the right end; the other is a fourth of the way from the left end, a "plan" view of a circular slug moving toward the eye of an observer.

The frequency of the vapor bursts is very high. For an over-all ΔT of 133°F., each inch of the photographed side of the tube exhibited 84 bursts per sec. A typical burst has a vigorous life of about 0.003 sec. A burst in profile shows that the surrounding liquid is shoved back about 0.16 in. in this time, or the average velocity of the interface is about 4.5 ft./sec.

B. Experimental Values

The boiling data of Drew and Mueller were exploratory in nature, and the authors did not claim high accuracy for the values. However, the general shape of the transition boiling curve was well illustrated. Figure 3 shows some of the data. Note that the ΔT includes the temperature drop across the steam film.

The data of Braunlich and of Kaulakis and Sherman, reported by McAdams, are shown in Figs. 4 and 5. These show the effect of modest changes in pressure. The important fact is that the individual curves remain apart. That is, increasing pressure (at least by small amounts)

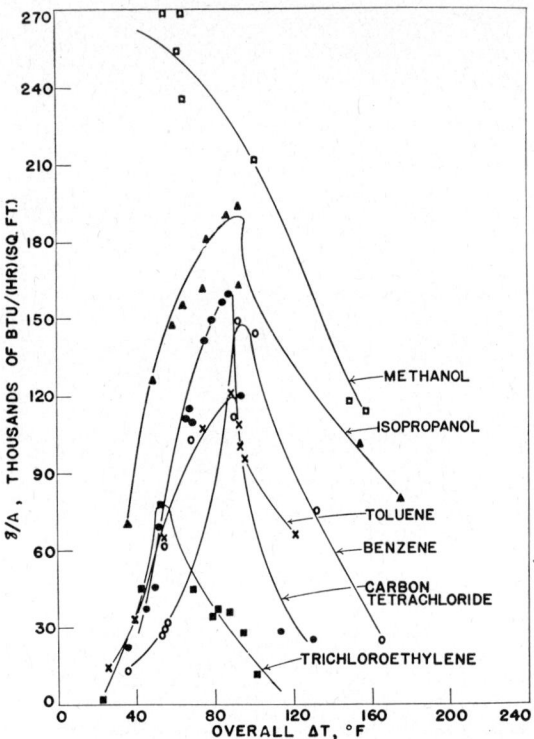

FIG. 3. Nucleate and transition boiling. Six liquids were boiled on a ¼-inch steam-heated, copper coil, at one atmosphere (D3).

improves the heat transfer in the transition region just as it does in the nucleate region.

In addition to pressure, certain other variables which are important during nucleate boiling are important during transition boiling. Figure 6 shows the results obtained when small amounts of a material which is known to be surface active with water was added to methanol. The change in the boiling curve is very real, although no proof exists that surface tension effects are the correct explanation. In fact, the liquid-vapor interfacial tension of methanol is not affected by the agent. Figure 7 shows the effect of surface texture on boiling methanol. The results are very similar to those obtained in nucleate boiling. Figure 8 shows that agitation improves the heat transfer in the transition region as well as in the nucleate region.

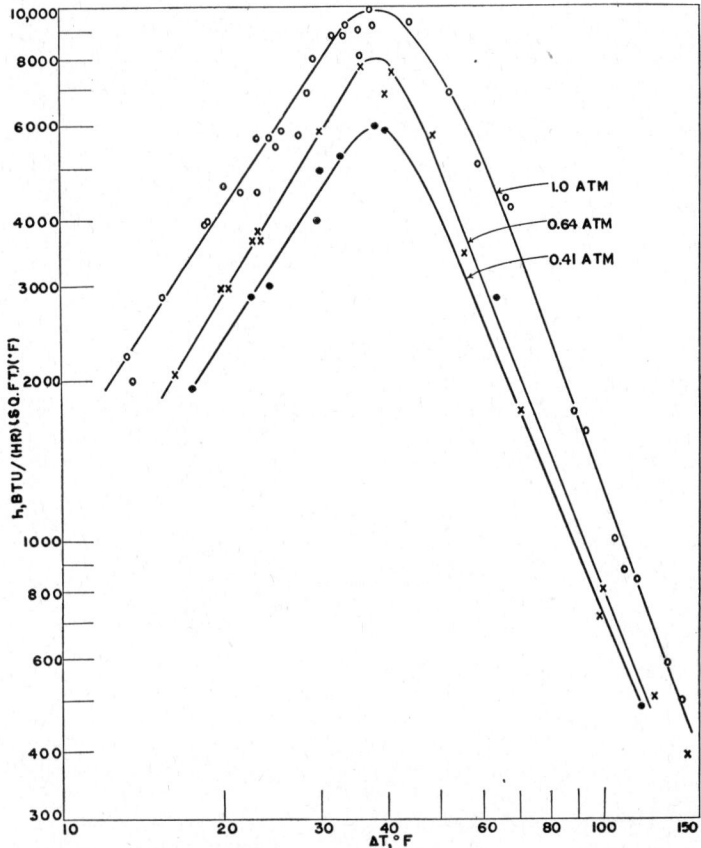

FIG. 4. Nucleate and transition boiling for water on a horizontal tube (M2).

To summarize the small quantity of data: the variables for transition-type boiling seem to be the same as those for nucleate boiling. The magnitude of each effect is not necessarily the same. As the film boiling region is approached, agitation remains an important variable, whereas surface texture becomes of negligible importance. The addition of surface active agents seems to result in an improvement in heat flux, even as the film boiling region is approached closely.

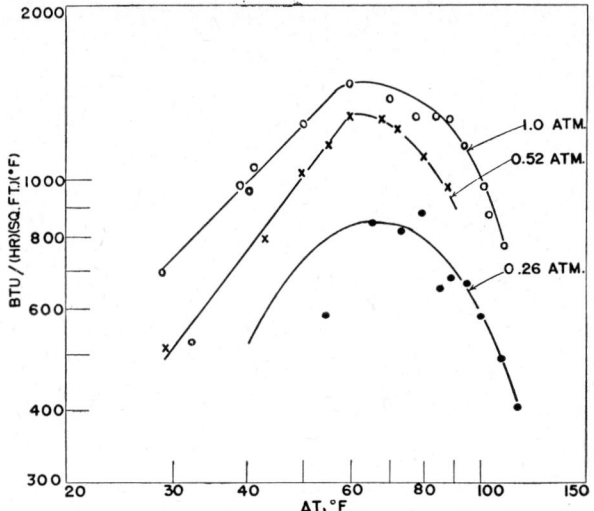

Fig. 5. Nucleate and transition boiling for isopropanol (M2).

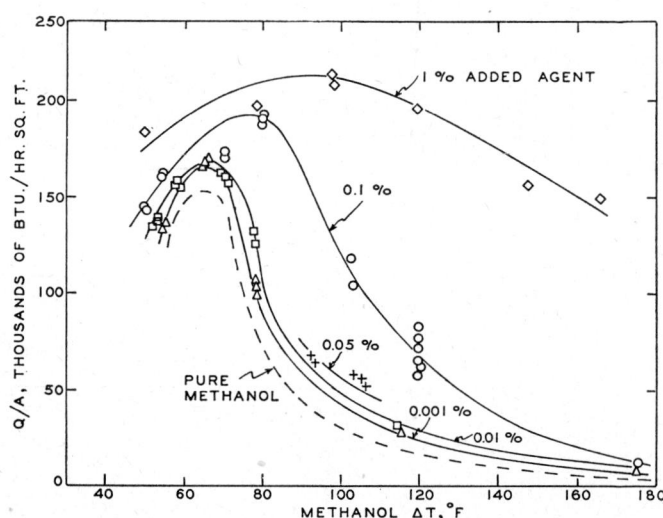

Fig. 6. Effect of surface-active agent on transition boiling. Rohm and Haas Hyamine 1622 (benzyldimethyl {2-[2(p-1,1,3,3-tetramethylbutylphenoxy)ethoxy]-ethyl} ammonium chloride: mol. wgt. = 466) was added to methanol boiling on a ¼-inch, horizontal steam-heated, copper tube at one atmosphere (L1).

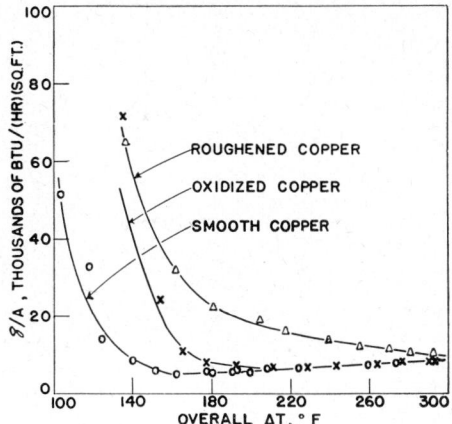

Fig. 7. Effect of surface texture on transition boiling. Methanol was boiled at one atmosphere on a ⅜-inch, horizontal, steam-heated, copper tube (D2).

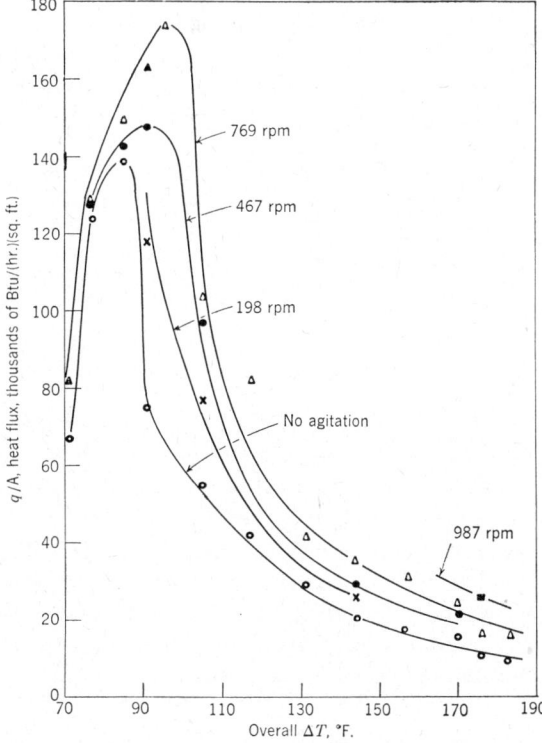

Fig. 8. Effect of agitation. Methanol was agitated with a 3-blade propeller while boiling at one atmosphere on a ⅜-inch horizontal, steam-heated, copper tube. The shaft speed is indicated (P1).

It is obvious that much more experimental work is needed. No theoretical treatment of transition boiling has ever been given. A convincing explanation of why there should be a smooth decrease in h with an increase in ΔT has never appeared. It is possible to "explain" the effect in terms of film thickness, but this is a superficial explanation based on the false assumption of a stable film. It is also possible to "explain" that transition boiling is a mixture of nucleate and film boiling, but this is contrary to photographic evidence.

At present it is impossible to design commercial equipment to operate in the transition region. From a practical standpoint, equipment should not be operated in this region anyhow. The performance would be superior at the critical ΔT.

II. Film Boiling

Film boiling has received more attention than the transition type, but still only a fraction of the attention given to nucleate boiling. The reasons are clear. Film boiling results in very low heat fluxes if conveniently small temperature differences are used; therefore the region of "easy" study is commercially unimportant. The region of high heat fluxes (and of practical interest) requires high-temperature techniques if ordinary liquids are used. Film boiling may be the convenient operating region when liquefied permanent gases are boiled, but this demands a knowledge of low-temperature techniques.

A. Description from Photographic Studies

Film boiling is characterized by action which is the slowest, most orderly, and best defined of the three main types of boiling. It is significant that film boiling is the first type to be attacked, with success, from a theoretical standpoint. Boiling heat transfer in most of the film region is described fairly well by the equations of Bromley (B3, B4, B5).

During film boiling no active centers exist. No explosive bursts occur. The hot solid is encased completely in a slowly moving film of vapor. No vapor is generated at the solid surface; it is all generated at the liquid-vapor interface. If a horizontal hot tube is used, the vapor flows up the tube in a series of ripples. In Fig. 9, the ripples are quite evident. For the conditions of Fig. 10, the ripples move up the tube at a rate of 74 each sec. on each side of the tube.

As vapor accumulates at the top of the tube a wavy, rodlike mass forms. At first the mass is fairly smooth and uniform, as shown in Fig. 10, although some movement occurs constantly. Eventually the vapor-rod becomes unstable and changes into a series of nodes, as shown in Fig. 9. Then the entire rod ruptures, between all the nodes at about the same instant. A horizontal row of bubbles, side by side, rises. After the rupture,

Fig. 9. Film boiling of methanol. Point E on the boiling curve. Overall ΔT = 148°F. Heat flux = 12,970 B.t.u./hr. sq. ft. (W2).

Fig. 10. Film boiling of methanol. Point F on the boiling curve. Overall ΔT = 181°F. Heat flux = 5,470 B.t.u./hr. sq. ft. (W2).

the whole procedure is repeated. For Fig. 10, one complete cycle required 0.06 sec.

The second vapor-rod usually ruptures at points between the preceding break-points. Thus even-numbered rows of bubbles are displaced sideways by a half-space from the odd-numbered rows. This is clear in

Fig. 9. The fact has been observed for film boiling on wires (C1) as well as tubes. No information is available for film boiling on plates.

Bubbles formed during film boiling are much larger than those from nucleate boiling. For the instant of release in Fig. 10, the diameters range from 0.20 to 0.36 in., averaging 0.30 in. The horizontal spacing, center to center, is about 0.5 in., with an occasional value as great as 0.75 in. Bubbles are released from the top only, never from the bottom or sides of the tube as in nucleate and transition boiling.

The history of these large bubbles after release is interesting. A "stretched" vapor filament is the last part of a bubble to break from the hot solid. The filament contracts rapidly, because of surface tension, and rams into the main body of the bubble. The bubble shape is distorted by the impact, becoming umbrellalike. The deformed surface then snaps back down, and the bubble vibrates as it rises. The true shape of rising, large bubbles is evident in Fig. 10.

B. Theoretical Treatment

Two types of condensation, drop-wise and film-wise, have been known for many years. As soon as the two types of boiling, nucleate and film, were described, certain similarities to condensation became evident. Nucleate boiling and drop-wise condensation were seen to be analogous. This is of little practical value, because no good theory of drop-wise condensation exists. However the analogy between film boiling and film condensation is fruitful, because a good theory of film condensation exists.

Colburn (C2), and probably other writers, suggested that Nusselt's excellent equation for film condensation could be modified for use with film boiling. Colburn gave as a guess that the proper form of the equation for boiling outside a horizontal tube would be the same as Nusselt's, except for certain changes from the physical properties of the liquid to those of the vapor and an adjustment in the dimensionless coefficient. Colburn suggested a coefficient of ½ for a liquid boiling outside a horizontal tube and ⅔ for a vertical tube.

In 1950, Bromley (B3) presented a complete derivation for film boiling outside a horizontal tube. The derivation is based on the fact that liquid is separated from the tube continuously. Important assumptions are involved: (1) The vapor film is smooth, and vapor escapes from the top of the tube as if issuing from a narrow slot. (Figures 9 and 10 show that this is not truly the case.) (2) The rise of vapor due to buoyancy is a viscous flow. (3) The kinetic energy of the vapor is negligible. (4) The liquid-vapor interface is at the liquid boiling point. (5) The temperature drop across the vapor film is assumed to be constant. (6) Heat transfer across the vapor film is by conduction and radiation, but not by convection. (7) Sensible heat transfer to the vapor is negligible.

The principal features of the derivation consist of a force balance and a heat balance for an element of vapor volume. These are also the principal features of Nusselt's derivation for film condensation. The forces acting on a tiny volume are those due to buoyancy and viscosity. The equation for the heat balance states that the increase in the flow rate of vapor which occurs along an increment of tube circumference corresponds to the heat flow rate in that region. Inasmuch as Nusselt's derivation is available in numerous books, Bromley's equivalent derivation will not be given here. The final equation is

$$h' = 0.62 \left[\frac{k_v^3 \rho_v (\rho_L - \rho_v) \lambda g}{D \mu_v \Delta T} \right]^{1/4} \quad (1)$$

(Bromley equation, horizontal tube)

The coefficient 0.62 is empirical. The theoretical value is either 0.512 (stagnant liquid around the vapor) or 0.724 (liquid moving with same velocity as the vapor). The average of these two numbers is 0.62.

Equation (1) omits the effect of radiation. Radiation is accounted for by use of the familiar expression for parallel plates,

$$h_R = \frac{\sigma' F_\epsilon (T_s^4 - T_L^4)}{\Delta T} \quad (2)$$

The combined film coefficient becomes $h = h' + h_R$.

Verification of Eq. (1) is shown in Figs. 11 to 13. The value of h_R has been subtracted from the observed h. Figure 13 shows the comparable magnitudes of h' and h_R and the agreement between the observed sum and the predicted sum.

For a vertical tube, Bromley predicts the equation,

$$h' = (\text{Const.}) \left[\frac{k_v^3 \rho_v (\rho_L - \rho_v) \lambda g}{L \mu_v \Delta T} \right]^{1/4} \quad (3)$$

No data are available to test this expression.

If the sensible heat content of the vapor is important, it can be shown (B4) that a correction factor $(1 + 0.4 C_v \Delta T/\lambda)^{1/2}$, should be included on the right side of Eq. (1).

Bromley's equations are consistent with reality in that they predict a minimum value of h'. As the solid temperature is increased, two counteracting effects occur. The values of k_v and ΔT both increase. At first the

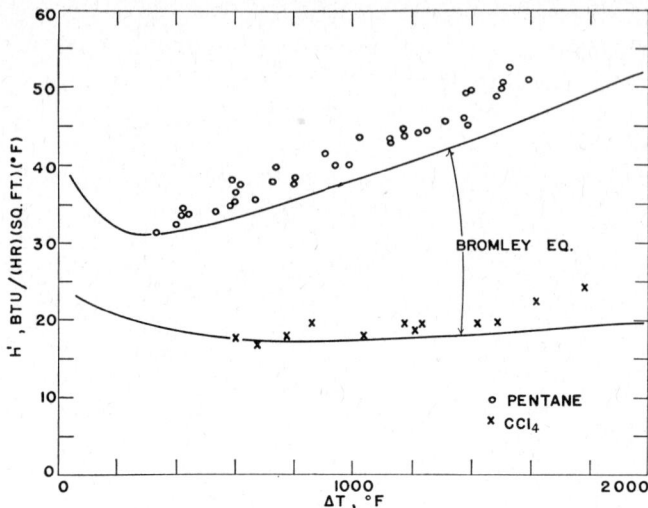

Fig. 11. Film boiling. The boiling heat transfer coefficients, corrected for radiation, are shown for two liquids, at one atmosphere, on a 0.352-inch carbon tube (B3).

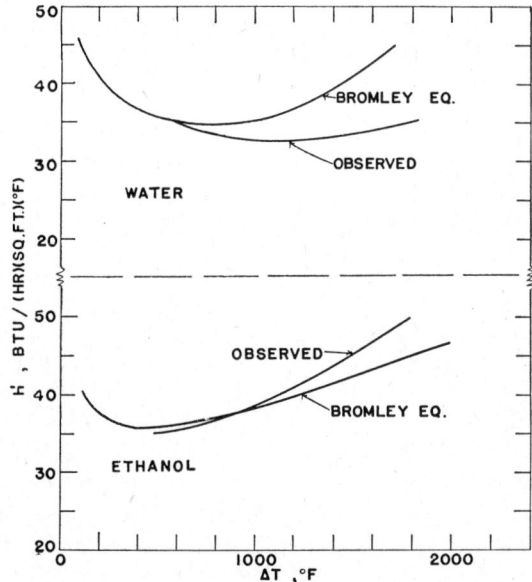

Fig. 12. Film boiling. The conditions were the same as for Fig. 11. The actual data scatter was similar to that in Fig. 11 (B3).

change in ΔT is predominant, but as the temperature level becomes greater, the change in k_v becomes more important. The other physical properties are temperature-sensitive, but their effects are not so great. Of course at very high temperatures, radiation is sufficient to account for increases in h.

Marx and Davis (M1) present the interesting thought that the film thickness during film boiling is some value greater than the mean free

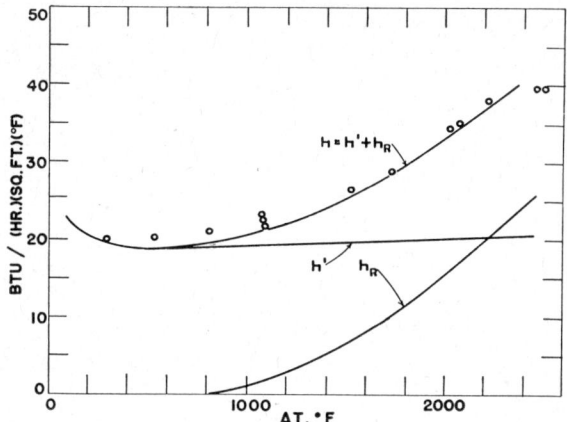

Fig. 13. Importance of radiation during film boiling. Nitrogen was boiled at the same conditions as for Fig. 11 (B3).

path of a vapor molecule. This idea leads to reasonable qualitative predictions as to the effects of pressure and temperature. These authors suggest also that transition boiling occurs if the film thickness becomes equal to the mean free path, but proof is not available.

C. Experimental Values

1. *Type of Liquid*

a. Water. Data for water in film boiling are available for horizontal wires and a tube. Bromley's results are shown in Fig. 12. The data of Rinaldo (M3) and Nukiyama (N1) are compared with Bromley's in Fig. 14. The agreement, for the wires, is good if consideration is given to the effect of diameter. Platinum was used by both Rinaldo and Nukiyama; Bromley used a carbon tube. All these tests were at atmospheric pressure.

Water in film boiling is normal in that it does not give unusual test values. The physical properties for which water is most different from common liquids are the latent heat of vaporization, heat capacity, and surface tension. According to Eq. (1), surface tension and heat capacity are not variables, and the other factor appears to a low exponent. Thus on theoretical grounds, water should be rather ordinary.

b. Other Common Liquids. Data are available for six organic liquids (benzene, methanol, ethanol, carbon tetrachloride, *n*-pentane, and diphenyl ether) and two permanent gases (nitrogen and oxygen). Data for some of these are shown in Figs. 11, 12, 13, 15, 16, and 17.

A striking fact is that at atmospheric pressure all these materials have roughly the same value of h', namely about 20 to 50 B.t.u./(hr.)(ft.^{2})(°F.).

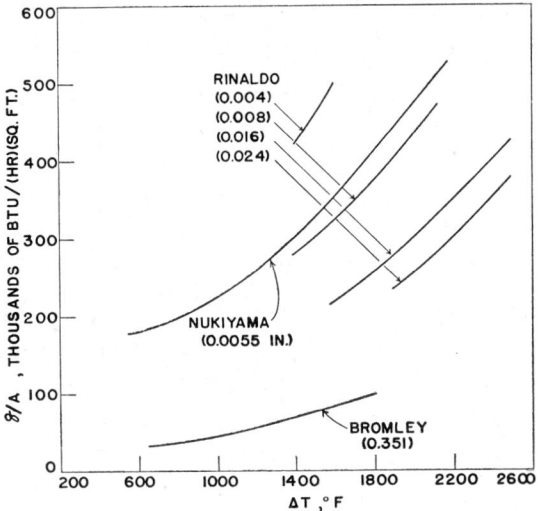

FIG. 14. Observed film boiling values for water at one atmosphere on horizontal tubes and wires.

If one were to use $h' = 35$, the error would be no worse than 50% for nearly every point in the graphs shown. One reason for this interesting occurrence is that h' is a weak function of all the variables except k_v, at atmospheric pressure. The thermal conductivities of different gases are not highly variable. At high pressures, the vapor density will become a relatively strong function, and the "uniformity" of different boiling liquids should no longer exist.

c. Liquid Metals. A major problem with boiling metals is the decision as to which type of boiling is occurring. Transparent vessels are not convenient, because of the high temperatures; thus rarely is the boiling visually observed.

Farmer (F1) obtained data for mercury boiling on a horizontal chromium plate at an absolute pressure of 0.25 in. mercury (340°F.).

Lyon and co-workers (L2) have published data for sodium, sodium-potassium alloy, cadmium, and mercury, at 1 atmos. The values of q/A vs. ΔT are shown in Fig. 18. In this graph the results all appear to be for nucleate boiling.

However, if the results are plotted as h vs. ΔT, the issue is confused. Figure 19 suggests that pure mercury boils in the film-fashion, even at

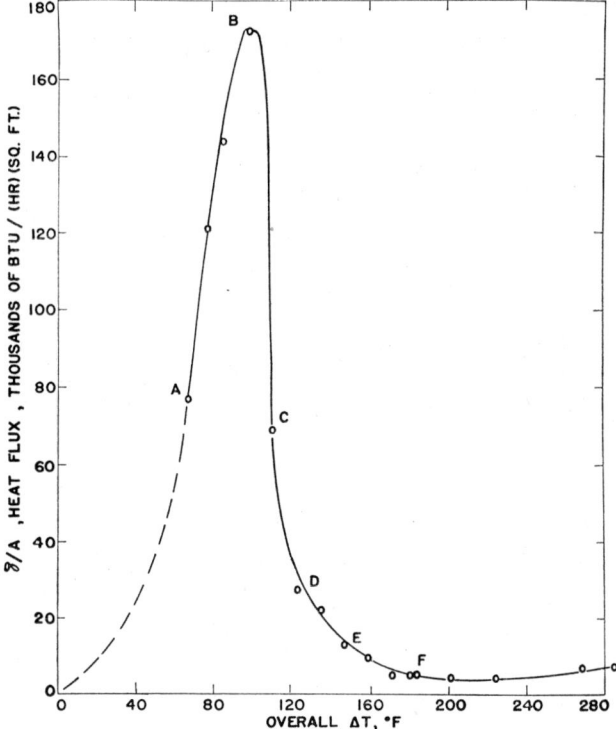

Fig. 15. Boiling curve for methanol. The liquid was outside a ⅜-inch, steam-heated, horizontal copper tube, at one atmosphere. The letters represent the conditions for the photographs, Figs. 1, 2, 9, and 10 (W2).

the surprisingly low ΔT of 15°F. The cadmium data may be for film boiling. Obviously, much more information is needed concerning boiling metals.

At temperature differences of 500 to 1000°F., pure mercury gives values of h of from 30 to 50. This is about the same as for water and other non-metals. The advantage of using boiling mercury in the film region therefore is not in a superiority in h at a stated ΔT, but rather in superiority from other standpoints such as discussed in Sec. II, E, 1 of Part I.

2. *Type of Hot Solid and Surface Texture*

Because the hot solid is covered completely with vapor during film boiling, the type of solid and its surface texture should be of little importance. Nucleation occurs at the vapor-liquid interface, and the process should not be influenced by conditions remote from that location.

Bromley's data for *n*-pentane boiling on carbon and stainless steel (Fig. 20) show no effects attributable to the surface. Banchero's data for

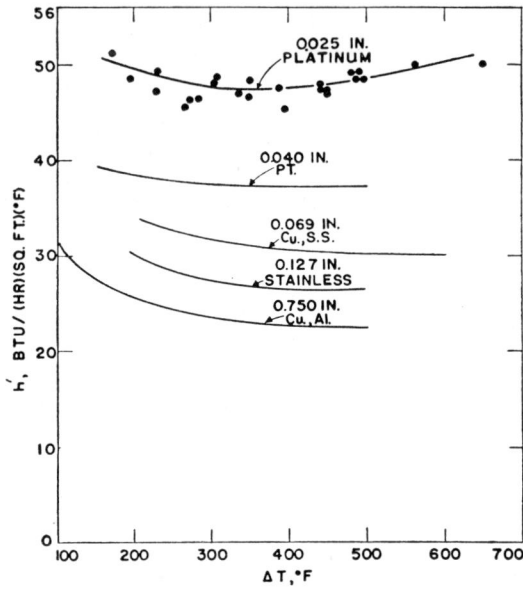

Fig. 16. Film boiling of oxygen. The effect of diameter of horizontal tubes and wires is demonstrated at one atmosphere. The type of metal has no effect. Data are shown for one line only (B1).

oxygen boiling on platinum, copper, aluminum, and stainless steel (Fig. 16) show no surface effects. Both fresh and oxidized surfaces were used.

Two imagined conditions for which the solid surface could become important are: (1) At very high temperatures radiation is important, and the emissivity of the solid should be significant. (2) For very coarse surfaces the flow of vapor on the solid could be disturbed, and a change in h could result. These conditions have not been studied experimentally.

3. *Geometric Arrangement*

The data for film boiling permit a discussion of the effect of diameter for horizontal tubes. Equation (1) predicts that h' is proportional to $D^{-¼}$.

Figure 20, based on diameters from 0.188 to 0.468 in., shows good agreement with this prediction for *n*-pentane at one atmosphere. However for smaller diameters, 0.004 to 0.024 in., a better correlation for water at 1 atmos. is $q/A = $ (const.) $D^{1/2}$, as shown by McAdams (M3). Note that the McAdams expression includes radiation. The Bromley expression excludes it. On the other hand for oxygen at one atmosphere, on diameters ranging from 0.025 to 0.750 in., Banchero (B1) states the best correlation

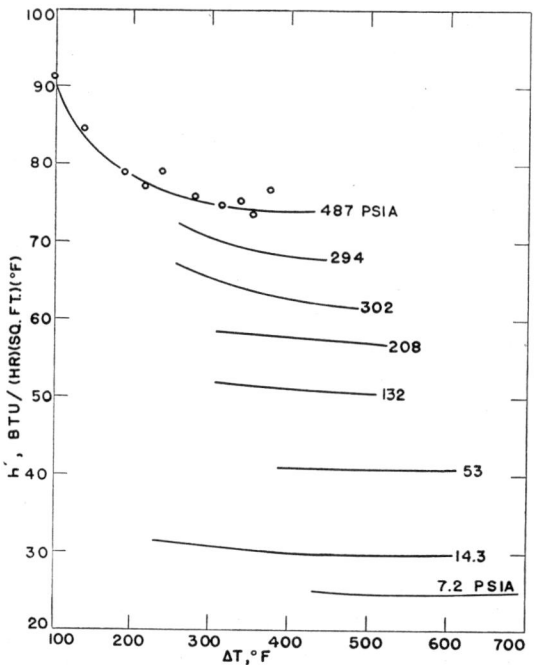

FIG. 17. Effect of pressure on film boiling. Oxygen was boiled on a 0.069-inch stainless-steel, horizontal tube. Data are shown for one line only (B1).

is $h' = $ (const.) $(D^{-1} + C)$, where C is a "constant" which is slightly temperature dependent.

The theoretical expression, Eq. (1), should be expected to fail for very small and very large diameters. For small diameters, the diameter and film thickness become comparable in size, a condition not allowed for in the derivation. For very large diameters, the equation predicts $h' = 0$. This prediction cannot be correct.

No data are available for demonstrating other geometric factors.

4. Pressure

According to Eq. (1), a change in pressure should affect h', because of changes in k_v, ρ_v, λ, and μ_v. Of these, ρ_v is the most sensitive. Figure 17 illustrates experimental results for the effect of pressure on boiling oxygen. The authors state that these results agree with predictions from Eq. (1).

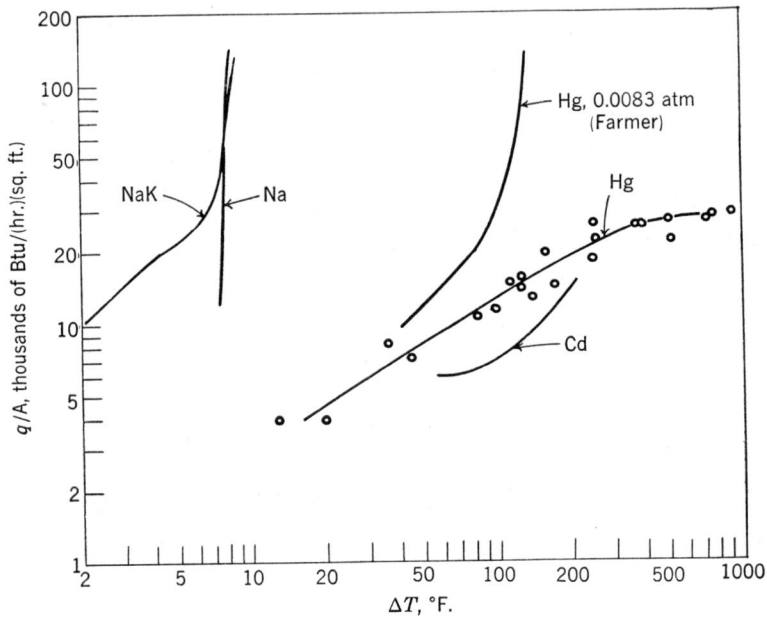

FIG. 18. Liquid metals boiling on a horizontal stainless-steel tube at one atmosphere. Data points are shown for mercury only (L2).

5. Surface Tension

Surface tension is not a variable in Eq. (1). The fact that water fits the expression as well as other liquids which have much lower surface tensions indicates that omission of σ in Eq. (1) is defensible.

Inasmuch as surface tension is an important factor for nucleation, one may conclude that nucleation is not rate-controlling during film boiling. The controlling mechanism must be the transfer of heat across the vapor film.

A theoretical treatment of the formation of bubbles from the gas rod that appears on the top of a tube is still in the future. Intuition suggests that surface tension should enter such a theory. If so, this bubble forma-

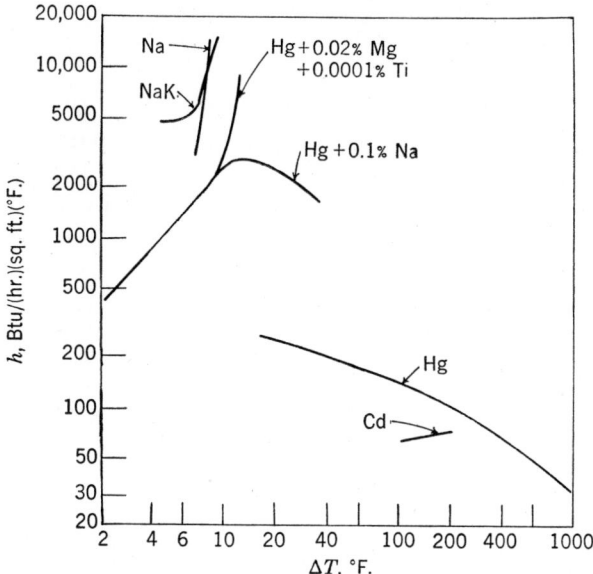

Fig. 19. Individual heat transfer coefficients for boiling metals. Data from Fig. 18 are included (L2).

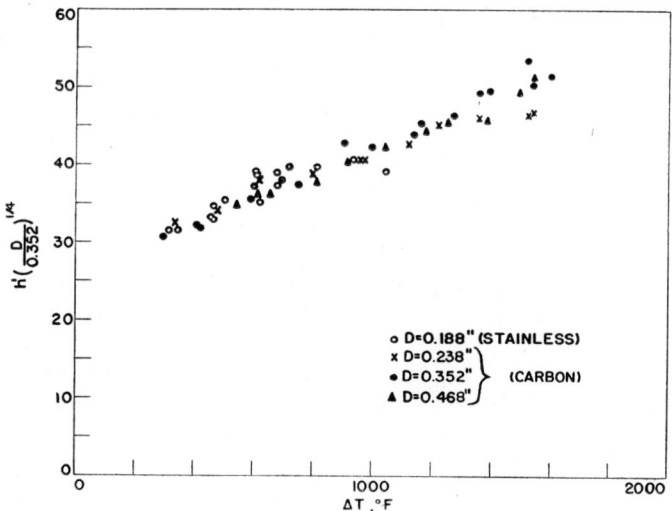

Fig. 20. Effect of diameter for film boiling of n-pentane on a horizontal tube at atmospheric pressure. The type of solid has no effect (B3).

tion may be the source of disagreement between Bromley's equation and observations for very small and very large tubes.

6. *Agitation*

Figure 8 includes data for methanol extending partly into the film-boiling region. It is clear that mechanical agitation increases the value of h in the film region as well as in the nucleate and transition regions.

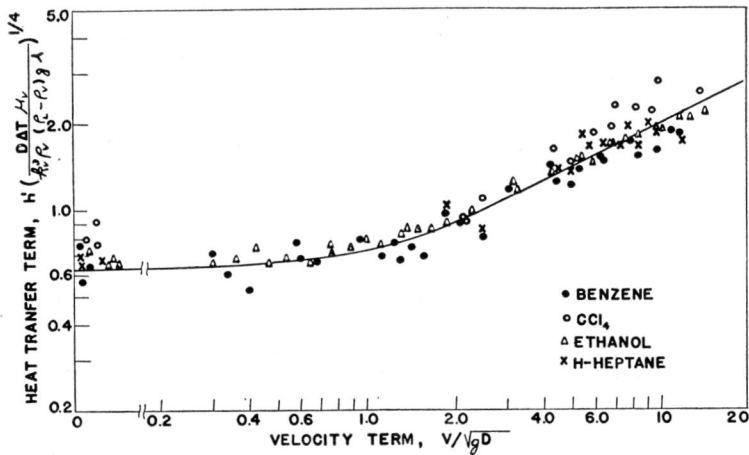

FIG. 21. Film boiling with forced convection. The liquid flow was normal to a horizontal tube at one atmosphere. Velocity = 0 to 14 ft./sec. Diameter = 0.387 to 0.637 inch (B5).

Bromley and co-workers (B5) derived an expression for film boiling with inclusion of the effect of forced flow upward normal to a horizontal tube. The main effects of the liquid flow are to produce changes in the velocity profile for the vapor film and in the thickness of the vapor film. The theoretical differential equation is too complex for a direct solution, but it does show that the variables occur in three groups. By use of experimental observations, it was possible to obtain a correlation of the groups,

$$h'\left[\frac{D^2\mu_v\,\Delta T}{v^2 k_v^3 \rho_v \rho_L \lambda}\right] = 0.88\left[\frac{gD(\rho_L - \rho_v)}{4v^2\rho_L} + \frac{3D\mu_v}{vk_v^2\rho_L}\left(\frac{\pi h'}{\theta'}\right)^2\right]^{1/4} \quad (4)$$

where v is the liquid velocity of approach and θ' is the working angle for heat transfer around the tube. For natural convection θ' is taken to be 180 degrees, for high velocities it is taken as 90 degrees.

A modified form of Eq. (4) is given in Fig. 21. The correlation covers four liquids, three tube diameters (0.387, 0.496, and 0.637 in.), and a velocity range of zero to 14 ft./sec. A single, horizontal, carbon tube was used in each test at atmospheric pressure. The correlation is satisfactory.

Figure 21 can be represented fairly well by two straight lines. In this case, simplified equations result. If $v/(gD)^{1/2} < 1$, the liquid flow is negligible and the usual film-boiling equation results, Eq. (1). If $v/(gD)^{1/2} > 2$, then

$$h' = 2.7 \left[\frac{v k_v \rho_v \lambda}{D \, \Delta T} \right]^{1/2} \quad (5)$$

In this case, h' varies directly as the square root of the velocity, and the vapor viscosity has ceased to be a variable.

7. *Impurities*

The role of impurities in a liquid during film boiling is not known. Because surface tension appears to be unimportant, at least for tubes of intermediate size, a first guess is that impurities should be of small importance also. Experimental study is indicated.

III. Boiling of Subcooled Liquids

The usual water-cooled, internal-combustion engine is designed for non-boiling in the heat-transfer sections. Engineers have realized for many years that the heat transfer rates occurring in such equipment sometimes become much greater than can be explained by the usual forced-convection equations. It is known now that a localized boiling on the hot solid can occur if the solid is sufficiently hot, even if the liquid bulk temperature is below the boiling point.

A. Description from Photographic Studies

Gunther and Kreith (G2) have photographed water during subcooled boiling at 1 atmosphere in the absence of forced convection. The liquid was kept cold by cooling coils in the upper portion of the boiling vessel. The important characteristics of the bubbles formed were their small size, large numbers, and rapid collapse. Some of the bubbles grew and collapsed at one location on the hot solid, while some succeeded in breaking away from the solid, only to collapse in the surrounding cold liquid.

Figure 22 is typical for the bubble behavior with no forced convection. The sizes of the bubbles decrease rapidly as the liquid temperature is decreased. For water at 60°F., the bubbles are about 12% of the size at 195°F. The lifetime of a bubble in water at 60°F. is only 2 or 3% of that in water at 195°F. There is good visual evidence that the rapid growth and collapse of tiny bubbles in greatly subcooled water causes strong convection near the hot solid.

If forced convection, the common practice, is used during subcooled boiling, some bubbles slide along the solid surface as they grow and collapse. Gunther (G1) found photographically that increases in the flow velocity result in a decrease in the average bubble size, bubble lifetime, and bubble population (at constant heat flux and liquid temperature). Decreases in the liquid temperature (or an increase in the degree

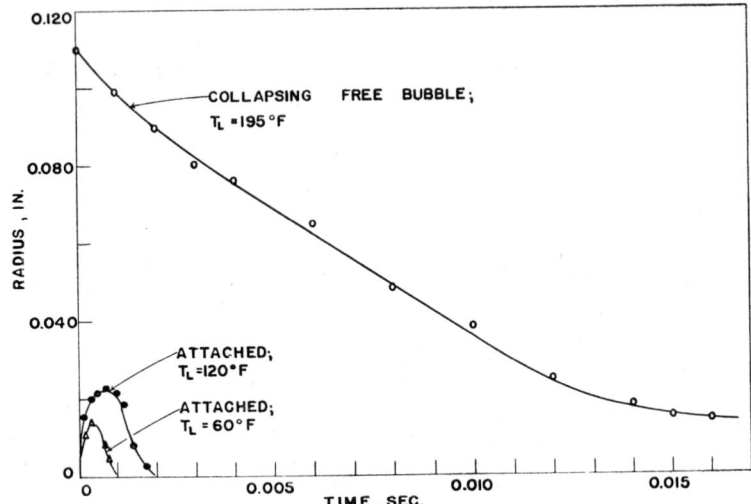

Fig. 22. Behavior of steam bubbles during subcooled boiling. No forced convection was used. The heat flux from the stainless-steel heating surface was 518,000 B.t.u./hr. sq. ft. (G2).

of subcooling), cause a decrease in bubble size and life and an increase in their population. The motion pictures of Dew (D1) confirm all these observations except one. Dew found that increased subcooling causes a decrease in the bubble population. Increases in the heat flux give results such as shown in Fig. 23. Note the tremendous increase in bubble population as the maximum heat flux is approached.

B. Data and Correlations

For subcooled boiling with no forced convection, Gunther and Kreith found that h vs. ΔT data were different from data obtained for ordinary nucleate boiling in at least three respects. The slope of the curve was steeper; the burnout heat flux was increased greatly, as demonstrated in Fig. 24; and lastly, it was necessary to use $T_s - T_{BP}$ for ΔT (rather than $T_s - T_L$) to get good correlations of the data for burnout. The amount of subcooling is not as significant as the temperature excess of

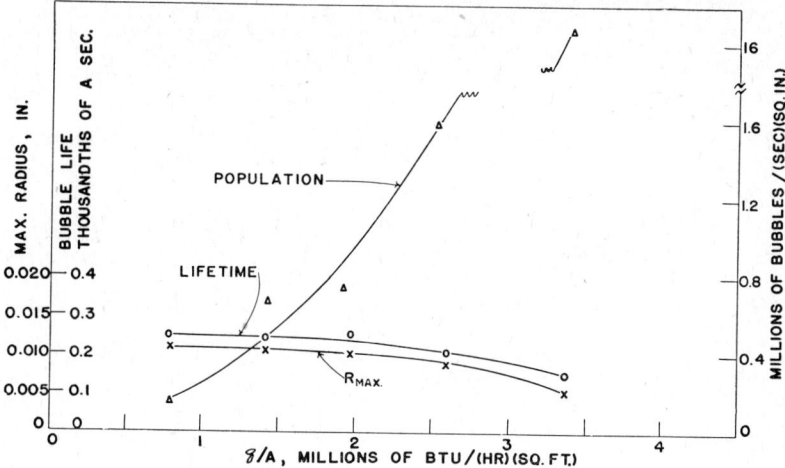

Fig. 23. Bubble histories for forced-convection, subcooled boiling. Water was boiled at 25 psi. on a stainless-steel strip. Velocity = 10 ft./sec. Subcooling = 155°F. (G1).

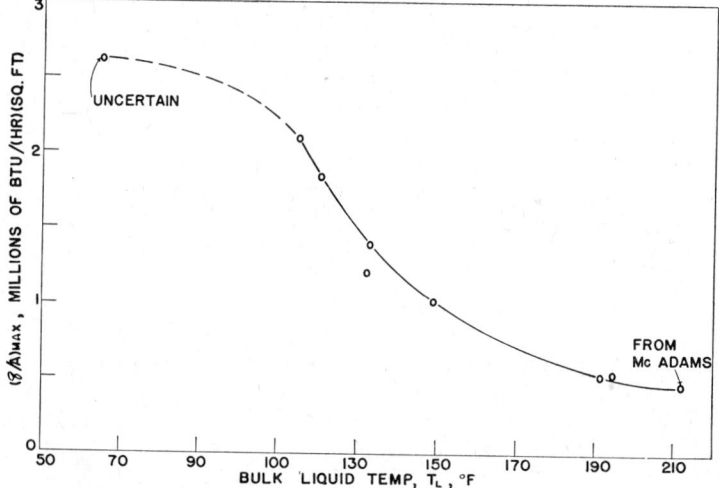

Fig. 24. Effect of subcooling on burnout point. Water was boiled on stainless-steel at one atmosphere with no forced convection (G2).

the solid above the saturation temperature of the liquid, if one wishes to express the critical ΔT. In fact for a given system at a fixed pressure, the critical $T_s - T_{BP}$ is practically constant, regardless of subcooling.

When forced convection is superimposed on subcooled boiling, the resulting heat transfer rate is some sum of the rates due to boiling and to forced convection. Figure 25 shows forced-convection, subcooled boiling observations for water under pressure (K2). Other workers report similar data. The lower portion of the graph indicates that no boiling at all occurs when ΔT is very small. This portion of the graph can be represented accurately by the usual forced-convection, heat transfer

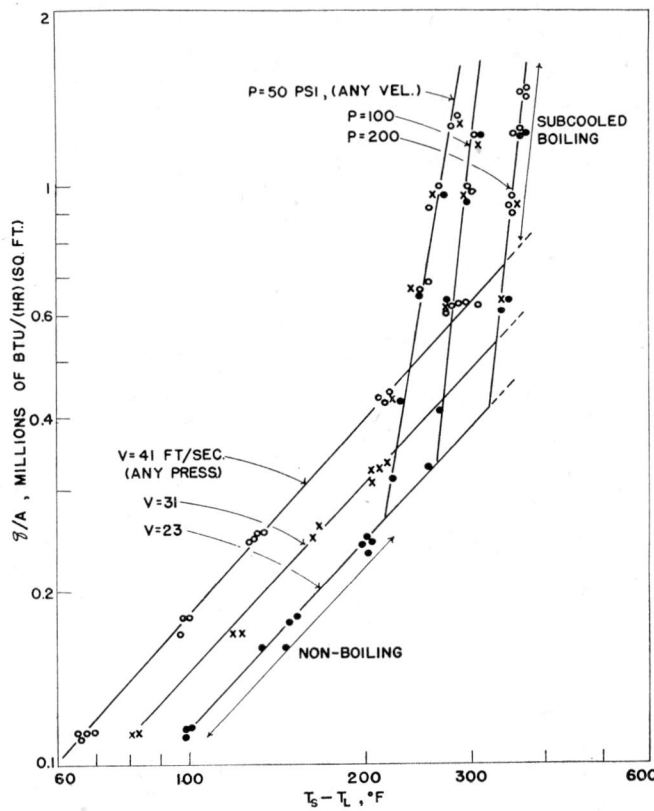

Fig. 25. Effect of pressure and velocity on subcooled boiling. Normal butanol was boiled inside a 0.587-inch stainless tube (K2).

correction such as used by Sieder and Tate or Colburn. In this region, the flow velocity is very important, and the pressure is not significant.

When the ΔT is large enough, subcooled boiling begins. After the boiling is well established, corresponding to the upper portion of Fig. 25, the pressure becomes very important and velocity is less significant. The reason pressure becomes important is that a vapor is created and vapor properties (such as density) are affected by pressure. The reason velocity becomes less important is that the agitation caused by the growth and collapse of bubbles is so great that it overshadows additional convection caused by forced fluid flow. At least this is true for liquid velocities up to about 40 ft./sec.

As the burnout point is approached, velocity again becomes important. It was pointed out (Part I, Sec. II, E, 7) that for ordinary nucleate boiling, agitation causes but a slight change in h until the critical ΔT is reached. The value of h at the critical ΔT is increased significantly by agitation. The same is true for subcooled boiling. Data from McAdams and coworkers (M4) show the effect of velocity and subcooling on the

Velocity ft./sec.	Subcooling °F.	Max. q/A B.t.u./(hr.)(ft.²)(°F.)
1	50	651,000
4	50	1,030,000
12	50	1,390,000
1	100	854,000
12	100	2,010,000

burnout heat flux for water at 60 lb./sq. in., abs. outside a vertical 0.25-inch, stainless tube.

An examination of Fig. 25 reveals an interesting point concerning pressure. Consider a forced convection, non-boiling system at a reasonable high temperature driving force, $T_s - T_L$, and a high pressure (for example: $\Delta T = 300°F.$, $v = 23$ ft./sec., $p_L = 200$ lb./sq. in., abs. in Fig. 25). An improvement in heat transfer can be achieved by *decreasing* the pressure. If the pressure is dropped to 100 lb./sq. in., abs., boiling occurs and the heat flux increases to about 2.5 times the original value.

No theoretical description in mathematical form is available for forced-convection, subcooled boiling. A number of correlations have been presented. McAdams (M4) gives

$$q/A = 0.378(T_s - T_{BP})^{3.86} \qquad (6)$$

for water on a ¼-in. stainless tube, at pressures below 100 lb./sq. in., abs., for velocities up to 12 ft./sec. Jens and Lottes (J2) give

$$q/A = 0.45 \cdot e^{p_L/225} \cdot (T_s - T_{BP})^4 \qquad (7)$$

for water in a tube at velocities up to 45 ft./sec., showing that pressure definitely influences the heat transfer. Neither expression can be general, since both omit the fluid velocity, tube dimensions, and other factors. The expressions show that the heat flux is very sensitive to the temperature difference between the solid and the boiling point (saturation temperature) of the liquid. In fact many writers have commented that

q/A seems to be independent of the solid temperature T_s. This cannot be true, strictly speaking. Rather it means that a modest change of q/A can occur with almost no observable change in T_s. At high pressures this becomes particularly noticeable, because the critical ΔT becomes very small. For example, with water at different amounts of subcooling, at 2000 lb./sq. in., abs., and $v = 10$ ft./sec., Rohsenow and Clark (R1) found that a change in $T_s - T_{BP}$ from 4.3°F. to 8.6°F. was sufficient to change the heat flux from 1,000,000 to 2,000,000 B.t.u./(hr.)(ft.^{2}).

Bernath (B2) has presented a correlation which is shown to fit the burnout points for water at various pressures and velocities, in systems of various geometries.

$$\text{Max. } h = 5710 \left(\frac{D_e}{D'}\right)^{0.6} + 48 \left(\frac{v}{D_e^{0.6}}\right) \tag{8}$$

Here D_e is the equivalent diameter of the flow passage in ft., D' is the heated perimeter divided by π, and v is the liquid velocity in ft./sec. The equation fits the data from six observers for pressures up to 3000 lb./sq. in., abs., velocities up to 54 ft./sec., and for tubes, annuli, and duct shapes. In spite of the good fit, the equation is not general. It predicts no effects for the type of metal or the pressure of the system.

Bernath also presents a correlation (for water) for the temperature of the solid at the burnout point.

$$\text{Burnout } T_s = 57 \ln p_L - 54 \left(\frac{p_L}{p_L + 15}\right) - \frac{v}{4(D_e/D')^{0.6}} \tag{9}$$

Here p_L is in lb./sq. in., abs., and T_s is °C. This predicts that pressure, velocity, and geometry are the only variables of significance.

The interesting case of a "chopped-cosine" heat flux distribution to a tube is discussed by Bernath. Expressions are developed for the burnout heat flux and the burnout location. An atomic reactor can have a heat flux of this type, so the problem is not trivial.

In addition to the recent Bernath correlation, a number of prior correlations for burnout have been published. Whereas Eq. (8) contains no temperatures, the older equations do. Jens (J1) reviews these equations and comments that they are not particularly reliable. It is obvious that theoretical work is needed badly. These expressions are for water only.

(1) McAdams (M4)

$$(q/A)_{\max.} = v^{1/2}[400,000 + 4800(T_{BP} - T_L)] \tag{10}$$

(2) Gunther (G1)

$$(q/A)_{\max.} = 7000 v^{1/2}(T_{BP} - T_L) \tag{11}$$

(3) Buchberg (B6)

$$(q/A)_{max.} = 520 G^{1/2}(T_{BP} - T_L)^{1/5} \qquad (12)$$

(4) Jens-Lottes (J2)

$$(q/A)_{max.} = C\left(\frac{G}{10^6}\right)^m (T_{BP} - T_L)^{0.22} \qquad (13)$$

$$C = f \text{ (Pressure)}, \qquad m = f \text{ (Pressure)}$$

Here q/A is in B.t.u./(hr.)(ft.^{2}), v in ft./sec., G in lb./hr. sq. ft., and ΔT in °F.

For a subcooled liquid in forced convection, boiling inside a tube, pressure drop considerations are important. As the burnout condition is approached, the volume of vapor formed becomes great and causes a back pressure. This back pressure can decrease the liquid flow rate so that the tube condition becomes serious. Burnout may occur because of this "choking." Kreith and Foust (K1) discuss this problem and present a graph for predicting choking for boiling water.

No data are available for subcooled boiling at temperature differences beyond the critical ΔT.

IV. Bumping during Boiling

Bumping during boiling has long been a scientific oddity and a laboratory nuisance. Scientific studies of the subject could throw light on an understanding of boiling. Unfortunately there is no practical incentive, for bumping is practically nonexistent in equipment of industrial size.

Bumping is a mode of boiling during which the liquid erupts suddenly, and a portion is ejected into the vapor space with considerable force. Bumping in a laboratory beaker can be severe enough to splatter the ceiling of a room.

The literature on bumping is curious. By the year 1919, the published methods for preventing bumping included the addition of chips or bits of glass beads, porcelain plates, bricks, granite, coal, talc, and pumice. One author noted that powdered coal was very effective, while solid chunks were valueless. Another author claimed the exact reverse.

Since 1919, progress in the prevention of bumping has not been great. Capillary glass tubes, hydrogen peroxide, platinum tetrahedrons, and other items have joined the list of additives. Air bubblers are claimed to prevent bumping; so are bare-wire electric heaters immersed in the liquid. The introduction of electrodes is a patented idea for prevention of bumping. The explanations of why these devices sometimes work are plentiful, but none has been proved. Adsorbed gas is suggested for things such as glass beads. Contact with trapped vapor is given as the explanation for capillary tubes. Local convection currents are claimed for the platinum, glass beads, and chips by one author. He states that these local currents give good mixing, intimating that bumping is caused by lack of mixing. Air bubbles and bubbles from peroxide or from electrolysis are supposed to furnish nuclei for vapor generation. No bibliography is included for this early work, since the tests were not designed carefully. Not one author "measured" bumping; it was "observed" only.

A. Torpidity Theory

Several papers discussing torpidity have been presented recently by Hickman (H1, H2, H3). A bumping liquid is stated to be one with a torpid surface. A torpid surface is one with an extremely thin layer of a surface-seeking impurity. Hickman reports, for example, that pure water at a high vacuum was boiled normally from an immersed nichrome heating wire. However, if the water was allowed to stand in glass for 16 hours, normal boiling was impossible. Violent bumping, breaking a flask on one occasion, was the result.

Hickman suggests that a tiny amount of something was leached from the glass. This material then formed a membrane-like layer on the surface. The membrane suppressed normal boiling and resulted in bumping.

These observations could be dismissed were it not for the plentiful evidence that boiling liquids sometimes exhibit a split surface. One part will be working, with a steady discharge of vapor. The other part will be torpid with no vapor discharge. Hickman has published numerous photographs of these schizoid surfaces. He has demonstrated also that it is possible to mechanically remove a torpid surface. However the torpid material has never been concentrated or chemically analyzed.

If this view of bumping is correct, agitation can decrease or prevent bumping. Sources of inert bubbles could also break the surface and be effective. Inert material such as clean chips of insoluble solids should have no effect.

B. Nucleation Theory

If a liquid is pure and its container is clean, considerable superheating is possible as discussed earlier. Under these conditions boiling could occur in a cyclic manner: first a quiet, non-boiling period as the liquid superheats; then an explosion when the nucleation rate becomes large; and finally a sudden drop in the liquid temperature after the superheat has disappeared as heat of vaporization. The process would then repeat.

If this view is correct, cleanliness and purity are necessary for bumping. Addition of most impurities would catalyze nucleation and decrease bumping. After bubbles are created, it is difficult to concede that a thin layer of molecules could prevent their escape.

Glaser's observations, Part I, Sec. II, C, 1b, that superheat (of particular liquids) is destroyed by particles of high energy radiation, are pertinent to bumping. Glaser calculated that hard cosmic ray particles struck his small laboratory apparatus at a rate of about one per half-hour. This would mean that a large-scale industrial boiler would be struck by cosmic particles continuously. Thus nucleation should be continuous, and bumping should be absent in industrial equipment. Of course an alternate explanation for the lack of bumping in large equipment is the great difficulty in obtaining sufficient purity and cleanliness.

C. Measurement of Bumping

Before bumping can be studied, it must be measured. A suggested method has been tried at the University of Illinois. Bumping in a flask creates sudden pressure fluctuations. These are used to vibrate a flexible diaphragm. An electrical scheme is used to measure the amplitude and frequency of the resulting deflections. Preliminary measurements with methyl alcohol show quantitatively that a decrease in pressure increases the severity of bumps, use of a smooth glass surface causes stronger bumps than an etched glass surface, and emanations from radium do not affect the course of bumping. The first two observations are consistent with nucleation theory. The later observation may mean that methyl alcohol does not ionize readily. Additional quantitative tests are needed.

Acknowledgment

Encouragement and assistance for the preparation of this portion of the writing were furnished by the National Science Foundation. Helpful suggestions and aid were given by H. B. Clark and J. C. Zahner.

Nomenclature

A Surface area, sq. ft.
C Constant
C_v Heat capacity of vapor, B.t.u./lb. (°F.)
D Diameter of tube or wire, ft.
D' Heated perimeter divided by π, ft.
D_e Equivalent diameter of flow passage, ft.
F_ϵ Radiation emissivity factor, dimensionless
g Acceleration of gravity, ft./hr.²

G Mass velocity, lb./hr. sq. ft.
h Individual heat transfer coefficient, B.t.u./(hr.)(sq. ft.)(°F.)
h' Heat transfer coefficient, neglecting radiation, B.t.u./(hr.)(sq. ft.)(°F.)
h_R Heat transfer coefficient for radiation, B.t.u./(hr.)(sq. ft.)(°F.)
k_v Thermal conductivity of vapor, B.t.u./(hr.)(ft.)(°F.)

L Length of tube, ft.
p_L Pressure imposed on a liquid, lb./sq. ft.
q Heat transfer rate, B.t.u./hr.
ΔT Temperature driving force, $T_L - T_v$ for homogeneous case; $T_s - T_\infty$ for heterogeneous case, °F.
T_{BP} Boiling point of the bulk liquid at the existing pressure; T_∞, °F.
T_L Temperature of the bulk liquid, °F, or °R.
T_s Temperature of a solid surface, °F, or °R.
T_v Temperature inside a bubble, °F.
T_∞ Saturation temperature of a liquid flat surface at the existing pressure, °F.
v Velocity, usually ft./hr.
θ' Working angle around a tube, radians
λ Latent heat of vaporization, B.t.u./lb.
μ_v Viscosity of vapor, lb./(ft.)(hr.)
ρ_L, ρ_v Density of liquid, vapor, lb./cu. ft.
σ Surface tension, liquid-vapor interface, lb./ft.
σ' Stefan-Boltzmann radiation constant, B.t.u./(hr.)(sq. ft.) (°R)^4

References

B1. Banchero, J. T., Barker, G. E., and Boll, R. H., *Chem. Eng. Progr. Symposium Ser.* **51**, No. 17, 21 (1955).
B2. Bernath, L., *Chem. Eng. Progr. Symposium Ser.* **52**, No. 18, 1 (1956).
B3. Bromley, L. A., *Chem. Eng. Progr.* **46**, 221 (1950).
B4. Bromley, L. A., Brodkey, R. S., and Fishman, N., *Ind. Eng. Chem.* **44**, 2966 (1952).
B5. Bromley, L. A., LeRoy, N. R., and Robbers, J. A., *Ind. Eng. Chem.* **45**, 2639 (1953).
B6. Buchberg, H., Romie, F., Lipkis, R., and Greenfield, M., "Heat Transfer and Fluid Mechanics Institute," p. 177. Stanford Univ. Press, Stanford, 1951.
C1. Castles, J. T., M.S. Thesis in Chemical Engineering. Massachusetts Institute of Technology, Cambridge, 1947.
C2. Colburn, A. P., referred to by Castles (C1); also in unpublished University of Delaware lecture material, 1941.
D1. Dew, J. E., M.S. Thesis in Chemical Engineering. Massachusetts Institute of Technology, Cambridge, 1948.
D2. Dougherty, E. L., M.S. Thesis in Chemical Engineering. University of Illinois, Urbana, 1951.
D3. Drew, T. B., and Mueller, A. C., *Trans. Am. Inst. Chem. Engrs.* **33**, 449 (1937).
F1. Farmer, W. S., in "Liquid Metals Handbook," 2nd ed., p. 205. Supt. of Documents, Washington, D. C., 1952.
G1. Gunther, F. C., *Trans. Am. Soc. Mech. Engrs.* **73**, 115 (1951).
G2. Gunther, F. C., and Kreith, F., "Heat Transfer and Fluid Mechanics Institute," p. 113. Am. Soc. Mech. Engrs., New York, 1949.
H1. Hickman, K. C. D., *Science* **113**, 480 (1951).
H2. Hickman, K. C. D., *Ind. Eng. Chem.* **44**, 1892 (1952).
H3. Hickman, K. C. D., and Torpey, W. A., *Ind. Eng. Chem.* **46**, 1446 (1954).
J1. Jens, W. H., *Mech. Eng.* **76**, 981 (1954).
J2. Jens, W. H., and Lottes, P. A., Report 4627, Argonne National Laboratory, 1951.
K1. Kreith, F., and Foust, A. S., *Ann. Meeting Am. Soc. Mech. Engrs.*, Paper 54-A-146 (1954).
K2. Kreith, F., and Summerfield, M., "Heat Transfer and Fluid Mechanics Institute," p. 127. Am. Soc. Mech. Engrs., New York, 1949.

L1. Lowery, A. J., Jr., M.S. Thesis in Chemical Engineering. University of Illinois, Urbana, 1955.
L2. Lyon, R. E., Foust, A. S., and Katz, D. L., *Ann. Meeting Am. Inst. Chem. Engrs., St. Louis,* Preprint No. 6 (1953).
M1. Marx, J. W., and Davis, B. I., *J. Appl. Phys.* **23,** 1354 (1952).
M2. McAdams, W. H., "Heat Transmission," 3rd ed., Chapter 14. McGraw-Hill, New York, 1954.
M3. McAdams, W. H., Addoms, J. N., Rinaldo, P. M., and Day, R. S., *Chem. Eng. Progr.* **44,** 639 (1948).
M4. McAdams, W. H., Kennel, W. E., Minden, C. S., Carl, R., Picornell, P. M., and Dew, J. E., *Ind. Eng. Chem.* **41,** 1945 (1949).
N1. Nukiyama, S., *Soc. Mech. Eng. Japan* **37,** No. 206, 367 (1934).
P1. Pramuk, F. S., and Westwater, J. W., *Chem. Eng. Progr. Symposium Ser.* **52,** No. 18, 79 (1956).
R1. Rohsenow, W. M., and Clark, J. A., "Heat Transfer and Fluid Mechanics Institute," p. 193. Stanford Univ. Press, Stanford, 1951.
W1. Westwater, J. W., *Sci. American* **190,** No. 6, 64 (1954).
W2. Westwater, J. W., and Santangelo, J. G., *Ind. Eng. Chem.* **47,** 1605 (1955).
W3. Westwater, J. W., and Santangelo, J. G., "A Photographic Study of Boiling" (Motion Picture). University of Illinois, Urbana, 1954.

AUTOMATIC PROCESS CONTROL

Ernest F. Johnson

Department of Chemical Engineering
Princeton University, Princeton, New Jersey

I. Introduction	34
A. Scope	34
B. Control as a Basic Concept in Chemical Engineering	34
C. Automatic Control in the Process Industries	35
II. Nature of Control and Control Problems	36
A. Definition	36
B. General Characteristics of Control Systems	36
1. Information Processing	36
2. Control System Components	38
3. Systems Concept	39
4. Feedback	39
5. Stability	40
C. Control Problems	40
1. Analysis	40
2. Synthesis	41
III. Treatment of Control Problems	41
A. History	41
B. Analysis	42
1. Process Elements	43
a. First Order Lag	44
b. Series of First Order Lags	50
c. Second Order Elements	51
d. Distributed Parameter Components	52
e. Time Delays	54
f. Nonlinear Process Components	55
2. Measuring Elements	55
a. Pressure	56
b. Temperature	56
c. Flow	57
d. Liquid Level	57
e. Composition	57
3. Controllers	57
a. Two-Position Control	58
b. Proportional Control	59
c. Integral Control	59
d. Derivative Action	60
e. Proportional Plus Integral Control	60
f. Three-Mode Control	61

	4. Final Control Element....................................	62
	a. Valve Types..	62
	b. Valve Characteristics...............................	62
	c. Valve Actuators....................................	63
	5. Transmission Lines......................................	64
	6. Loop Behavior..	64
	a. Linear Systems.....................................	64
	b. Nonlinear Systems..................................	67
	c. Process Systems....................................	68
	d. Sampled-Data Systems..............................	69
	7. Complex Systems..	70
	8. Effect of Disturbance Location............................	70
C. Synthesis...		70
	1. Criteria of Control.......................................	70
	2. Control System Synthesis from Frequency Response Characteristics.	71
	3. Empirical Methods of Controller Specification...............	73
	a. Reaction Curve (Signature Curve).....................	73
	b. Optimum Controller Adjustments from Ultimate Gain and Frequency..	74
	4. Complex Systems..	75
IV. Present Status and New Directions................................		76
A. Summary of Present Status...................................		76
B. Future Directions...		77
Nomenclature...		78
References..		79

I. Introduction

A. Scope

The field of automatic process control has only recently come to the attention of the chemical engineer as an area wherein a quantitative treatment offers interesting possibilities for technological advance. This brief survey undertakes to orient automatic process control in the chemical engineering picture and to identify and evaluate the possibilities of intelligent development of the field. The important broad concepts are presented in a general description of the nature of control. These concepts are applied to the solution of typical process control problems in an elucidation of available techniques.

A brief history of the development of modern control theory is included to provide some indication of what kind of help may be obtained from developments in other control areas such as the field of servomechanisms. The final section of the chapter deals with the future trends in the field of automatic process control and how chemical engineers can contribute materially to the acceleration of these trends.

B. Control as a Basic Concept in Chemical Engineering

Chemical engineering deals with the design, development, and operation of processes and plants in which bulk raw materials are converted by chemical and physical means into bulk products of varying degrees of finish. The purely technical problems of design and analysis of processes and equipment are solved readily on the basis of a few fundamental concepts. If we regard smooth and stable operation of process plants as an important quality of a successful design, we must regard the general concept of feedback control as one of our basic concepts, coequal in importance with such concepts as those of equilibrium, conservation, and rate processes.

The technical side of chemical engineering may be contemplated most broadly in terms of the most immediately creative aspect of the field, namely, the design of a plant. We may regard such design as involving a succession of integrating steps in which first of all the basic properties of matter are used in conjunction with basic rate expressions to describe the component rate processes; next the rate equations are combined with the constraints imposed by thermodynamics, e.g., conservation of mass, momentum, and energy, and chemical and physical equilibrium to permit sizing the process units; and finally the process units are fitted into the overall plant sequence with such control instrumentation as is necessary for satisfactory operating performance. In this picture of a plant design there are four distinct basic concepts: *conservation, equilibrium, rate process,* and *control*. Each must be considered in maximizing the economic gain to be realized from the plant. A sound understanding of these concepts and how they are applied is essential to successful design.

C. Automatic Control in the Process Industries

A typical characteristic of much of the chemical process industry is that the operating plants appear to require little human supervision. The process units are often extensively instrumented and the desired operating conditions are maintained rigidly by arrays of control instruments. Automatic control is clearly an important feature of modern processing.

The process industries were among the first to make widespread use of automatic controls with the attendant advantages of continuous operation, improved yields and product quality; and the general adaptability of the many processes and operations to automatic regulation has contributed greatly to the rapid growth of the industry.

Because of this ready adaptability of the processes and process equipment to automatic control there has been little incentive to develop a sound quantitative theory of automatic process control. Chemical engineering ingenuity has been applied to more immediately profitable process problems. Processes involving difficult control problems have been rejected in favor of easily controllable processes without any concerted effort to solve the control problem. As a result, chemical engineers have made little contribution to control theory, and even today most chemical engineers regard control problems as lying a little outside their field.

However it is obvious in many operations that improved overall performance can result from a clearer understanding of the control problems. Furthermore, the trends toward greater integration of operations and the exploitation of faster processes require that the control problems be handled in a quantitative manner by process engineers. The nature of control problems and what is involved in handling them, are discussed in Parts II and III.

II. Nature of Control and Control Problems

A. DEFINITION

In the broadest sense *control* is *regulation for some purpose*. Automatic process control is the mechanical (or other non-manual) regulation of processes for the purpose of producing high quality products at high rates and low cost. Generally this regulation involves holding process conditions relatively constant in the face of external disturbances. By contrast, the control action in the field of servomechanisms is designed to maintain a close agreement between a varying desired behavior and the actual behavior, as for example, in steering devices.

Control, of course, is not limited to machines and process units but is a significant feature of living organisms, corporate enterprises, and sociological organizations. Regardless of the nature of the control system the basic characteristics are the same, and the broad principles of handling control problems are equally applicable. Only the terminology and degree of sophistication of treatment are different.

B. GENERAL CHARACTERISTICS OF CONTROL SYSTEMS

1. *Information Processing*

All control systems whether they be process plants, steering systems, living organisms, or industrial corporations may be regarded as information processing systems. They obtain information, analyze it, and on the basis of their analysis generate new information to take some kind of action.

Consider the simple heat exchange process shown in Fig. 1 wherein a reaction kettle is heated by steam condensing in the jacket of the kettle. The temperature in the kettle is measured by a thermometer which is connected to a controller. The controller in turn actuates a control valve in the steam line thereby ultimately causing changes in the reactor temperature. In this example information flows from the kettle to the thermometer; from the thermometer to the controller; from the controller to the control valve; from the control valve to the kettle; and so on around the control loop. The information flows as various kinds of signals, all of which may be regarded as measures of kettle temperature either actual or potential. Thus by virtue of heat exchange between the kettle

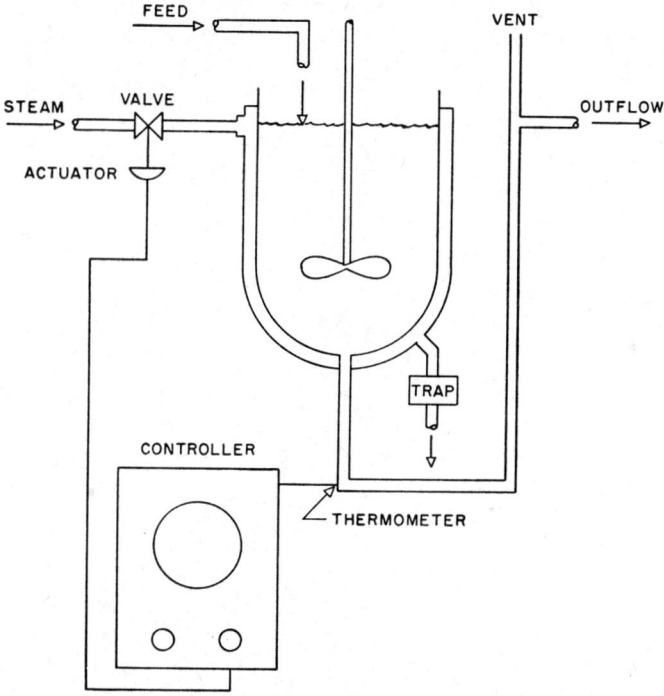

Fig. 1. Jacketed kettle control system.

contents and thermometer, the thermometer attains a temperature indicative of but not necessarily equal to the kettle temperature. Depending on the nature of the thermometer, it sends to the controller a pressure variation, a volume change, or an electrical impulse, which in the controller may be converted to mechanical displacements, air pressures or electrical signals. The controller sends out air pressures or electrical signals which become valve positions. These positions in turn moderate flows which affect steam pressures in the jacket and ultimately affect temperatures in the kettle. The signals flowing from the kettle to the controller are all measures of the actual kettle temperature; the signals flowing from the controller to the kettle are all *potential* kettle temperatures.

Since the information processing involved in control systems is the propagation of signals, the appropriate pictorial representation of such systems is the *signal flow diagram*. Such a diagram indicates not only where signals go but how they are related to each other. A simple kind of signal flow diagram for the heat exchange system just described is shown

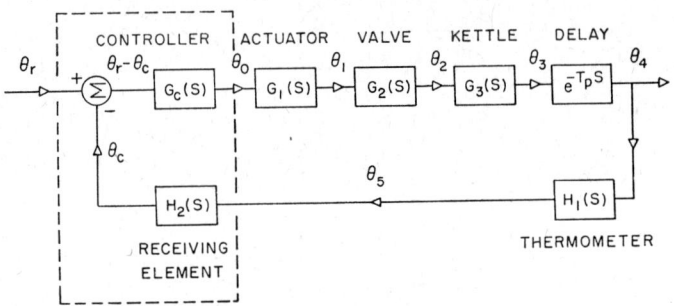

Fig. 2. Block diagram of jacketed kettle control system.

in Fig. 2. Campbell (C1) shows how primitive signal flow diagrams can be derived for process control systems and how they reveal weak process designs.

2. *Control System Components*

All process control systems comprise (a) measuring elements, (b) controlling elements, and (c) the controlled system or *plant*, as it is frequently called. The measuring elements produce signals related to the plant performance, and the controlling elements on the basis of these signals regulate the plant performance. In the example of Fig. 1, the measuring elements are shown as two blocks in the feedback path, one block for the primary measuring element which in this case is the temperature-sensing device in the kettle, and the other for the receiving element or secondary measuring element, which usually is located in the controller housing and which converts the signal from the primary element into something intelligible to the controller.

The controlling elements include the controller and its accessories and the final control element, which in this case and in most process cases is a valve. A single block represents the controlled system, which is the jacketed, agitated kettle.

In more complicated systems each of the blocks of Fig. 2 might be expanded to various multi-block diagrams.

3. Systems Concept

Figure 2 emphasizes an important characteristic of control systems, namely, that every component in the control circuit is important to the overall behavior of the system. It is not the size of a component, nor the amount of the flows of matter and energy to or from a component, that determines the contribution of the component to the system behavior. Rather it is the dynamic or time-dependent behavior of the component and how the various components interact with each other. An overall appraisal of the control system as an integrated unit is necessary in control technology.

Systems engineering is the name given to engineering activity which considers the overall behavior of a system, or more generally which considers all factors bearing on a problem, and the systems approach to control engineering problems is correspondingly that approach which examines the total dynamic behavior of an integrated system. It is concerned more with quality of performance than with sizes, capacities, or efficiencies, although in the most general sense systems engineering is concerned with overall, comprehensive appraisal.

4. Feedback

Control systems may be classified from their signal flow diagrams as either *open-loop* systems or *closed-loop* systems depending on whether the output of the primary control circuit is fed back to the controlling component. As Fig. 2 suggests, the typical control circuit consists of sequential arrays of components deployed about the process under control. If the controller is not apprised of the behavior of the controlled variable, the control system is an open-loop one. Conversely, if the measuring means on the controlled variable sends its signals back to the controller so that the behavior of the controlled variable is always under the scrutiny of the controller, the system is a closed-loop or feedback control system.

A simple example of an open-loop control system would be a steam-jacketed resin kettle very much like that in Fig. 1 except that the steam *pressure* is regulated automatically by the behavior of the measured jacket pressure but not by the actual temperature of the resin batch in the kettle. In the corresponding closed-loop system the steam pressure is regulated by the temperature of the resin batch as in Figs. 1 and 2. The only way open-loop control can be precise is through a close calibration between steam pressure and batch temperature. Since this close calibration can be maintained inexpensively only in the absence of load changes of any kind, it is obvious that the field of application of open-loop control is limited. In the example of Fig. 1, load changes would result from changes in flow rate or temperature of the feed to the kettle. Changes in either of these quantities would cause changes in the kettle temperature which would have no effect on the controller.

With feedback control, on the other hand, the controller action is dictated by the behavior of the controlled variable, and precise control is generally possible even though load changes occur.

An elaboration of open-loop control which sometimes can be competitive with closed-loop control is *feed-forward control*, in which the controller is apprised of factors which affect the key process variable to be controlled, but is not directly apprised of the behavior of this variable. Thus for the kettle of Fig. 1, it would be possible to measure the flow rate and temperature of the feed and have a simple computer analyze the information and on the basis of this information set the steam pressure to be maintained by the controller. So long as the computer considered all pertinent factors, the control would be satisfactory, but an unaccounted factor, such as a change in agitator speed, might make the control ineffectual.

5. *Stability*

Most processes or operations of importance in chemical engineering are by themselves stable, that is, in response to a small perturbation they do not run away or oscillate, but, rather, tend to level out at some new condition. When fitted into a feedback control system, however, any real process can be made to oscillate merely by sufficient amplification of the signals going around the loop. Since instability cannot be tolerated in a control system, the condition at which the system just becomes unstable is a critical limit in operating performance, and the degree of amplification required to produce the condition is a prime control characteristic of the system. This characteristic, called the *ultimate proportional gain*, provides a convenient basis for approximating a good control system design. In operating plants it is used frequently to make the final adjustments to control instruments.

C. Control Problems

1. *Analysis*

The analytic control problem deals with the appraisal of existing control systems. What is the overall dynamic behavior of the system, and what individual contributions are made to the overall behavior by the various components in the system? How may processes be characterized as to their dynamic behavior in terms useful for control system analysis? What are the control characteristics of typical instruments such as the measuring devices, controllers, and regulating units?

2. Synthesis

On the other hand the synthetic problem is the design of the whole control system, including in its broadest implication the design of process, as well as the specification of, control instruments. Before the synthetic problem can be tackled intelligently, the criteria of satisfactory control must be identified. These criteria are different for different systems, but most usually they are described in terms of the response of the system to certain stimuli. Having established the criteria of control, the problem of synthesis is one of optimizing the selection of control system components and their disposition in the control loop, so that the criteria are met.

III. Treatment of Control Problems

A. History

Although automatic control has existed as long as living matter has existed, control problems were not treated quantitatively and successfully much before the early 1920's.

The obvious formal approach to a quantitative treatment of automatic control is to write rigorous, general differential equations for the dynamic behavior of all the components in the system; combine these equations into a single description of the system; solve this single equation for the system response to a typical disturbance or stimulus; compare this response with the desired response; and make such adjustments in manipulable parameters as will produce the desired response. This kind of approach was applied successfully to the design of steering systems for ships and to the design of positioning systems for naval guns as early as 1922. It was also applied in the 1930's and 1940's to a variety of simple process systems. Unfortunately in the general case, this classical approach breaks down because rigorous general equations cannot be written for any but the simplest processes, or if they can be written, the overall equation becomes unwieldy and usually unsolvable. Furthermore the matching of actual response and desired response must be a trial-and-error procedure, since there are no simple means of gauging the total effect of changing the characteristics of individual components.

By the start of World War II, a new approach to control system synthesis was being developed from Nyquist's theoretical treatment (N1) of feedback amplifiers in 1932. This approach utilized the response of components and systems to steady-state sinusoidal excitation or *frequency response* as it is more usually called. The frequency response approach provides an important basis for present-day methods of handling control problems by affording a simply manipulable characterization which avoids the need for obtaining the complete solutions of system equations.

Since World War II, frequency response techniques have been applied to an increasing variety of control problems, and although the bulk of progress prior to the war came from treating servomechanisms, there has been heightened effort in process control problems since the war. The works of Rutherford (R3), Aikman (A1), and Young (Y1) in Great Britain have been notable in this respect.

A comprehensive summary with extensive bibliographies of what has been done with frequency response techniques is contained in the book edited by Oldenburger (O3), which includes in addition to a few solicited articles, all the papers presented at the international symposium on the subject sponsored by the Instruments and Regulators Division of the American Society of Mechanical Engineers in 1953. The use of frequency response in process control problems is described in an introductory fashion by Johnson in a series of two articles (J2, J3). Ceaglske's book (C4) is an elementary exposition of frequency response principles for chemical engineers. More practical but with somewhat alien terminology, is the book of Young (Y1) which describes the methods developed and used by Imperial Chemical Industries in Great Britain. Also alien and more theoretical is the short but surprisingly comprehensive book by Farrington (F1).

All of the foregoing deal with automatic process control. The great bulk of books in the field, however, are concerned more with servomechanisms of one kind or another. Among such books which are helpful for automatic process control are Brown and Campbell (B4), Chestnut and Mayer (C5), and Draper *et al.* (D1), to mention but a few. More advanced texts are those of Truxal (T3) and Tsien (T4).

Despite the growing body of knowledge of process control theory most process plants are instrumented empirically today. Some useful empirical procedures are described in Section III, C, 3.

B. Analysis

The analysis problem in control can be divided into an analysis of individual components and an analysis of integrated loop behavior, the latter depending on the former. Typical process control systems may be regarded as combinations of process elements, measuring elements, controllers and computing elements, final control elements, and transmission lines. They may also be regarded as combinations of time lags, time delays, amplifiers, summers, differentiators, integrators, and other simple functional units.

In analyzing individual components, what is required is the relationship between signal inputs (forcings) and outputs (responses). In the analysis of a whole process control loop what is required is the behavior of the controlled process variable in response to the typical disturbances imposed on the system. Trimmer (T2) gives an interesting and general treatment of the response of physical systems.

1. Process Elements

Typical process elements are flow systems, heat exchangers, contacting systems for diffusional operations, and chemical reactors. Each of these elements has to some extent the properties of storing energy (including matter) and of amplifying or attenuating signals.

a. First Order Lag. Consider the simple jacketed kettle of Fig. 1. At constant feed rate and feed temperature, the relationship between outlet liquid temperature and jacket steam temperature is obtained readily by combining the overall rate equation and a heat balance on the liquid. For a well stirred kettle, the bulk liquid temperature in the kettle and the outlet temperature of the liquid are the same and

$$W_f C_{p_f} \frac{d\theta_o}{dt} = UA(\theta_i - \theta_o) \tag{1}$$

where W = mass
C_p = heat capacity at constant pressure
θ = temperature
U = overall heat transfer coefficient
A = heat transfer surface area related to U
d = differential operator
t = time

Subscripts:
f = liquid feed in the kettle
i = steam in kettle jacket
o = liquid leaving kettle

This equation is a description of the dynamical behavior of the kettle in that it relates temporally the input or forcing temperature of the steam jacket to the response or output temperature of the kettle contents. It is an oversimplification in that the heat capacity of the kettle wall has been neglected.

Since many equations are involved in a control system analysis, it is desirable that each equation be written as simply as possible. Operational calculus provides a useful notation, and in particular the Laplace transformation permits a very simple treatment if the differential equations are linear. A further simplification results if the same types of initial conditions are taken for all problems, or if only steady-state sinusoidal behavior is considered. Churchill (C6) and Carslaw and Jaeger (C2) give particularly useful presentations and applications of the Laplace transformation.

Fortunately, process control problems are most usually concerned with maintaining operating variables constant at particular values. Most disturbances to the process involve only small excursions of the process variables about their normal operating points with the result that the system behaves linearly regardless of how nonlinear the descriptive equations may be. Thus Eq. (1) is a nonlinear differential equation since both C_{p_f} and U are functions of θ_o; but for small changes in θ_o average values of C_{p_f} and U may be regarded as constants, and the equation becomes the simplest kind of first order linear differential equation.

When the Laplace transform of a differential equation is taken, a term must be included for the initial conditions, i.e., for the conditions at $t = 0$. This term is zero for the special case when the dependent variable and all its derivatives with respect to time are zero at $t = 0$. Now the typical process control situation is that the process variables are normally constant with time except for occasional disturbances which temporarily derange the system. Thus the period of time which is of especial interest in control analysis is the time from the start of a disturbance until the system returns to its normal controlled condition. At $t = 0$ for this period all the time derivatives of the dependent variable are zero since all conditions are steady, and for convenience the steady initial value of the variable can be taken as zero. Hence for control purposes the initial conditions terms in the Laplace transformation may be eliminated.

The Laplace transform of Eq. (1) then, is

$$W_f C_{p_f} s \theta_o(s) = UA(\theta_i(s) - \theta_o(s)) \tag{2}$$

where s = Laplacian complex variable and the $\theta(s)$ terms are transformed variables. This equation can be rearranged to give the ratio of output to input transforms,

$$\frac{\theta_o(s)}{\theta_i(s)} = \frac{1}{1 + Ts} \tag{3}$$

where $T = (W_f C_{p_f})/UA$, the *time constant* of this stage (element). The operational expression for the ratio of output to input, in this case $1/(1 + Ts)$, is called the *transfer function* of the stage.

Process elements which are describable by Eq. (3) are often called *first order time lags* or *first order RC stages*, since they cause the output signal to lag behind the input signal and since the time constant in each case is the product of a resistance term and a capacitance term. For the jacketed kettle, the resistance $R = 1/(UA)$ and the capacitance $C = W_f C_{p_f}$. Among the process elements which may behave like first order time lags are mixers, single stage contactors, liquid storage tanks, gas pressure reservoirs, and stirred reactors.

Although the transfer function gives a complete dynamic description of process elements, some interpretation is necessary. The response of the element to any kind of forcing function can be determined from the transfer function, but only a very few kinds of forcing are of any importance in control problems. These important forcing functions are: (a) step; (b) pulse; (c) ramp; (d) steady state sine wave; and (e) random.

Step Forcing. In step forcing, the input which has been constant for all $t < 0$, suddenly jumps to a new value at $t = 0$ and for all time thereafter remains at the new value. The response to a step function is called the *indicial response* or the *response to a constant* or the *transient response* or merely the *step response*. Since step forcing is as severe a kind of forcing as can be imposed on a process, there is considerable advantage in handling control problems in terms of the indicial response insofar as is practicable. Furthermore this response is useful in specifying criteria of control performance. It is disadvantageous, however, in that its computation requires a total solution of the differential equations describing the element. Such a solution becomes difficult and cumbersome for all real elements except the very simplest.

The step response of a first order lag may be obtained readily by substituting in Eq. (3) the transform of $\theta_i = A$, which is $\theta_i(s) = \dfrac{A}{s}$, where A is the magnitude of the step [see tables of transforms in references (B4), (C2), and (C6)].

Thus

$$\theta_o(s) = \frac{A}{s(1 + Ts)} \qquad (4)$$

and the solution in the time domain may be written directly from the same tables of transforms referred to previously. For the conditions that the process element is initially at rest, the response to a rise of A in the input signal expressed as a departure from the initial condition is

$$\theta_o = A(1 - e^{-\frac{t}{T}}) \qquad (5)$$

The important characteristics of this response are that the slope at $t = 0$ is A/T, and at $t = T$ the response is 63% of its final value. Large values of the time constant T correspond to sluggish response. Obviously, large time constants are undesirable in measuring instruments, whereas in process elements the sluggishness arising from large time constants may in special cases be advantageous in providing a stabilizing inertia in the control system.

Pulse Forcing. In pulse forcing, a shot of energy of some kind is put into the element. The simplest kind of pulse to visualize (but not the simplest to generate) is the *rectangular* pulse which consists of a succession of two step inputs of equal magnitude but opposite sign so that the input returns to its initial condition at the end of the pulse. Other pulses are the *cosine pulse,* which is half a cosine wave, and the *unit impulse* which has infinite magnitude but a time lapse such that the input is a unit of energy. Pulse responses are harder to interpret than step responses, but they can be obtained experimentally with less total upset to the system

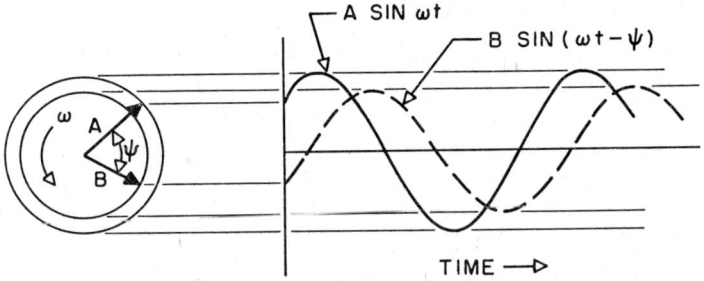

Fig. 3. Vector representation of sine waves.

than step responses. Lees and Hougen (L1) describe the successful use of a displaced cosine pulse to determine the dynamic characteristics of a heat exchange process.

Ramp Forcing. In ramp forcing the input changes at constant rate. This kind of forcing is of greater pertinence to servomechanisms (steering controls, for example) than to process systems.

Sinusoidal Forcing. From the standpoint of control analysis the most useful forcing function is the steady state since wave. If a steady-state, low amplitude sinusoidal variation is imposed on some property of an inlet process stream, the same property of the corresponding outlet stream will also vary sinusoidally and at the same frequency. For most process components the output wave will lag behind the input wave, and the output amplitude will be less.

The *magnitude ratio* (amplitude ratio, modulus) of the output wave to the input wave and the *phase angle* (argument) between the output and the input waves, both given as functions of frequency, constitute the frequency response characteristics of the component. Figure 3 gives a vector representation of the input and output sinusoids. The two sine waves may be regarded as being generated by the ordinate projections of the vectors as they rotate counterclockwise at an angular velocity of ω radians per unit time and displaced ψ radians with respect to each other. For these two waves the magnitude ratio is B/A and the phase angle is $-\psi$, the negative sign indicating that the output lags the input. If amplitudes are small, the amplitude ratios of a series of connected components are multiplicative and the phase angles are additive, providing there is no interaction between components. Thus the overall frequency response characteristics of a system of components can be determined readily from the frequency response characteristics of the individual components. The limitation regarding small amplitudes arises from the fact that the simple relationship between overall behavior and individual behavior is only valid for linear systems, i.e., for systems describable by linear differential equations. Fortunately, as has been intimated, although most real systems are nonlinear, their behavior for small perturbations is linear.

The frequency response characteristics of a process element or a group of elements can be computed readily from the corresponding transfer function merely by substituting $j\omega$ for s, where j is the imaginary number, $\sqrt{-1}$, and ω is the angular velocity. Thus the frequency response characteristics of a simple first order lag are given by

$$\frac{\theta_o(j\omega)}{\theta_i(j\omega)} = \frac{1}{1 + j\omega T} \tag{6}$$

Equation (6) may be split into real and imaginary terms by multiplying the numerator and denominator by $1 - j\omega T$ to get

$$\frac{\theta_o(j\omega)}{\theta_i(j\omega)} = \frac{1}{1 + \omega^2 T^2} - j\frac{T}{1 + \omega^2 T^2} \tag{7}$$

As shown in Fig. 4, this equation can be plotted for all frequencies on a complex plane having an imaginary vertical axis and a real horizontal axis. The result, which is a semicircle in the fourth quadrant, is the vector locus of the ratio of output to input. At any frequency ω the corresponding sine waves may be produced by rotating the input vector (magnitude 1.0) and the output vector counterclockwise at ω radians per unit time at constant ψ and taking the imaginary parts in a manner similar to that suggested in Fig. 3. Note that the magnitude ratio is merely the square root of the sum of the squares of the coordinates,

$$\left|\frac{\theta_o(j\omega)}{\theta_i(j\omega)}\right| = \frac{1}{(1 + \omega^2 T^2)^{1/2}} \tag{8}$$

and the phase angle is

$$\psi = -\tan^{-1} \omega T \tag{9}$$

Note further that the magnitude ratio is 1.0 at zero frequency and zero at infinite frequency and that the corresponding phase angles are zero and −90 degrees respectively.

Figures like Fig. 4, called Nyquist diagrams, are widely used in frequency response characterization.

Farrington (F1) hangs most of his penetrating treatment of control fundamentals on the *inverse Nyquist diagram*, which is a complex plane plot of the ratio of input to output. On this diagram the first order time lag is given by a straight vertical line through Imaginary = 0, Real = 1.0.

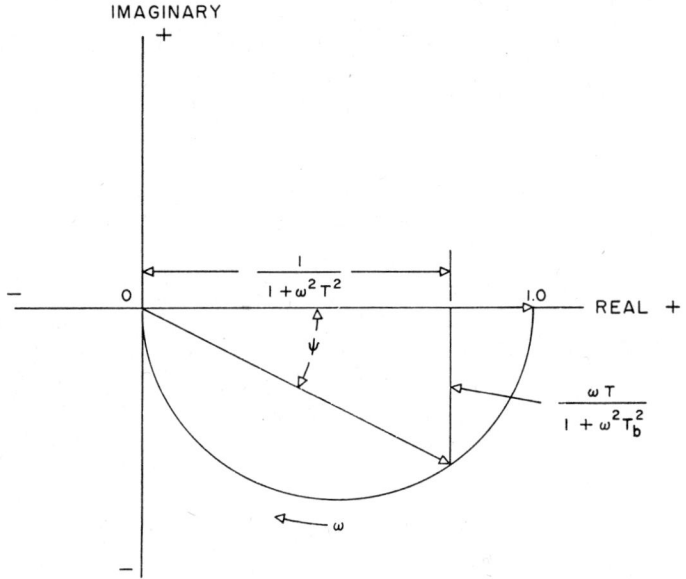

FIG. 4. Vector locus, output/input, first order lag.

A more convenient representation than either of the foregoing is the *Bode diagram*, in which the logarithm of the magnitude ratio and the numerical value of the phase angle, usually in degrees, are plotted against the logarithm of the frequency. The convenience of this plot lies in the fact that the frequency response characteristics of a series of coupled components can be determined readily by a simple graphical addition of the characteristics of the individual components. Furthermore for many kinds of loop components the characteristics can be approximated by a few straight lines. For example, Fig. 5 shows Bode diagrams for a single first order time lag and also for two lags in series. The straight line approximations, which are shown as dashed lines, are found readily from the following facts, which in turn are readily derivable from Eqs. (8) and (9):

(a) At low frequencies the magnitude ratio is unity, and at high frequencies it decreases with increasing frequency at a slope of $-n$ on a log-log plot where n is the order of the lag (or the number of first order lags in series).

(b) For a first-order time lag the high frequency approximation, with a slope of -1, intersects 1.0 magnitude ratio at a frequency of $1/T$

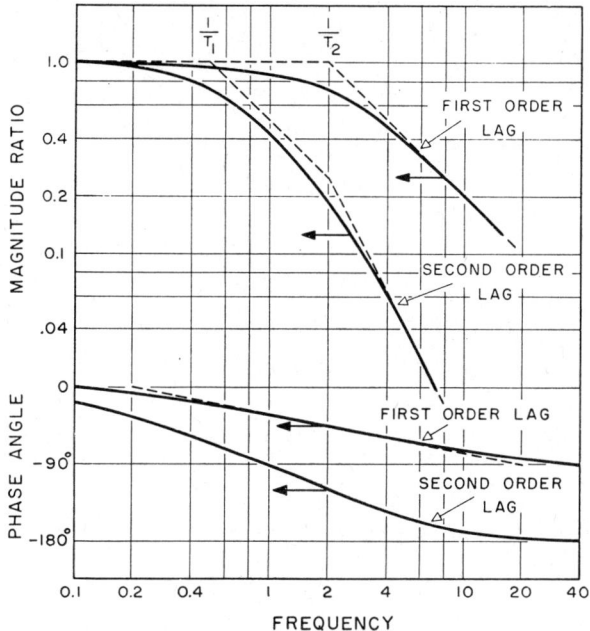

FIG. 5. Frequency response characteristics of time lags.

radians per unit time, where T is the time constant of the lag. This frequency is called the *corner frequency* or *break frequency*. The actual magnitude ratio at this frequency is $1/\sqrt{2} = 0.707$.

(c) For a series of first order lags in which the individual components are non-interacting, that is, in which the behavior downstream of a given lag has no effect on that lag, the break frequencies occur at each $1/T$ beginning at the one corresponding to the largest time lag (lowest frequency). At each break the straight line approximation increases in negative slope by one.

(d) The straight line approximation for the phase angle of a single, first order time lag is drawn through $-45°$ at $\omega = 1/T$ from 0 at $\omega = 1/(10T)$ and ending at $-90°$ at $\omega = 10/T$. At frequencies below

$1/(10T)$ the phase angle is zero, and above $10/T$ the phase angle is $-90°$. For a series of lags the individual straight-line approximations may be added graphically.

b. Series of First Order Lags. There are no real control systems so simple that they behave like a single first-order lag. Most systems behave like three or more lags coupled in series.

For three first order lags connected so that the output of the first lag becomes the input of the second lag and the output of the second becomes the input to the third, the overall transfer function relating the output of the third stage to the input to the first is

$$\frac{\theta_4(s)}{\theta_1(s)} = \frac{K_1}{1 + T_1 s} \cdot \frac{K_2}{1 + T_2 s} \cdot \frac{K_3}{1 + T_3 s} \tag{10}$$

where the individual $K/(1 + Ts)$ terms are the respective transfer functions of the lags. This equation may be derived simply by writing Eq. (3) for each stage and combining the resulting expressions algebraically. If the stages interact, that is, if what happens in stage 2 affects stage 1, etc., the individual time constants in Eq. (10) must be replaced by effective time constants which are related to the interacting resistances and capacitances (F1).

Interaction occurs in most systems of coupled stages. Thus heat transfer through a number of resistances in series may be treated as involving a series of interacting RC (resistance-capacitance) stages. An equation similar to Eq. (10) can be written for this system, and by equating the coefficients on like powers of s after expanding the equations, effective time constants can be determined which permit using the simpler equation of the non-interacting system.

Non-interacting, coupled stage systems are called *cascaded* systems. A simple example would be a series of liquid level tanks arranged so that the discharge of one is the feed to another and so that the flow from a given tank depends only on the liquid level in that tank. Systems of serial RC stages may be effectively non-interacting if the downstream capacitances are very small relative to upstream capacitances. Thus, for all practical purposes, a system composed of a small thermocouple immersed in a large bath would be non-interacting since the bath temperature would be insensitive to changes in the thermocouple temperature.

By substituting $j\omega$ for s in Eq. (10), the overall frequency response characteristics of a third order lag are found to be

$$\left|\frac{\theta_4(j\omega)}{\theta_1(j\omega)}\right| = \left|\frac{\theta_2(j\omega)}{\theta_1(j\omega)}\right| \cdot \left|\frac{\theta_3(j\omega)}{\theta_2(j\omega)}\right| \cdot \left|\frac{\theta_4(j\omega)}{\theta_3(j\omega)}\right| \tag{11}$$

for magnitude ratio and

$$\psi = \psi_1 + \psi_2 + \psi_3 \tag{12}$$

for phase angle. The convenience of the Bode diagram derives from these two equations and Eqs. (8) and (9). Note that on a Bode diagram at high frequencies the magnitude ratio of an n-th order lag decreases at a slope of $-n$ and the phase angle approaches $-90n$ degrees as a limit.

Random Forcing. All of the foregoing kinds of forcing whether step, pulse, or periodic are arbitrary functions. Yet in a typical plant or process under automatic control, the usual forcing function is a random one. Theoretically the output from a process component or an array of components must be related to the input via the transfer function of the component even though the input is purely random and the output seemingly random. Because of the high level of noise in the random signals it is not practical to attempt to relate input and output by performing Fourier analyses on input and output records. However the random input signals from a fairly long record, e.g., 2 minutes for a flow process, can be correlated statistically (*auto-correlation*), and a *cross correlation* for the same period of time can be obtained between input and output. The transfer operator relating the cross-correlation function and the auto-correlation function, is the same as the operator relating output and input in real time. Goodman and Reswick (G1) describe the application of their delay line synthesizer in determining the relation between input and output from auto-correlation and cross-correlation functions.

Up to the present time it has not been possible to demonstrate the ultimate reliability of characterizations based on random disturbances. However, the use of random disturbances offers great potential advantage in studying existing process control systems where upsets like step disturbances cannot be tolerated. Because of the extensive calculation required to reduce the random operating records to statistical-correlation functions, high speed digital computation is essential in this treatment.

c. Second-Order Elements. The transfer function of a second-order element may be written as

$$\frac{\theta_o(s)}{\theta_i(s)} = \frac{1}{As^2 + Bs + 1} \tag{13}$$

where the coefficients A and B depend on the nature of the element. Because some generalizations may be made regarding the behavior of second order elements, it is customary to write Eq. (13) in the form

$$\frac{\theta_o(s)}{\theta_i(s)} = \frac{1}{\frac{s^2}{\omega_n^2} + \frac{2\zeta}{\omega_n}s + 1} \tag{14}$$

where ω_n is the natural angular frequency of the element in radians per unit time and ζ is the damping ratio, dimensionless.

The two roots of the auxiliary equation of the differential equation corresponding to the above expression are

$$s = \omega_n[-\zeta \pm \sqrt{\zeta^2 - 1}] \qquad (15)$$

If the damping ratio, ζ, is zero, the roots are purely imaginary, and the response to a step disturbance would be a steady state cycling (*zero damped oscillation*) at a frequency of ω_n. If the damping ratio is unity, the element would be critically damped, and the transient response would be the swiftest possible recovery without overshoot. At damping ratios intermediate between 0 and 1.0, the transient response is a damped sinusoid, and at damping ratios greater than 1.0, the response is the aperiodic response of coupled RC stages.

d. Distributed Parameter Components. Systems which may be regarded as a series of time lags are *lumped parameter systems* in that the properties, such as resistance and capacitance which determine their dynamic characteristics, are presumed to operate at specific points in the system. The jacketed kettle, or a thermometer in a protecting well, or a compressed air system involving tanks separated by flow restrictions, may be treated adequately as lumped parameter systems. On the other hand, systems like a double-pipe heat exchanger, or a transmission line, or a tubular flow reactor, cannot be treated reliably as lumped systems, and consequently they are *distributed parameter systems*. Strictly speaking all systems are distributed and only under certain conditions may they be approximated as lumped systems.

There are two methods of tackling the distributed parameter case: (1) by arbitrarily dividing the system into n lumped parameter stages, where n is taken large enough (based on experience) to ensure a good approximation; and (2) by using operational calculus directly. Oldenbourg and Sartorius (O2) describe both the lumping approach and the direct approach. A straightforward general treatment of the direct approach is presented by Cohen and Johnson (C9) in an analysis of double-pipe heat exchangers. Takahashi (T1) has derived and tabulated transfer functions for a wide variety of heat exchange systems.

As an example of what is involved in the direct approach, consider a double-pipe heat exchanger in which condensing steam in the outer jacket heats up cold water flowing at constant mass rate in the inner pipe. It is desired to find the transfer function relating the temperature of the water leaving the inner pipe (as output) and the temperature of steam in the jacket (as input). Assuming for the purposes of this exposition that the steam pressure is constant throughout the length of the exchanger (but not necessarily constant with time), and that the metal walls have negligible thermal resistance, the combined enthalpy balances and heat transfer equations for a differential length of the exchanger are

$$\frac{\partial \theta_{fl}}{\partial t} = -u_{fl} \frac{\partial \theta_{fl}}{\partial x} + \frac{1}{T_1} (\theta_w - \theta_{fl}) \qquad (16)$$

and

$$\frac{\partial \theta_w}{\partial t} = \frac{1}{T_{22}} (\theta_{st} - \theta_w) - \frac{1}{T_{12}} (\theta_w - \theta_{fl}) \qquad (17)$$

where u_{fl} = fluid velocity
x = distance in the direction of fluid flow
$$T_1 = \frac{A_{fl} C_{p_{fl}} \rho_{fl}}{h_{fl} \pi D_i}$$
$$T_{12} = \frac{A_w C_{p_w} \rho_w}{h_{fl} \pi D_i}$$
$$T_{22} = \frac{A_w C_{p_w} \rho_w}{h_s \pi D_o}$$
A = cross-sectional area normal to flow
C_p = heat capacity at constant pressure
D = diameter
h = film coefficient for heat transfer at wall surface
ρ = density

Subscripts:

fl = fluid
i = inside
o = outside
st = steam (condensing)
w = wall

These equations are partial differential equations rather than ordinary differential equations, but through use of the Laplace transformation they may be converted to ordinary differential equations. Thus, Eq. (16) transformed is

$$s\theta_{fl}(s) = -u_{fl} \frac{d\theta_{fl}(s)}{dx} + \frac{1}{T_1} (\theta_w(s) - \theta_{fl}(s)) \qquad (18)$$

If Eqs. (16) and (17) are each written as the sum of a steady-state part and a transient part, and only the varying part is transformed, the resulting equations may be solved simultaneously to give

$$\theta_{fl,L}(s) = \left[\theta_{fl,0}(s) - \frac{b}{a} \theta_{st}(s) \right] e^{-\frac{L}{u_{fl}} a} + \frac{b}{a} \theta_{st}(s) \qquad (19)$$

where
$$a = s + \frac{1}{T_1} - \frac{T_{22}}{T_1(T_{12}T_{22}s + T_{12} + T_{22})}$$
$$b = \frac{T_{12}}{T_1(T_{12}T_{22}s + T_{12} + T_{22})}$$
L = length of exchanger

Subscripts:
  0 = point in exchanger where $x = 0$, i.e. inlet
  L = point in exchanger where $x = L$

For constant inlet water temperature ($\theta_{fl,0}$ = constant), Eq. (19) may be written

$$\frac{\theta_{fl,L}(s)}{\theta_{st}(s)} = \frac{b}{a}[1 - e^{-\frac{L}{u_{fl}}a}] \qquad (20)$$

This ratio is the transfer function between the outlet water temperature and the steam temperature in the jacket. For constant steam temperature, the transfer function between outlet water temperature and inlet water temperature is

$$\frac{\theta_{fl,L}(s)}{\theta_{fl,0}(s)} = e^{-\frac{L}{u_{fl}}a} \qquad (21)$$

As shown by Cohen and Johnson, Eq. (20) leads to frequency response characteristics which on a Bode diagram exhibit resonances both in the magnitude ratio and phase angle. The first resonance occurs at a period approximating the residence time of a slug of water in the inner pipe.

In general, distributed process systems are characterized by magnitude ratios and phase angles which decrease without limit as frequency increases. If the sinusoidal forcing is applied in a distributed manner, the magnitude ratios and phase angles decrease in a periodic or resonating manner.

e. Time Delays. Equation (21) describes a kind of *time delay* or *dead time* in that the effect of a change is not felt or observed until a finite time, L/u_{fl}, has elapsed. These time delays appear frequently in process systems as sampling lags in cases where the analytical or measuring means cannot be located near the process stream, and as distance-velocity lags in cases where the measuring point in the process flow stream must be located an appreciable distance from the point of real interest. The term *distance-velocity lag* is applied to delays for which the time constant or time lapse is given by the ratio of distance to velocity as in Eq. (21).

Consider a measuring point situated in a reactor effluent line a distance L from the reactor. If the fluid velocity in the line is u_{fl}, the time constant for the distance-velocity lag is $L/u_{fl} = T_D$, and the transfer function is

$$\mathcal{L}\left[\frac{\theta(t-T_\mathrm{D})}{\theta(t)}\right] = \frac{\theta_\mathrm{o}(s)}{\theta_\mathrm{i}(s)} \tag{22}$$

where $\mathcal{L}$ = Laplace transform, θ = measured variable, e.g., concentration or temperature; and subscripts o and i refer to outlet and inlet respectively, i.e. to $L = L$ and $L = 0$ respectively.

The transformation in Eq. (22) gives

$$\frac{\theta_\mathrm{o}(s)}{\theta_\mathrm{i}(s)} = e^{-T_\mathrm{D}s} \tag{23}$$

Substituting $s = j\omega$ gives the frequency response characteristics, which are that the magnitude ratio is unity for all frequencies, and the phase lag (negative phase angle) increases with increasing frequency without limit. Since the characteristic of unlimited phase angle promotes system instability, time delays are undesirable and should be minimized whenever possible.

f. Nonlinear Process Components. All the components discussed above have been treated as linear systems. Some components which are definitely nonlinear may also be treated as linear systems on the basis that controlled behavior necessarily results in minor fluctuations about mean operating points. No techniques are available for dealing with nonlinear elements generally, but there are two quite different methods which have had some success in dealing with nonlinear elements within certain limitations. These methods are described in Section III, B, 6.

2. *Measuring Elements*

Although the process elements are the heart of the process, they are only a part of the control system, neither more important nor less important than the other elements of the loop. Measuring elements are certainly as important as process elements, for no control is possible if there is no measure of what is going on in the process. In what follows, the emphasis will be on primary sensing elements, although strictly speaking the measuring elements in typical process control systems include both the sensing elements at the point of measurement and the receiving element in the controller.

The great majority of automatic process control systems involve one or more of only five process variables, namely, *pressure, temperature, flow rate, composition,* and *liquid level*. Many of these variables are measured by the same kind of instrument, and indeed, all of them under certain circumstances can be evaluated in terms of pressures. Thus temperature can be measured by the pressure exerted by a confined gas in the gas thermometer; the differential pressure across a restriction in a flow line is a measure of flow rate; the pressure exerted by a boiling liquid mixture at constant temperature is a measure of liquid composition; and the hydrostatic pressure at the bottom of a tank is a measure of the height of liquid in the tank.

A comprehensive summary of measuring elements appears in Perry (P1); and Eckman (E1) gives general descriptions of measuring instruments and their characteristics. *Industrial and Engineering Chemistry* devotes a monthly column to instrumentation, which deals primarily with new developments in the measurement of process variables.

The desired characteristics in measuring elements—including both the primary sensing element and secondary receiving elements—are accuracy and response speed consistent with low overall cost. Absolute accuracy is seldom a requirement, but it is important that the measuring means provide a reliable measure of the process variable. The speed of response should be high relative to the response speed of the process being monitored. A low overall cost requires that the first cost and also all associated operating and maintenance costs be low. What constitutes a low cost for any given problem will depend of course on a total economic appraisal.

Some of the more important measuring methods are catalogued briefly below:

a. Pressure. Pressures are most usually measured by balancing the pressure against a column of liquid as in the *manometer* or against a *pressure spring* of some sort such as a *bourdon tube, diaphragm, bellows,* or *helix.* Where it is desirable to have an electrical signal as the measure of pressure, the *strain gage* and differential transformer is useful.

Manometers and pressure springs may be described dynamically to a first approximation by second-order differential equations for which the roots of the characteristic equation are conjugate complex. As shown in Section III, 8, 1c, since the roots are complex, these systems have an oscillatory mode, and the response of the system to step forcing, for example, is a damped sinusoid.

b. Temperature. Temperature is perhaps the most widely measured process variable, yet it is in principle the most difficult to measure since it cannot be measured directly.

The three most important types of thermometers are expansion-type thermometers (pressure thermometers), electrical thermometers, and radiation thermometers. In expansion-type thermometers the primary sensing element is a bulb containing an expansible fluid. The bulb is connected to a pressure spring through capillary tubing. Expansion of the thermometric fluid with rising temperature causes expansion of the pressure spring, which in turn is converted to a mechanical displacement as the final measure of temperature. The response of these thermometers is determined by the dynamics of the sensing bulb and usually can be regarded as involving a pair of coupled RC stages: two resistances given by the external and internal fluid films, and two capacitances given by the bulb wall and the bulb fluid. Transmission lags in the capillary tubing and the behavior of the pressure spring are of secondary importance.

It is frequently desirable to protect the sensing bulb by sheathing it in a thermowell with the result that at least another RC stage is added to the measuring system and the response becomes more sluggish.

Because of the ease with which electric signals can be transmitted and manipulated it is not surprising to find that electrical thermometers are the most widely used thermometers in control systems. Both the thermoelectric type thermometer (thermocouple) and the resistance thermometer can be made small, and hence quite high response speeds are realizable. Even sheathed couples can be obtained which have time constants as low as three seconds. In these instruments the hot junction of the couple is welded directly to the sheath.

c. Flow. Most process flow streams are metered for continuous automatic control by orifice meters. The pressure drop across the orifice is sensed either by an enlarged leg manometer or by a pneumatic differential pressure cell. In both cases the response is rapid, usually far more rapid than is required for typical flow control problems.

d. Liquid Level. Liquid level can be measured directly by floats or indirectly by measuring hydrostatic head or by measuring the buoyancy of a submerged mass (displacement float). Thus the problem is essentially one of measuring either a mechanical displacement or a pressure.

e. Composition. There are many ways of monitoring composition including measuring properties such as density, viscosity, refractive index, thermal conductivity, absorption and emission spectra, dielectric constant, and the like. An increasingly popular method is mass spectrometry, despite greater expense than conventional methods.

Most of these methods are rapid in their response to the particular property which is the measure of composition, but frequently their overall behavior is sluggish because of the time delays in leading the stream samples to the test cells.

In some cases, for example where mass spectrometers are used in analyzing complex mixtures, high speed computing equipment is necessary for interpretation, since a large number of simultaneous equations must be solved. The total time lapse for such cases may run into minutes.

3. *Controllers*

The signal from the measuring elements in control systems goes to the controller, where it is compared with some measure of the command signal to the controller. In process control terminology the command signal is called the *set point*. Related but differing quantities are the *desired value* and the *control point*. The former is the value at which it is desired to hold the process variable being controlled; the latter is the point at which the variable is actually controlled.

Most usually the controller takes the difference between the set point and the controlled variable and on the basis of this difference or *deviation*, generates a control action. The relationship between control action and deviation provides a convenient means of classifying controllers.

In process control only a few types of control action (control modes) are important, namely: (1) on-off or two-position control; (2) proportional control; (3) integral control or automatic reset; (4) derivative or rate action.

With the exception of derivative action any of these control modes may be used alone in certain applications. Integral and derivative actions are most usually combined with proportional control to give *proportional plus integral control (proportional control with automatic reset); proportional plus derivative control;* or *three-mode control*, which is *proportional plus integral plus derivative*.

These various control actions may be generated electrically, mechanically, pneumatically, or hydraulically. Pneumatic and electrical controllers are most widely used. The former are often preferred in petroleum refining and similar industries where fire hazards make electrical devices dangerous unless specially protected. Present trends are toward increased use of electrical controls, particularly where signals must be transmitted more than 200 ft.

a. Two-Position Control. The least expensive and most widely used mode is two-position or on-off control. As the name implies the controller output is one value for all positive deviations from the set point and another value for all negative deviations. Usually these output values are the extremes of full output and zero output. Since this type of control is discontinuous and produces cycling in the controlled variable, it is most applicable to stable systems characterized by large energy capacities and small load changes. Household heating systems and constant temperature baths are familiar examples.

A mathematical treatment of discontinuous automatic control systems with special attention to high-speed position control problems is given by Flügge-Lotz (F2).

To avoid excessive wear due to too frequent action of the on-off controller, it is customary to provide a fixed dead zone or differential around the set point of perhaps 3% of full variable range. The effect of the dead zone is to decrease the frequency of cycling and to increase the phase lag between controller output and input to slightly more than 180°.

b. *Proportional Control.* In proportional control the controller output is proportional to the deviation, that is

$$\theta_o - \theta_a = K_p(\theta_r - \theta_c) \qquad (24)$$

where θ_o = controller output
θ_a = controller output for zero deviation
K_p = controller proportional gain
θ_c = measured variable as seen by the controller, per cent of full scale
θ_r = set point, per cent full scale

Since the controller output must counteract the measured variable, these two quantities, θ_o and θ_c, are of opposite sign and hence are inherently 180° out of phase. In commercial proportional controllers this 180° phase shift and also any set gain K_p, are constant for all practical ranges of frequency. Thus the frequency response characteristics of a proportional controller are a magnitude ratio of K_p and a phase lag of 180°.

Instrument manufacturers sometimes present the scales on proportional action as *per cent proportional band* which is $100/K_p$.

Proportional control is the basic control mode. Indeed on-off control may be regarded as a limiting case of proportional control at very high K_p. Proportional control produces an immediate and opposing particular output for every particular deviation. It has the disadvantage that it will tolerate an *offset*, i.e., a steady-state error or sustained deviation, since in general, the particular deviation at which the corresponding controller output balances the process, will not be zero. The offset produced by a given load change can be shown to be equal to $1/(1 + K_p)$ times the deviation that would result if no control action were taken to correct for the load change. This computation is valid only if K_p is expressed in terms of the potential corrections resulting from the actual controller outputs.

It is clear from the above expression that offset can be minimized with proportional controllers by setting the proportional gain at high values. At high gains, however, control systems become increasingly unstable, and for some systems instabilities occur at relatively low gains, so that proportional control alone is unsatisfactory. In order to eliminate offset without promoting system instability, an additional mode of control such as integral control must be used.

c. *Integral Control.* Integral control has also been called *proportional speed floating control*. The former name derives from the fact that the controller output is proportional to the time integral of the deviation;

the latter from the fact that the control action is always changing, hence floating.

In differential form the expression for integral control is

$$\frac{d\theta_o}{dt} = \frac{1}{T_{int}} (\theta_r - \theta_c) \qquad (25)$$

where T_{int} is the integral time (reciprocal reset rate). Because of the floating action the integral controller takes increasing action so long as there is any deviation, hence offset is impossible.

From the Laplace transformation of Eq. (25), the transfer function for an integral controller is $1/(sT_{int})$ and by substituting $j\omega$ for s the corresponding frequency response characteristics are found to be a phase angle of $-90°$ and a magnitude ratio of $1/\omega T_{int}$.

d. Derivative Action. Offset may also be reduced by adding a stabilizing control mode which permits the use of higher gains in the proportional mode. Such a stabilizing mode is derivative action, or *rate* action as it is sometimes called.

In derivative control, the controller output is proportional to the rate of change of the input with time, or in differential form

$$\frac{d\theta_o}{dt} = T_d \frac{d^2}{dt^2} (\theta_r - \theta_c) \qquad (26)$$

where T_d = derivative time.

Unlike integral control this action by itself provides no control, since it gives a sustained controller output for a constantly changing input. The output does not depend on the amount of deviation but only on how fast the deviation is changing.

Derivative action is used occasionally with proportional control alone, but more frequently it is used with proportional plus integral control.

The transfer function for derivative control is sT_d, and the frequency response characteristics are a phase angle of $+90°$ and an amplitude ratio of ωT_d. Control stability results from the leading phase angle.

e. Proportional Plus Integral Control. The most important control mode combinations are proportional plus integral and three-mode control. Proportional plus derivative control finds occasional use, but the combination of integral plus derivative control has limited practical application because at certain frequencies the two actions cancel each other and no control results.

Proportional plus integral control (also called proportional plus reset) has the transfer function

$$\frac{\theta_o(s)}{(\theta_r - \theta_c)(s)} = K_p \left(1 + \frac{1}{sT_{int}}\right) \qquad (27)$$

The frequency response characteristics are sketched in Fig. 6 at low frequencies. Note that the integral action provides infinite gain at zero frequency.

f. Three-Mode Control. The transfer function for ideal three-mode control is

$$\frac{\theta_o(s)}{(\theta_r - \theta_c)(s)} = K_p \left(1 + \frac{1}{sT_{\text{int}}} + sT_d\right) \tag{28}$$

Although proportional plus integral controllers follow quite closely the idealized behavior of Eq. (27), most commercially available three-mode

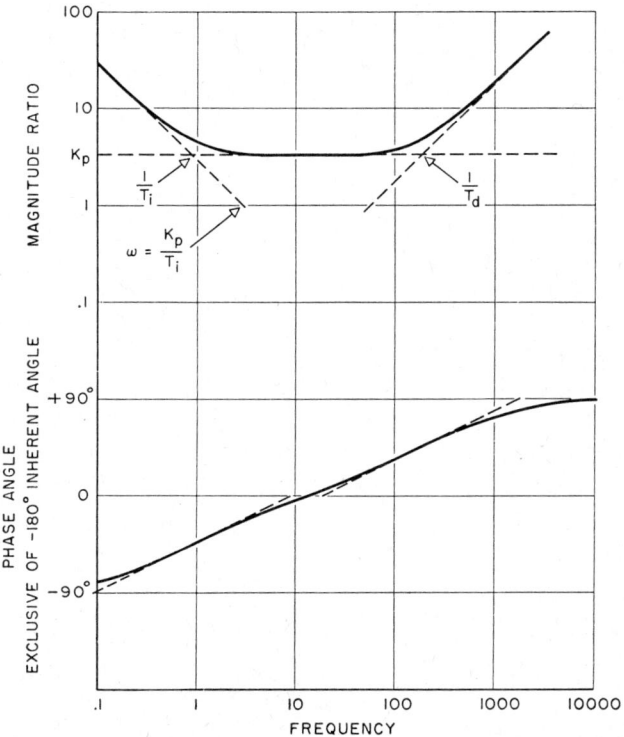

FIG. 6. Frequency response characteristics of three-mode controller.

controllers only very approximately follow Eq. (28). The reason for this departure from ideal behavior is that the derivative action cannot readily be generated independently of the other actions, and particularly in pneumatic controllers the resulting interaction alters the overall performance of the controller. Typical equations for various commercial three-mode controllers are given by Farrington (F1) and Young (Y1).

Figure 6 gives the frequency response characteristics of ideal three-mode control. The chief contributions of the derivative action are to decrease the phase lag of the controller and provide high gain at high frequencies.

It can be seen from the ideal equation that the reset rate, $1/T_{int}$, is the number of times per minute that the integral action repeats the proportional action, and the rate time, T_d, is the time that the derivative mode advances the control action over that of the proportional mode alone.

4. *Final Control Element*

The controller output becomes the input to the final control element or *regulating unit* as it is often called.

a. Valve Types. For most process control problems the final control element is a valve. Depending on whether the controller output is an air pressure or a voltage the drive for the valve will be a diaphragm motor or a reversing electric motor (or for on-off control, a solenoid).

The most widely used control valves for liquid and high pressure gas streams are the various sliding stem valves such as the *bevel plug, V-port, parabolic plug,* and for small flows the *needle* valve. For corrosive materials and slurries, pinch-type valves like the *Saunders patent* valves are used. Rotary stem valves include plug types and for gas flows *butterfly valves, dampers,* and *louvers.*

The bevel plug valve, which is essentially a globe valve, is used with on-off and narrow band proportional control. For more difficult control systems, *characterized valves* like the V-port and parabolic plug are preferred because the flow-lift (valve position) behavior is more uniform over the full range of the valve.

Where there are large pressure drops across the valve, *double-seated* valves are used to balance the pressures on the plug and ensure easy operation. To avoid erratic control action due to valve sticking the controller output may be applied to a *valve positioner,* which is a servomechanism (position controller) designed to fix the valve position as required by the controller.

b. Valve Characteristics. In characterizing valves for control purposes it is necessary to examine both the valve itself and the valve motor or actuator. Valves are usually characterized by their manufacturers in terms of the relationship between valve stem position (lift) and flow through the valve at constant pressure drop. Thus a *linear valve* is one for which the flow at constant pressure drop is linearly dependent on the lift. A *semilogarithmic* characteristic produces a straight line when the logarithm of the flow is plotted against lift. The latter characteristic is also called *equal-percentage*, since equal increments of valve lift produce equal percentage flow changes (percentage of previous flow). Equal-percentage valves are desirable in control circuits which are subject to large load changes because precise adjustment of flow rate is possible whether the valve is almost fully open or almost fully closed. With linear valves on the other hand, a given shift in valve position will cause a very high percentage change in flow rate at low flows and negligible percentage change at high flows.

For many process control applications it is possible that the pressure drop across the valve will have more effect on flow rate than will the valve position. The relationship between pressure drop and flow rate at constant valve position is the square-root expression for flow through an orifice. This relation introduces a nonlinearity in the control system which complicates the analysis of the system except insofar as it can be linearized. For equal-percentage valves the effect of changing pressure drop in the lift-flow characteristic of the valve is small, but for linear valves the effect is large.

The selection of valve size is something of an art in that a compromise is usually required. In general, the valve should be small enough to take up most of the pressure drop in the supply line but large enough to pass the maximum possible control agent flow that might be needed by the process. An incorrectly sized linear valve can seriously affect the stability of the control loop, whereas the size of an equal-percentage valve has little effect on stability.

Flow ranges of control valves frequently are characterized as *rangeability*, which is the ratio of maximum to minimum flow rates, and *turndown*, which is the ratio of normal maximum operating flow rate to minimum flow rates. The minimum flow rate for control valves is usually not zero to preclude hysteresis effects that might result from total closing.

c. Valve Actuators. Valve motors in pneumatic systems are most usually spring-loaded diaphragms. Their behavior can be characterized by second-order equations, but in most process control systems their response is sufficiently rapid to permit characterizing them as simple first-order lags.

Electric motors may be variable-speed, reversible motors or simple solenoids (for on-off control). Because the former are expensive, it is a common practice to transduce the electric signals to pneumatic signals and use inexpensive diaphragm valves.

Valve actuators may be classified broadly as *directional*, such as solenoids and electric motors; *force-balanced*, such as spring-loaded diaphragms, and *position-balanced*, such as valve positioners. On the basis of this classification, Close (C7) describes the important types of actuators in commercial use, their dynamic characteristics, and how they are integrated into process systems.

5. Transmission Lines

Transmission lines in process control systems rarely make any significant contribution to the overall loop characteristics. Where signals are transmitted electrically, there is no detectable signal attenuation for any frequencies characteristic of the process components, and even for pneumatic transmission lines as long as 200 feet there is little loss of signal. Transmission lines have distributed properties, but according to Bradner (B3) who has studied pneumatic transmission lines extensively, they can be approximated as second-order systems.

6. Loop Behavior

a. Linear Systems. If the dynamic characteristics of all the components in the control system are known, it is possible in principle to predict how the whole system will behave. Although the closed-loop can be analyzed by formal solution of the system equations or by means of the Nyquist diagram or the inverse Nyquist diagram (F1) or by the root-locus method of Evans (E2), for process control systems it is usually more convenient to use the Bode diagram, particularly for systems involving distance-velocity lags.

Consider the simple control loop of Fig. 2. The open-loop characteristics for a typical system exclusive of the controller, are a magnitude ratio and phase angle which fall off with increasing frequency. When the loop is closed through a controller, the magnitude ratio of the closed loop has a resonance peak if the controller gain is sufficiently great.

If the transfer function for all the components in the process control loop lumped together exclusive of the controller is $G(s)$, and the transfer function of the controller (exclusive of the summer which takes the difference between set point and measured variable) is $G_c(s)$, the transfer function of the closed loop is

$$\frac{\theta_o(s)}{\theta_r(s)} = \frac{G_c(s)G(s)}{1 + G_c(s)G(s)} \tag{29}$$

The closed-loop frequency response characteristics may be obtained by replotting the open-loop characteristics on a graph having contours of constant closed-loop magnitude ratios and angles. Points for the closed-loop characteristics are obtained from the intersections of the open-loop locus and the contours. Either polar diagrams such as the transfer function plot (Nyquist diagram) and the inverse Nyquist diagram, or a plot of log magnitude versus phase (Nichols or Black chart) (B4) may be used.

It can be shown by a simple algebraic manipulation that the loci on a Nyquist diagram of constant closed-loop magnitude ratio, M, are eccentric circles of radius $M/(1 - M^2)$ having centers along the real axis at $-M^2/(M^2 - 1)$ from the origin. These M circles are for the closed-loop magnitude ratio of controlled variable to set point, corresponding to the transfer function given by Eq. (29).

In an analysis of the closed-loop behavior of an automatic process control system three items are of particular interest: (a) How stable is the system in recovering from upsets? (b) How quickly does it recover? (c) What is its steady-state deviation after recovery?

The question of stability can be considered most readily from the Bode diagram for the open-loop characteristics of the whole system excluding the controller. Because of the summing mechanism in the controller which gages the deviation, there is inherent in any controller a basic phase lag of 180°. Taking the deviation involves reversing the sign of the signal going around the loop, and this reversal corresponds to a phase shift of exactly 180°.

If a sinusoidal input signal of sufficient frequency is impressed on the process components in the loop so that the corresponding output signal from the process (input to the controller summer) lags by 180°, the addition of the controller to the loop will bring the total phase shift to 360° and the signals at this particular frequency will be in phase around the loop. If the gain in the controller is just sufficient to offset the attenuation in the process, the signals will cycle around the loop continuously without abatement. Higher gains in the controller lead to instability and runaway, while lower gains result in damped response. Thus the frequency at which the process phase lag is 180°, is the critical frequency or *ultimate frequency* of the system. Similarly the proportional control gain which just produces steady-state sinusoidal oscillation is the *ultimate proportional gain*. At this controller gain, the total loop gain, which is the product of controller and overall process gains, is unity. A step disturbance anywhere in the loop will cause sustained oscillations of the controlled variable at the ultimate frequency since the step disturbance excites all frequencies.

The ultimate frequency response properties of a control loop are given special names. *Gain crossover* is the point at which the overall open-loop magnitude ratio is unity. *Phase crossover* is the point at which the overall open-loop phase angle is $-360°$. Since the controller has an inherent phase angle of $-180°$, it is a common practice to disregard this angle and handle the analysis on the basis of the overall open-loop characteristics exclusive of the additional 180° lag. Thus, *phase crossover* occurs at the frequency for which the total phase angle of the open-loop system—exclusive of the −180° inherent shift in the controller—is −180°, i.e., at the ultimate frequency. *Phase margin* is the difference between the phase angle at gain crossover (exclusive of the −180° in the controller) and −180°. For example, if the phase angle at gain crossover is −120°, the phase margin is 60°. Phase margin may be regarded simply as the margin of stability in terms of phase angle that the system has at the critical condition of unity magnitude ratio. Similarly, *gain margin* is the measure of stability in terms of magnitude ratio that the system has at the critical condition of ultimate frequency. Gain margin is expressed as a factor. Thus for a control loop with an overall magnitude ratio of 0.4 at phase cross-over, the gain margin is 1/0.4 or 2.5.

Control loops with gain margins exceeding 2.0 and phase margins not less than 30° are reasonably stable, although for some cases somewhat higher minima may be desirable, At these margins the loop response to disturbances usually will be oscillatory with fairly rapid damping.

How much oscillation and how much damping there will be can be determined reliably only by solving the descriptive equations of the system for the case of a typical disturbance. Since such solutions are tedious at best and impossible at worst, it is desirable to make use of approximations derived from the frequency response characteristics insofar as practicable. It is generally true for simple systems, that the initial overshoot of the controlled variable in response to a step change will be approximately equal to the resonance peak of the corresponding closed-loop magnitude ratio, provided the peak is about 1.2–1.3. That is, a 20–30% overshoot (based on the amount of the step change) will occur if the peak in the closed-loop magnitude ratio is 1.2–1.3. For higher magnitude ratios the peak overshoot will be higher but not proportional to the closed loop peak. Furthermore the frequency of the damped oscillation will approximate the resonance peak frequency, and for typical stable gain margin and phase margin, e.g., 2.5 and 45° respectively, the subsidence ratio (ratio of a succeeding peak to the peak of the preceding cycle) will approximate one-third.

From the foregoing it is possible to sketch in the transient response except for the final steady value of the controlled variable. The speed of recovery is most conveniently expressed in terms of the *settling time*, which is the time required to come within some limit of the final steady value, usually within 5%. A system with a 1/3 subsidence ratio would have a settling time for 5% limit of roughly two times the period of oscillation of the transient.

The steady-state error of the system is the deviation at infinite time after the disturbance. For a unit change in set-point the error is

$$\left| \frac{\theta_r - \theta_c}{\theta_r} (j\omega) \right|_{\omega=0} = 1 - \left| \frac{\theta_c(j\omega)}{\theta_r(j\omega)} \right|_{\omega=0} \tag{30}$$

But from Eq. (29)

$$\left| \frac{\theta_c(j\omega)}{\theta_r(j\omega)} \right|_{\omega=0} = \left| \frac{G_c(j\omega)G(j\omega)}{1 + G_c(j\omega)G(j\omega)} \right|_{\omega=0} \tag{31}$$

and since all θ are actual or potential values of the controlled variable, $|G(j\omega)|_{\omega=0} = 1.0$ and therefore

$$\left| \frac{\theta_c(j\omega)}{\theta_r(j\omega)} \right|_{\omega=0} = \left| \frac{G_c(j\omega)}{1 + G_c(j\omega)} \right|_{\omega=0} \tag{32}$$

With proportional control $G_c(j\omega) = K_p$ and $|G_c(j\omega)| = K_p$ so that Eq. (32) becomes

$$\left| \frac{\theta_c(j\omega)}{\theta_r(j\omega)} \right|_{\omega=0} = \frac{K_p}{1 + K_p}$$

and Eq. (30) becomes

$$\left| \frac{\theta_r - \theta_c}{\theta_r} (j\omega) \right|_{\omega=0} = \frac{1}{1 + K_p} \tag{33}$$

as stated in Section III, B, 3b.

With the addition of integral control to give proportional plus reset control, $G_c(j\omega) = K_p \left(1 + \frac{1}{j\omega T_{int}} \right)$ and the magnitude ratio at zero frequency becomes infinite with the result that $\left| \frac{\theta_r - \theta_c}{\theta_r} (j\omega) \right|_{\omega=0} = 0$ and no sustained deviation is possible.

The same result obtains with three-mode control. On the other hand the combination of proportional and derivative control gives the same steady state error as proportional alone, since the derivative contribution disappears at low frequency.

b. *Nonlinear Systems.* The foregoing discussion of closed loop behavior has been limited for the most part to linear systems. For such systems the problem of analysis is relatively simple because frequency response techniques are applicable, and the individual component characteristics can be combined readily to give the overall system characteristics.

With nonlinear systems, however, all simplicity disappears. No general methods of solving even the simplest, nonlinear differential equations are known. Frequency response characterization is useless since sinusoidal forcing will not produce sinusoidal response. The only recourse other than arbitrary linearization of the equations is to utilize approximation methods like those of describing function analysis and phase-plane analysis.

The describing function, first suggested in this country by Kochenburger (K1), is merely the ratio of the fundamental component of the output of the nonlinear element to the amplitude of the sinusoidal input. If there is only one nonlinear element in the control system, all harmonics higher than the fundamental will be attenuated in the linear parts of the system, and the describing function may be used as the frequency response characterization of the nonlinear element. The only other restriction is that any analyses based on frequency response must be at particular signal levels in the nonlinear element.

A serious limitation on describing function analysis is that there is no means of gaging its reliability. For example, it might be expected that describing function analysis would be particularly useful in identifying the existence of limit cycles in nonlinear systems. (Limit cycles are the conditions of sustained oscillation that obtain in nonlinear systems. The familiar condition of ultimate frequency and ultimate proportional gain in a process control system, represents a limit cycle since it is not possible to adjust the system exactly to the ultimate gain and as a result a truly linear system would run away.) Yet there are frequent examples of cases where describing function analysis has failed to indicate limit cycles which actually existed (T3).

In general, the transient response of stable nonlinear systems cannot be inferred from the frequency response characteristics derived from describing function analysis.

Phase-plane analysis of nonlinear systems, suggested by MacColl (M1), is based on the characteristics of the equation

$$\frac{d^2\theta}{dt^2} + A\left(\theta, \frac{d\theta}{dt}\right)\frac{d\theta}{dt} + B\left(\theta, \frac{d\theta}{dt}\right)\theta = 0 \tag{34}$$

The phase-plane representation is a plot of $d\theta/dt$ vs. θ as families of curves for a given system with initial conditions $d\theta/dt(0)$ and $\theta(0)$ as parameters. This plot or phase portrait, provides a useful indication of the transient response of a nonlinear system. It cannot be applied to sinusoidal or other continuing forcing functions. Furthermore, the method is limited to second-order systems or systems that can be handled as second-order systems.

A succinct appraisal of phase-plane analysis, its possibilities and limitations, is given by Truxal (T3).

c. *Process Systems*. In recent years a number of process control systems have been studied theoretically, experimentally, and by analogy. No attempt will be made here to list all these studies, but a few of the more quantitative studies will be cited by way of illustration.

Williams *et al.* (W1), describe the results of studies of the automatic control of continuous fractional distillation. These studies were made on an analog computer which could simulate a five-plate tower. The effects of column design, varying feed rate, imperfect sampling, and quality of feed and reflux on controllability were evaluated. An earlier article by Rose and Williams (R2) on the same system compares various schemes for controlling fractionation columns. One interesting conclusion is that derivative control action cannot improve the control for any of the various combinations of measurement and regulation that were studied.

Bilous and Amundson (B1, B2), also using an analog computer, examined the stability and sensitivity of continuous, stirred reactors and tubular reactors. For the former type of reactor the methods of small perturbations used in nonlinear mechanics are directly applicable, but for the latter the partial differential equations necessary for the basic description make a rigorous solution impossible. Both the methods of small perturbations and of frequency response were used in the study. For large perturbations, the solutions for the stirred reactor were obtained on the analog computer. It was concluded that the natural behavior of stirred reactors is not to approach unstable states. The parametric sensitivity, i.e., the sensitivity to changes in operating conditions, of tubular reactors was predicted semi-quantitatively from frequency responses and transient responses approximated by linearizations about fixed operating points.

In one of four articles on the design of gas flow systems Schwent *et al.* (S2), present a theoretical analysis of a wind tunnel installation and a comparison of theoretical and observed performance. All possible simplifying assumptions and approximations are used to advantage.

d. Sampled-Data Systems. The analysis up to this point has considered systems in which the flow of signals is continuous. Virtually all process systems involve continuous measurement and recording of process variables and continuous regulation and control of processes. However, there are a few process systems in which the controlled variable is measured intermittently. In some fractional distillation operations, for example, an expensive process stream analyzer may be shared between two separate units, analyzing first one unit stream and then the other. Such systems are called sampled-data systems, because the flow of signals is carried in samples at regular time intervals. Because of the time lapse between signals, the sampled-data systems are potentially less stable than the corresponding continuous systems.

Fortunately, frequency response techniques are applicable with slight modification to sampled-data systems. Ragazzini and Zadeh (R1) describe the analysis of these systems. Applications in process control with large distance-velocity lags are considered by Oldenbourg (O1) and Sartorius (S1).

7. *Complex Systems*

Complex, multiple-loop, process control systems do not pose any new problems of analysis. The closed-loop characteristics of the smaller loops become the individual component characteristics of successively larger loops.

8. *Effect of Disturbance Location*

Although it is common practice to analyze control loops in terms of the response of the controlled variable to changes in set point, the usual disturbances in process control systems occur at various points in the process rather than at points in the controlling instruments. No special techniques of analysis are required to determine the response of the controlled variable to disturbances applied anywhere in the loop. It is only necessary to manipulate the component transfer functions algebraically, until the ratio of controlled variable to disturbance is found.

Little is known about the nature of process disturbances, so some arbitrary upset like a step change must be used in the analysis. In general, the resulting upset in the controlled variable will be more severe, the closer the point of disturbance is to the point of measurement of the controlled variable. For the usual types of process components, the effect of interposing components between the disturbance and the controlled variable is to attenuate the disturbance. A convenient control system property, which gives a clear picture of the effect of disturbance location, is the *deviation reduction factor*. This factor is the ratio of the change in the controlled variable when the system is disturbed to the change that would have resulted had there been no control action. European engineers (J1, Y1) have made greater use of this quantity than have American engineers.

C. Synthesis

1. *Criteria of Control*

Before an engineer can undertake a design he must have a clear picture of the purpose of the design. In the case of process control systems, the ultimate objective is an operating process plant which will maximize the return on the allowed investment. A more narrow specification might be to design an operating process plant of given capacity and maximized return on investment. An even narrower specification but one more typical of the kind given to the process control system designer, would be to design a plant of a given capacity with an operating performance specified as to tolerable limits of critical process variables.

Since process plants most usually operate continuously under steady-state conditions, the criteria for good control may be specified realistically in terms of the allowable behavior of the process variables following typical upsets in operations. Hence the transient response of process variables to step changes in loads or step changes in set point provides a convenient basis for fixing control performance specifications.

For some processes no cycling of the controlled variable is desirable, for example, because the oscillations might be amplified in process units downstream. In other processes, continuous cycling is not objectionable so long as the controlled variable always lies within a specified range. In still other processes a temporary oscillation which damps out quickly may be required.

For some processes, it is not necessary for the controlled variable to line out right on the set point. In most cases, however, little if any offset is permissible.

A generally used set of criteria for good control is that the controlled variable in response to a unit step change in set point (a) overshoot by not more than 20 per cent of the step and (b) damp out with a subsidence ratio of about one-third. This behavior is approximated by many systems if the closed-loop frequency response and the corresponding open-loop frequency response have certain simple characteristics. Since the closed-loop frequency response characteristics can be determined readily from the open-loop frequency response, the latter characteristics of simple control systems can be used as a convenient basis for design.

2. *Control System Synthesis from Frequency Response Characteristics*

Generally recommended frequency response characteristics for automatic control systems are given by Oldenburger (O3) as follows:

(a) Phase margin should be at least 30 degrees and the gain margin at least 2.5.

(b) Maximum closed-loop magnitude ratio should be less than 2.0, and preferably about 1.3. The resonant frequency at the peak magnitude ratio should be as high as possible.

In the absence of other specifications, the problem of control system synthesis is one of adjusting the characteristics of the various loop components until the recommended frequency response characteristics are obtained. All the control components including measuring elements should have negligible lagging and attenuation characteristics over the range of frequencies which are of importance for the control system. The process components should have as little phase lag as possible. A large attenuation in the process magnitude ratios, however, is not objectionable and in many cases makes for simple and precise control.

Minimizing the overall phase angle of a typical process system, maximizes the frequency at phase crossover and hence maximizes the response speed of the system. Large phase lags in process components, for example resulting from distance-velocity lags, can be counteracted by the phase lead characteristic of derivative control action.

To determine the controller gain required for a particular closed-loop resonance peak in magnitude ratio a simple graphical proportionality can be applied on the Nyquist diagram. The procedure is to draw the open-loop vector locus of the ratio of the controlled variable to the command for some arbitrary controller gain, e.g., one equivalent to a gain margin of 2.5. If the desired resonance peak in the corresponding closed-loop magnitude ratio is 1.3, draw in the $M = 1.3$ circle. Estimate where the open-loop locus would be just tangent to the M circle. From the point of expected tangency, draw a vector to the origin through the arbitrary open-loop locus. The necessary controller gain for a 1.3 resonance peak, will be given by multiplying the arbitrarily chosen controller gain by the ratio of the length of the vector between the tangent and the origin to the length between the origin and the intersection of the vector with the open-loop locus.

If the proportional gain of the controller for reasonably stable control is too low so that there is excessive steady-state deviation, integral control action must be added. Ordinarily the reset rate is kept low enough to avoid changing the phase margin of the system. A commonly used rule of thumb (Y1), is that the integral time should be set equal to the ultimate period (reciprocal ultimate frequency). In cases where the initial phase margin is reduced by the integral action, it is necessary to reduce also the proportional gain to ensure the stability of the control system.

Flow systems have very small capacitances and as a result are most effectively controlled by proportional plus integral controllers having low proportional gains and very high reset rates. Such controllers are essentially proportional speed floating controllers (pure integral) and usually the proportional gain is non-adjustable at a value near unity. Catheron and Hainsworth (C3) show that for liquid flow systems the process response is so rapid that the controlling components are limiting and hence the controller characteristics can be preset and never adjusted further.

Improvement of the speed of response of control systems, requires that the frequency at gain crossover be increased by increasing the controller proportional gain. Since such an increase in gain crossover frequency decreases both the phase margin and the gain margin, the stability of the system will decrease. What is needed is the phase-leading characteristic of derivative action to increase the ultimate frequency, and consequently increase both the gain margin and the phase margin of the system at increased proportional gain. The maximum derivative time is fixed by the allowable gain margin, since the derivative action contributes increasing gain with increasing frequency.

3. *Empirical Methods of Controller Specification*

The more usual problem in the synthesis of automatic process control systems is simply the selection of appropriate controlling components

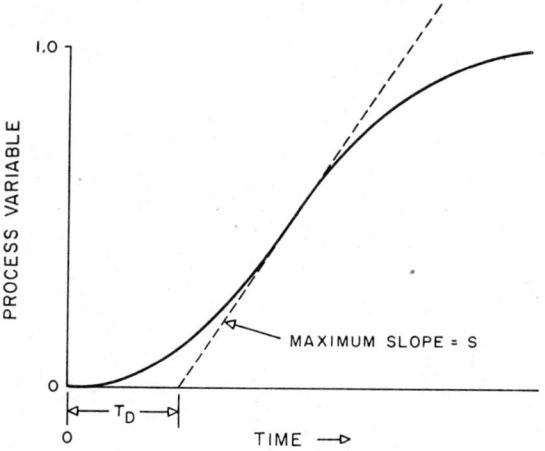

FIG. 7. Process reaction curve.

including measuring and regulating instruments which, when hung on the process equipment, will provide an adequate control performance in response to the normal disturbances imposed on the process.

There are a couple of simple empirical approaches for estimating the optimum controller settings for a particular process. Both approaches require data on the response of the existing process to simple stimuli: one the open-loop response to a step; the other the behavior of the closed-loop at the condition of ultimate gain.

a. Reaction Curve (Signature Curve). Theoretically, the transient response of a system to a step-forcing function contains all possible information about the dynamic characteristics of the system, since the step contains the whole spectrum of frequencies. Unfortunately this information cannot be elicited reliably because much of it is contained in the response near zero time.

Ziegler and Nichols (Z1) on the basis of studies of a variety of process systems, proposed that systems generally can be characterized by the *apparent dead times* and the *maximum reaction rates* of their transient responses (reaction curves or signature curves as they are often called in this context). The reaction curve can be obtained readily if the process may be subjected to a step input with the loop open, and the two quantities can be taken from the curve as indicated in Fig. 7.

The maximum reaction rate is the slope at the point of inflection, S, and the apparent dead time is the abscissa intercept of the maximum reaction rate line, T_D.

Optimum controller settings based on these quantities are as follows:
For proportional control only

$$K_p = \frac{1}{ST_D} \tag{35}$$

For proportional plus integral

$$K_p = \frac{1}{1.1 ST_D}$$

and $\qquad\qquad\qquad\qquad\qquad\qquad\qquad\qquad\qquad\qquad\qquad\qquad\qquad$ (36)

$$T_{int} = \frac{T_D}{0.3}$$

For three-mode control with action similar to the Taylor Fulscope Controller,

$$K_p = \frac{1}{0.83 ST_D}$$
$$T_{int} = 2T_D \tag{37}$$
$$T_d = \frac{T_D}{2}$$

where all times are in minutes, T_D is the apparent dead time, and S is the maximum slope of the reaction curve.

Cohen and Coon (C8) present a similar set of rules based on a theoretical treatment of signature curves, assuming that a reasonable representation of real processes is a time delay plus a single RC stage.

b. *Optimum Controller Adjustments from Ultimate Gain and Frequency.* Ziegler and Nichols also suggested setting controller adjustments on the basis of the ultimate proportional gain and the frequency of cycling at that gain as follows:
For proportional control only

$$K_p = \frac{K_0}{2} \tag{38}$$

corresponding to a gain margin of 2.0.

For proportional plus integral control

$$K_p = 0.45 K_0$$
$$T_{int} = \frac{1}{1.2 f_0} \qquad (39)$$

For three-mode (Taylor Fulscope) control

$$K_p = 0.6 K_0$$
$$T_{int} = \frac{1}{2 f_0} \qquad (40)$$
$$T_d = \frac{1}{8 f_0}$$

where K_0 is the ultimate proportional gain and f_0 is the ultimate frequency in cycles per minute. These expressions are based on the following relations between the reaction curve constants and the ultimate gain and frequency:

$$T_D = \frac{1}{4 f_0}$$
$$S = \frac{8 f_0}{K} \qquad (41)$$

The ultimate characteristics of a process control system are obtained readily by closing the loop through a proportional controller, and increasing the gain on the controller to the minimum proportional gain at which the system oscillates steadily. Johnson and Bay (J4) describe tests of both the ultimate gain approach and the reaction curve approach applied to pneumatic analog systems.

4. *Complex Systems*

The techniques for treating the synthesis of simple control systems are generally applicable to complex systems if the complexity arises from multiplicity of loops rather than intransigence of loop elements.

A common kind of complicated system is the *cascaded control system*, in which a master controller reacts to disturbances in the controlled variable by manipulating the set point of a secondary controller, which in turn manipulates a control agent affecting the process. Franks and Worley (F3) present a quantitative analysis of a cascade control system simulated on an analog computer.

Among other kinds of complex systems are those which incorporate high speed computers in the control loop to generate the command signals for the controllers on the basis of broad scale appraisals of operating performance, output requirements, and other pertinent variables. Such systems are not yet completely in being in the process industries, but studies of the possibilities are in progress both in industry and in academic laboratories.

IV. Present Status and New Directions

A. Summary of Present Status

Automatic control has long been a significant feature of the process industries, but only very recently has there been any attempt to treat process control problems quantitatively. It is now clear that the theoretical developments in the field of servomechanisms and governors are generally applicable to process control problems. The great simplification in treatment that results from frequency response characterization, makes possible a quantitative appraisal and design of process control systems.

Although most chemical engineers are aware that feedback control theory has important applicability to industrial process control problems, the number of engineers who make use of this applicability is small.

One limitation on using control theory in process control problems has been a lack of data on process characteristics, but now an increasing volume of effort is being applied to the dynamic characterization of processes. The fact that many typical processes of importance in chemical engineering are much harder to describe dynamically than the simple positioning operations in servomechanisms, may have been a deterrent to any earlier development of this area. Up to now most process characterization has focused on heat transfer processes, but every other type of process or operation has been studied to at least some extent. A number of process companies, particularly certain petroleum companies, have conducted extensive frequency response studies of their process units. Also a few process companies have undertaken dynamic studies of various kinds of instruments.

At the practical level, the status of some recent past trends in the automatic control of process plants has crystallized. Centralization of control and monitoring functions in compact and protected control centers has long been an important feature of process instrumentation. In some cases the centralization was overdone; today the practice is to mount as much instrumentation as possible at the points of application, leaving only critical items at the control center. By recording only really significant variables and by using miniaturized instruments, modern control centers are much less cluttered and confusing than their predecessors. Graphic panels, wherein the control center instruments are mounted on a flow diagram of the process, are widely used particularly for well-proved processes which are not likely to undergo appreciable revision after initial start-up.

A great variety of measuring devices for determining the chemical composition of process flow streams is now available, and many of these instruments are being incorporated in automatic control systems.

Although automatic computers are used widely in economic control of process plants and organizations, the incorporation of computers in the process control loop to obtain fully automatic control (automation) has been explored only in preliminary fashion.

At the theoretical level three different developments extend the possibilities of treating control problems. These developments are: (1) dynamic characterization of processes from normal operating records without need for disturbing process operations; (2) methods of attacking problems involving nonlinear systems; and (3) improved techniques for sampled-data systems. This last development is of particular importance in systems which utilize digital computation within the control loop.

B. Future Directions

The preceding sections have summarized briefly what is involved in automatic process control and how control problems may be attacked quantitatively. In the course of the presentation some soft spots have been manifest. For one thing it is clear that we do not know how to specify the kind of controlled behavior that is optimum for a given plant or process. General or approximate specifications can be given but they are not optimal. Actually, control quality specification can be made only on the basis of the disturbances affecting the process, and at the present time there is virtually no reliable knowledge about process disturbances and where they apply.

We are only beginning to characterize chemical engineering processes in ways useful for control system synthesis. Only very simple processes have been analyzed with success, and much more work is needed not only in analyzing complex processes but also in developing improved and simplified methods of analysis. Most of the important industrial processes involve distributed parameters, interacting components, and nonlinearities, all of which make the applicability of servomechanism techniques at best an approximation.

Because research and development engineers are not able to deal effectively with control system design, it is likely that many potentially profitable chemical processes never see the light of day. Processes which cannot be managed by conventional control instrumentation hardware are just not studied or developed, although a little imagination and an appreciation of what is involved in controlled loop behavior might make the processes commercially feasible.

The most important need for the immediate future is the continuing education on an increasingly broad scale of process engineers as to the practical possibilities of automatic control. Techniques developed by servomechanism engineers for handling control system analysis and design must be exploited insofar as they are applicable to process control systems, and new techniques immediately pertinent to process control problems must be developed by process-oriented chemical engineers. In particular the chemical engineer must offer improved methods of characterizing his processes and process systems. He must be quick to adapt new measuring techniques to process needs and quick to recognize possibilities in new instruments.

In the long run economic advantage is the only justification for automatic control in the process industries. Chemical engineers bear the primary responsibility for demonstrating such economic advantages, since they are the engineers most directly concerned with process performance. They must be alert to the possibilities of radically revamping processes and procedures, and in particular they must not be bound by preconceived notions. For example, if there is a potential economic advantage in harnessing certain unwieldy chemical reactions by means of carefully integrated control systems and reactors of special design, the chemical engineer should willingly explore the possibilities. Similarly, the whole problem of maintaining large inventories of products and stocks in process operations to provide stabilizing ballast against fluctuating demands and relatively inefficient plant control, should be reappraised. These and other problems cannot be assessed properly unless there is a clear understanding of the possibilities and limitations of automatic control.

Nomenclature

a	intermediate coefficient	j	$\sqrt{-1}$
A	constant; heat transfer area; area normal to process flow	K	gain
		L	length
		$\mathcal{L}$	Laplace operator
b	intermediate coefficient	M	closed loop magnitude ratio
B	constant	n	number of stages
C	capacitance	R	resistance
C_p	heat capacity at constant pressure	s	Laplacian complex variable
		S	maximum slope of reaction curve
d	differential operator		
D	diameter	t	time
e	base of natural logarithms	T	time constant
f	frequency, cycles per unit time	u	fluid velocity
		U	overall heat transfer coefficient
$G(s)$	transfer function		
h	film coefficient for heat transfer	W	mass
		x	distance

y intercept of maximum slope of reaction curve

GREEK LETTERS
∂ (small delta) partial differential operator
ζ (small zeta) damping ratio
Θ (capital theta) process variable
ρ (small rho) density, mass per unit volume
τ (capital tau) time constant, time
ψ (small psi) phase angle, radians
ω (small omega) frequency, radians per unit time

SUBSCRIPTS
a zero deviation
c controlled variable; controller
d derivative
D delay, dead time
f feed
fl fluid
i inlet, input; inside
int integral
L point where $x = L$
n natural
o outlet, output
p proportional control; constant pressure
r reference, set point
s steam
w wall
0 (number) point where $x = 0$; ultimate

References

A1. Aikman, A. R., *in* "Frequency Response" (R. Oldenburger, ed.), p. 141. Macmillan, New York, 1956.
B1. Bilous, O., and Amundson, N. R., *A.I.Ch.E. Journal* **1,** 513 (1955).
B2. Bilous, O., and Amundson, N. R., *A.I.Ch.E. Journal* **2,** 117 (1956).
B3. Bradner, M., *Instruments* **22,** 618 (1949).
B4. Brown, G. S., and Campbell, D. P., "Principles of Servomechanisms," Wiley, New York, 1948.
C1. Campbell, D. P., *Ind. Eng. Chem.* **47,** 409 (1955).
C2. Carslaw, H. S., and Jaeger, J. C., "Conduction of Heat in Solids," Chapter XIV, Oxford Univ. Press, London and New York, 1947.
C3. Catheron, A. R., and Hainsworth, B. D., *Ind. Eng. Chem.* **48,** 1042 (1956).
C4. Ceaglske, N., "Automatic Process Control for Chemical Engineers," Wiley, New York, 1956.
C5. Chestnut, H., and Mayer, R. W., "Servomechanisms and Regulating System Design," Vol. 1. Wiley, New York, 1951.
C6. Churchill, R. V., "Modern Operational Mathematics in Engineering," McGraw-Hill, New York, 1944.
C7. Close, C. D., *Control Eng.* **2,** No. 9, 97 (1955).
C8. Cohen, G. H., and Coon, G. A., *Trans. Am. Soc. Mech. Engrs.* **75,** 827 (1953).
C9. Cohen, W. C., and Johnson, E. F., *Ind. Eng. Chem.* **48,** 1031 (1956).
D1. Draper, C. S., McKay, W., and Lees, S., "Instrument Engineering," Vols. 1 and 2. McGraw-Hill, New York, 1952, 1953.
E1. Eckman, D. P., "Industrial Instrumentation," Wiley, New York, 1950.
E2. Evans, W. R., "Control-System Dynamics," McGraw-Hill, New York, 1954.
F1. Farrington, G. H., "Fundamentals of Automatic Control," Wiley, 1951.
F2. Flügge-Lotz, I., "Discontinuous Automatic Control," Princeton Univ. Press, 1953.
F3. Franks, R. G., and Worley, C. W., *Ind. Eng. Chem.* **48,** 1074 (1956).
G1. Goodman, T. P., and Reswick, J. B., *Trans. Am. Soc. Mech. Engrs.* **78,** 259 (1956).

J1. Janssen, J. M. L., *in* "Frequency Response" (R. Oldenburger, ed.), p. 131. Macmillan, New York, 1956.
J2. Johnson, E. F., *Chem. Eng. Progr.* **51,** 353 (1955).
J3. Johnson, E. F., *Chem. Eng. Progr.* **52,** 64 (1956).
J4. Johnson, E. F., and Bay, T., *Ind. Eng. Chem.* **47,** 403 (1955).
K1. Kochenburger, Ralph, *Trans. Am. Inst. Elec. Engrs.* **69,** 270 (1950).
L1. Lees, S., and Hougen, J. O., *Ind. Eng. Chem.* **48,** 1064 (1956).
M1. MacColl, L. A., "Fundamental Theory of Servomechanisms," Appendix. VanNostrand, Princeton, New Jersey, 1945.
N1. Nyquist, H., *Bell System Technical J.* **11,** 126–147 (1932).
O1. Oldenbourg, R. C., *in* "Automatic and Manual Control" (A. Tustin, ed.), p. 435. Academic Press, New York, 1952.
O2. Oldenbourg, R. C., and Sartorius, H., "Dynamics of Automatic Control" (translated and edited by H. L. Mason), Am. Soc. Mech. Engrs., New York, 1948.
O3. Oldenburger, R., ed., "Frequency Response," Macmillan, New York, 1956.
P1. Perry, J. H., ed., "Chemical Engineers' Handbook," 3rd ed., Section 19. McGraw-Hill, New York, 1950.
R1. Ragazzini, J. R., and Zadeh, L. A., *Trans. Am. Inst. Elec. Engrs.* **71,** 225 (1952).
R2. Rose, A., and Williams, T. J., *Ind. Eng. Chem.* **47,** 2284 (1955).
R3. Rutherford, C. I., *Inst. Mech. Engrs. (London) Proc.* **162,** 334 (1950).
S1. Sartorius, H., *in* "Automatic and Manual Control" (A. Tustin, ed.), p. 421. Academic Press, New York, 1952.
S2. Schwent, G. V., McGregor, W. K., and Russell, D. W., *ISA Journal* **3,** 274 (1956).
T1. Takahashi, Y., *in* "Automatic and Manual Control" (A. Tustin, ed.), p. 275. Academic Press, New York, 1952.
T2. Trimmer, J. D., "Response of Physical Systems," Wiley, New York, 1950.
T3. Truxal, J. G., "Automatic Feedback Control System Synthesis," McGraw-Hill, New York, 1955.
T4. Tsien, H. S., "Engineering Cybernetics," McGraw-Hill, New York, 1954.
W1. Williams, T. J., Harnett, R. T., and Rose, A., *Ind. Eng. Chem.* **48,** 1008 (1956).
Y1. Young, A. J., "An Introduction to Process Control System Design," Instruments, Pittsburgh, 1955.
Z1. Ziegler, J. G., and Nichols, N. B., *Trans. Am. Soc. Mech. Engrs.* **64,** 759 (1942).

TREATMENT AND DISPOSAL OF WASTES IN NUCLEAR CHEMICAL TECHNOLOGY

Bernard Manowitz

Brookhaven National Laboratory, Upton, L. I., New York

I. Introduction	82
A. Magnitude of the Problem	82
B. Sources of Waste	84
II. Waste Disposal as a Consideration in Site Selection	87
A. Regulations and Practices in Discharge of Radioactivity to the Environment	87
B. Geology and Hydrology of Site	88
C. Meteorological Problems	88
III. Waste Treatment and Disposal Practices	90
A. Liquid Wastes	90
1. Collection and Pretreatment	90
a. Uncontaminated Waste	90
b. Low-Level Waste (μ curie/gal. range)	90
c. Low-Level Waste (millicurie/gal. range)	90
d. High-Level Waste	90
2. Storage of Liquid Wastes	91
a. Storage Tanks—High-Level Waste	91
b. Storage Tanks—Low-Level Waste	92
3. Concentration of Liquid Wastes	92
a. Evaporation	92
b. Ion Exchange	96
c. Precipitation	97
4. Dispersal of Liquid Wastes to the Environment	99
B. Gaseous Wastes	100
1. Stack Dispersal	100
2. Equipment for Particulate Removal	101
3. Equipment for Vapor Removal	103
C. Disposal of Solid Wastes	103
1. Collection and Packaging	103
2. Incineration	106
3. Burial on Site	107
IV. Recovery of Fission Products from Radiochemical Wastes	108
A. Processes for Recovery of Specific Fission Products	108
B. Processes for the Recovery of Mixed Fission Products	113
V. Other Waste Problems	113
VI. Future Problems	114
References	115

I. Introduction

A. Magnitude of the Problem

Man has always been exposed to radiation from cosmic rays and other natural sources. Present developments in nuclear energy are such, that with the passage of time, the radiation background will be raised significantly by radiation sources of man's own making, if these are not controlled. From what is already known about the biological effects of radiation, an intensification of the radiation background is likely to lead to somatic and genetic effects in man. The former will occur in the population exposed, while the latter will accumulate and affect future generations. Although the National (N1) and International (I1) Committees on radiation protection have agreed upon maximum permissible concentrations of radioisotopes in air and water for continuous exposure, it is important to note their conclusion: "While the values proposed for maximum permissible exposures are such as to involve a risk that is small compared to the other hazards of life, nevertheless, in view of the unsatisfactory nature of much of the evidence on which our judgments are based, coupled with the knowledge that certain radiation effects are irreversible and cumulative, *it is strongly recommended that every effort be made to reduce exposures to all types of ionizing radiations to the lowest possible level.*"

Thus, an important task, responsibility, and challenge to the nuclear energy industry and its progress is a constant improvement in methods and facilities for control of radioactive waste. Simultaneous with an improved waste technology, cost reductions must be effected if the nuclear industry is to be competitive with other sources of energy.

A number of attempts have been made to assess the magnitude of the disposal problem the world may face during the next half century. Estimates have been made by Hatch (H1), Glueckauf (G2), Rodger (R4), and others. These estimates all tend to agree that the maximum plausible use of nuclear energy will result in the burnup of approximately 1000 tons of fissionable material (and the production of 1000 tons of fission products) per year. This is equivalent to approximately 2×10^6 megawatts, installed reactor capacity. Assuming this production rate, Rodger has calculated the equilibrium quantity of eight significant isotopes given in Table I. Also given are the maximum permissible concentrations of these isotopes in water and in air, and from these are calculated the volumes of water and air required to dilute the isotope in question to tolerance. Note that strontium-90 is the controlling isotope and that the dispersal volume of water needed, 2.6×10^7 cubic miles, is about 5% of the entire world ocean volume.

It should be noted that radioactive contaminants are not to be compared with their stable chemical counterparts. For example, because of the unchangeable character of the nuclear disintegration phenomenon, the natural purification processes which restore chemical and biological balances in the environments of living things, would have no effect on the radio-toxic properties of the radioactive elements. The time required for accumulations of strontium-90 to decay to background levels, would

TABLE I
50-Year Accumulation of Long-Lived Isotopes and Required Dispersal Volumes
Basis: 2.2 × 10⁶ MW Installed Reactor Capacity—3 Tons of Fission Products per Day

Isotope	Accumulated Quantity in 50 years, curies	Maximum Permissible[c] Concentrations (μ curies/ml.)		Volume Required to Dilute to Tolerance (cubic miles)	
		Water	Air	Water	Air
Zr^{95}	1.3×10^{11}	4×10^{-3}	4×10^{-7}	7.8×10^{3}	7.8×10^{7}
Ce^{144}	1.1×10^{11}	4×10^{-2}	7×10^{-9}	6.6×10^{3}	3.8×10^{9}
Ru^{106}	1.0×10^{11}	0.1	3×10^{-8}	2.4×10^{2}	8×10^{8}
Pm^{147}	5.1×10^{10}	1	2×10^{-7}	12	6×10^{7}
Sr^{90}	8.6×10^{10}	8×10^{-7}	2×10^{-10}	2.6×10^{7}	1×10^{11}
Cs^{137}	8.1×10^{10}	1.5×10^{-3}	2×10^{-7}	1.3×10^{4}	9.7×10^{7}
$Tc^{99a}\perp$	2.0×10^{7}	3×10^{-2}	3×10^{-6}	0.2	1.6×10^{3}
$Pu^{239b}\perp$	2.8×10^{6}	1.5×10^{-6}	2×10^{-12}	4.5×10^{2}	3.4×10^{8}

[a] Decay neglected.
[b] Based on a loss of 0.1 % in processing.
[c] From National Bureau of Standards Handbook **52** (1953).

be measured in centuries. It is clear that an amount of radio-toxic materials such as is indicated in Table I, should never be introduced into our world, except under the most carefully planned and controlled conditions.

The problem of what to do with these large quantities of waste has received serious attention. Glueckauf (G2) has suggested separating out strontium-90 and cesium-137, storing them separately, and storing the remainder of the waste in tanks for 13 years, after which time these could be discarded into the sea. He calculates that with strontium and cesium removed, the other activities after 13 years would be negligible compared with the potassium-40 activity, which is naturally present in the sea. Rodger (R4) points out that the separation of strontium and cesium must be truly quantitative. Leaving as little as 0.001% of the strontium-90, will cause the storage requirement to be raised from 13 years to a century.

The general conclusion is that wastes should be reduced to solids and confined. If the most optimistic assumption is made, namely, that the inert salts can be held to a weight equal to the fission products themselves, at the 1000 ton/year rate, six tons per day of concentrate would be formed. This is not an impractical amount to consider storing permanently.

Of course, permanent storage implies facing a set of opposing forces tending toward dispersal. Corrosion and other forms of chemical and physical disintegration of containers would be unknown factors, involving the consideration of very long periods of time. All sorts of possible disturbances of the balance of natural forces would have to be taken into account, as well as the faculty of man himself to depart from principle in order to follow short-term objectives.

B. Sources of Waste

As our nuclear technology develops, the major sources of radioactive wastes will be reactor sites and fuel recovery sites.

Most of the present nuclear reactors have been burning solid fuel elements of either normal or enriched uranium. Thus far, it has been necessary to reprocess fuel in order to recover valuable fissionable or fissile material. It is possible that fuel elements will be developed for future reactors which can be burned to the point where it is not economically justifiable to recover fissionable materials. Obviously, this depends upon the value of these materials. Such a procedure would provide an optimum solution to the major part of the waste disposal problem. The fission products would still be locked in the fuel element, simple disposal techniques could be employed, and in fact, spent fuel elements would probably have secondary uses as radiation sources.

A second and more likely possibility is that fuel from future reactors will have to be reprocessed, resulting in the dilution and dispersal of the fission products into many media.

A third possibility is that fluid fuel reactors, such as the ORNL aqueous reactor, or the BNL liquid metal reactor, will supply some fraction of our future nuclear economy. These reactors are usually designed as completely self-contained systems including continuous reprocessing of fuel.

Various sources of waste in a complete reactor cycle are shown in Fig. 1.

The processes associated with fluid fuel reactors are too varied to be considered in detail here. The reader is referred to references (F1) and (D3) for further information.

There are three sources of contamination connected with the operation of solid fuel reactors, excluding uncontrolled nuclear reactions. The first source is an imperfection or failure of the jacket which protects metal fuel elements. If the coolant contains oxygen or an oxidizing agent, uranium will readily oxidize and particles of uranium containing fission products, will be carried out by the coolant.

The second source of contamination in normal reactor operation is induced activity by neutron irradiation of the coolant. The amount of

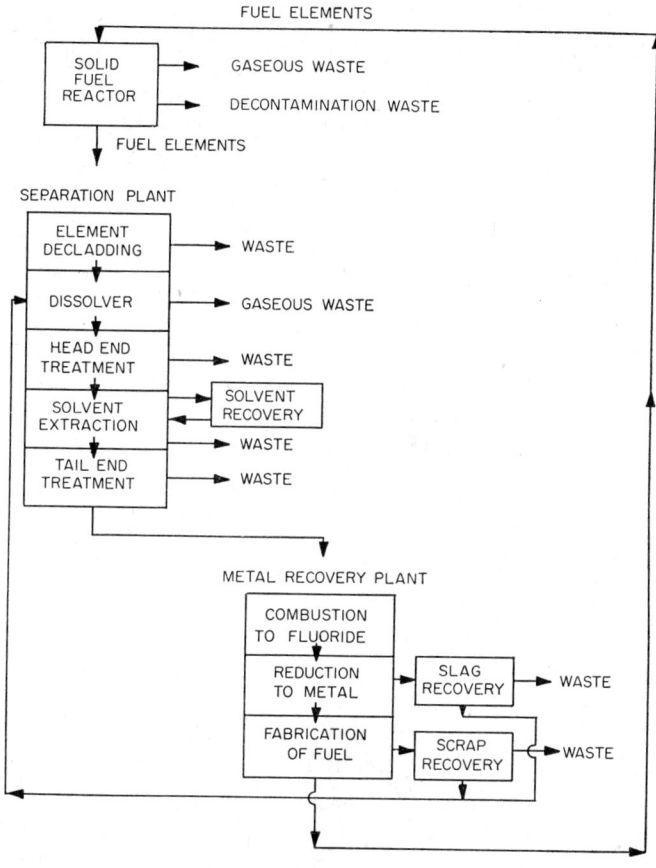

Fig. 1.

activity formed in the coolant may be determined for each element. It depends upon the average neutron flux, the absorption cross-section for the element, the number of atoms of the element irradiated, the concentration of the element in the coolant, and the half-life of the isotope formed. In air coolants, A^{41} is the principal activity formed. Particulate matter may also contribute to air pollution. If water is the coolant, the dissolved impurities can be a source of appreciable radioactivity.

Leaks and auxiliary equipment failure, can be a third source of release of radioactivity.

The major source of radioactive waste is likely to be the separation plants designed to treat solid fuel elements and to separate and purify uranium and plutonium from the fission products. Detailed descriptions of such plants are given in reference (R1).

The first thing that will happen to the fuel sent to such a plant will be the removal of its cladding. This may be done mechanically or chemically, resulting in either a solid or a liquid low-level waste. Next the fuel itself must be put into solution. This may involve the use of some hazardous chemicals, depending upon the nature of the fuel. As long as the fuel is composed of uranium metal, it can be dissolved in nitric acid. However, enriched fuels in alloy form require solvents which will dissolve the fuels. For instance, hydrofluoric acid may be required for zirconium-uranium alloys. Typical separation processes employ solvent extraction as the primary means for separating uranium and plutonium from each other and from the fission products, accompanied by head-end treatments (usually precipitation) and tail-end treatments (usually adorption or ion exchange), to improve the purity of the product. These processes result in both high-level and low-level liquid waste streams, and both must be properly disposed of.

For a typical plant large enough to process fuel from five to ten reactors, the liquid waste generation picture is likely to be as follows:

The high-level waste stream may consist of from 10^5–10^6 gallons per year of an aqueous solution of various inert salts, ranging from 0 to 20% of total solids contaminated with 10 to 100 parts per million of fission products. The activity level of such a waste would be in the hundreds of curies per gallon range, so that even small samples of such waste would be dangerously radioactive.

The low-level waste stream may consist of 10^6–10^7 gallons per year of an aqueous solution of various inert salts ranging from 0 to 1% total solids contaminated with 0.01 to 0.1 parts per million of fission products. The activity level of such waste would be in the millicurie per gallon range.

Gaseous waste will be generated during the fuel-dissolving stage. Exit gases from the dissolver will contain contamination in the form of undesirable vapors (iodine), noncondensable gases (Xe, Kr), and particulate matter on which radioactivity has been adsorbed. Cell ventilation gases may also contain gaseous and particulate contamination.

There will be an inevitable accumulation of solid wastes at such a plant. In general, these may be categorized into contaminated equipment, contaminated burnable trash, and contaminated nonburnable trash.

II. Waste Disposal as a Consideration in Site Selection

A. Regulations and Practices in Discharge of Radioactivity to the Environment

Relatively few states have as yet adopted regulations governing the disposal of radioactive wastes. A considerable amount of work on determining the biological hazards of various radioisotopes, has been done by Morgan, Stone, and others, leading to the concept of the mean permissible concentration of radioisotopes in air and in water. Specifications proposed depend upon:

(1) The chemical and radiochemical nature of the fission products: for example, Ac^{227}, At^{211}, Po^{210}, Am^{241}, Pb^{210} and Sr^{90}—Y^{90} are generally among the more hazardous isotopes, the first four being among the 12 most hazardous in every case.

(2) The site in which they are deposited; for inhalation, the lung or gastrointestinal tract is usually the critical body organ.

(3) The length of time during which they are retained by the body, this factor being termed the biological half-life. (The longest biological half-lives are associated with Sr^{90}—Y^{90}, Sm^{151}, Pu^{239}, Ra^{226}, and Th^{232}.)

(4) The method by which they are adsorbed, i.e., whether as dissolved ions, particulates, or gases.

General tolerance levels for unknown mixtures of fission products are roughly as follows (as taken from "Maximum Permissible Amounts of Radioisotopes in the Human Body and Maximum Concentrations in Air and Water" [N1]):

Kind of Uptake	Type	Maximum Weekly Dose
1. Total body irradiation	$\beta\gamma$	0.3 r/week
2. Ingestion in water, food, etc.	$\beta\gamma$	3×10^{-7} μ curie/cc., water, based on Sr^{90}
3. Inhalation of particulates	$\beta\gamma$	2×10^{-10} μ curie/cc., air, based on Sr^{90}
4. Inhalation of fission gases	$\beta\gamma$	3×10^{-9} μ curie/cc., air, based on I^{131}

It has been common practice to set the mean permissible concentrations for environmental exposure at $1/10$th of the above occupational exposure values. It is likely that these values will be incorporated into state and local regulations restricting the *concentration* of radio-nuclides that may be discharged to the local environment. It is probable, however, that some legislation will be developed that will also regulate the total amount of radioisotopes per year that may be discharged from a site. Agreements have been made between the AEC (Atomic Energy

Commission) and Brookhaven National Laboratory, for instance, that in addition to the above concentration requirements, no more than 1.5 curies per year of mixed fission products will be discharged to the local environment. Using this level as a guide, it will probably be permissible to discharge no more than one or two curies per year to the environment in populated areas. In relatively unpopulated areas, these requirements may be relaxed to perhaps 100–1000 curies per year. In cases where specific nuclides of short half-life are released, as for example, A^{41} from air-cooled reactors, the amounts permissible for release even in populated areas, are likely to be greatly in excess of 1000 curies per year.

B. Geology and Hydrology of Site

In any consideration of waste disposal, it is of prime importance to first obtain as good a picture as possible of the geology and hydrology of the site. Some information should be obtained on shallow and deep geology and on surface and ground-water hydrology. The presence of faults and the history of seismic disturbance should be determined. The chemical and physical nature of the soils should be ascertained.

It is important that these geologic data be determined both from the point of view of evaluating potential hazards, and from the point of view of evaluating the maximum amount of activity that may be safely discharged to the local environment.

C. Meteorological Problems

Weather factors are of importance in site selection, site construction and site operation. Preliminary advice on the climatological picture will often point out potential atmospheric pollution hazards and help in making a decision on selecting one particular site out of several.

The conventional use of meteorology will be most important during the construction period at a site. On some projects it may be necessary to furnish climatological data for prospective bidders, seasonal work scheduling, and on information not commonly available, such as freeze-thaw cycle frequency, and annual wet-bulb temperature frequencies and ranges.

After an atomic energy site begins operations involving the use of radioactive materials, meteorological services may be required under the following conditions: (1) if there is a continuous or intermittent source of radioactive effluent present in the area, or if there is a source potentially present to the extent that an atmospheric monitoring program is required; (2) if there is a daily or occasional requirement for weather forecasts, which may be required for the control of stack effluent; (3)

if there is a need for meteorological observations for documentation and historical purposes.

These records may be used in conjunction with testing and experimentation, or in the event of legal action arising from alleged radiation exposure.

The need for meteorological control should be carefully evaluated before being resorted to. "Meteorological control" implies stopping or slowing down plant operations for the duration of meteorological conditions, which are estimated to produce above-tolerance, ground radioactivity.

Two atomic energy sites where experience has been obtained in the routine forecasting for the control of radioactive effluents, are Brookhaven (B2) and Hanford (C2). The scheme for estimating probable ground concentrations at Brookhaven was based on O. G. Sutton's diffusion theories, on empirical relations, and on certain micrometeorological variables at the 355-foot level. A series of templates were prepared for a wide range of weather conditions, which gave the mean value of the dose rate averaged for 1 hour over an area 1 kilometer radius by 10 degrees, extending from the stack base outward for a distance of about 15 kilometers.

Since the radiation tolerance for both on and off-site locations was on a 7-day basis, the stack output for the previous 6 days was converted by means of the appropriate templates into radiation-dosage values. A decision could then be made as to whether continuation of reactor operations would result in dosages higher than the specified maximum.

As the experience at Brookhaven grew, it became evident that the curtailment of reactor operations due to meteorological conditions was sufficiently infrequent to warrant discontinuation of the regular forecast service. Accordingly, routine 24-hour forecasts of radiation dosages were discontinued on November 30, 1952. Since then, the 6-day dosages have been computed from the past meteorological and stack output records, but a forecast is prepared only when there is a chance of exceeding the permissible radiation limits.

At Hanford, a somewhat different forecasting procedure has been used. Teletype reports, records of instruments from a 410-foot tower, and pilot balloon observations four times daily, have been used to arrive at a "least dilution" factor (C2) which is likely to occur during a particular plant operation.

In planning for any new atomic energy area, the decision on the scope of the required meteorological program is fundamental to economic operations and safety. This is especially true where meteorological control of the operation is being considered. In most instances the use of reliable air-cleaning devices, sufficiently high stacks, or isolated sites will be more economical and much more satisfactory than operating a program only during favorable meteorological conditions.

III. Waste Treatment and Disposal Practices

A. LIQUID WASTES

1. *Collection and Pretreatment*

Liquid wastes at an atomic energy plant can, in general, be categorized as follows:

  a. Uncontaminated Waste. These are usually cooling waters or process waters that are normally clean, but that may potentially be contaminated. Careful consideration should be given to finding ways in which to discharge washes known to be uncontaminated, into drains other than the special waste system to prevent unnecessary overloads on sampling and holdup equipment.

  b. Low-Level Waste (μc./gal. range). At some reactor sites and at research laboratories, considerable amounts of liquid waste in the μc./gal. to mc./gal. range will be generated. These wastes are normally neutralized, sampled, and held up prior to disposal in site storage tanks, or in some cases, are finally disposed of by incorporation into concrete and burial at sea. Neutralizers for these "warm" wastes can be of several types—chemical, injection, or leaching tanks. Holdup tanks should have provisions for sampling, radioactive level recording, as well as convenient access for possible decontamination. Frequently a small diversion tank is placed just before the holdup tank, specially instrumented so that a potential slug of highly active waste will be diverted before being added to a large volume "warm" holdup tank. These tanks should be located so that no pumping has to be done until after the wastes reach them and have been neutralized. At least two holdup tanks should be present, so that the contents of one tank may be neutralized and disposed of, while the other is filling.

  c. Low-Level Waste (mc./gal. level). These wastes result from second-cycle processes and cell washings at separations plants. Frequently, a portion of these are highly acid wastes which first go through an acid recovery process. At plants processing natural uranium, these wastes are neutralized and concentrated before storage. Neutralization and concentration are remotely controlled operations. Stainless steel equipment is used. Transfer of liquids from neutralization to concentration to storage is by jet. At plants processing enriched uranium, low-level wastes are concentrated but are stored in the acid state.

  d. High-Level Waste (multicurie level). These wastes result from head-end treatments, in which a large fraction of the activity in the feed to an extraction process, is removed by a scavenge cake, and which result from first-cycle raffinates. Again, in plants processing natural uranium these wastes are neutralized and concentrated before storage, whereas at plants processing enriched uranium they are concentrated but stored as acids.

2. *Storage of Liquid Wastes*

  a. *Storage Tanks—High-Level Waste.* Neutralized high-level wastes can be stored in reinforced concrete tanks lined with steel. Acid wastes are stored in thin-walled, stainless-steel tanks inserted in a secondary concrete tank. For volumes to be handled at separation plants, the tanks are very large ranging in capacity from 300,000 gallons (acid waste) to over 1,000,000 gallons (neutralized waste). For the purpose of shielding, the tanks are usually placed underground. An annulus between the stainless-steel and concrete tanks, provides a means for prompt detection of leaks. The concrete and steel construction for these facilities, requires the highest quality of materials and workmanship.

  Means should be provided for temperature, pressure, liquid-level, and radiation measurement, as well as sampling. In some cases, provisions for cooling, liquid transfer, and agitation, are necessary.

  At production plants, the delivery of wastes to individual tanks is made through a diversion box. The diversion box is the junction point for a manifold pipe system leading to the various tanks. Routing of wastes to specific tanks, is accomplished by means of jumpers which are placed in the diversion box and attached and detached remotely.

  All long runs of pipe which carry high-level waste, are placed in concrete troughs with suitable provisions for expansion and movement of pipes caused by temperature changes. The troughs drain to the diversion box, which in turn may drain into a catch tank from which contaminated leakage can be readily delivered to the storage tanks.

  The major problems to be overcome in storing high-level wastes are the following:

  (1) *Cooling.* High-level waste may generate heat of up to 10 B.t.u./hr./gal., due to absorption of nuclear radiation. Since tanks are in the million-gallon capacity range, millions of B.t.u./hr. must be removed. In addition, acid wastes must frequently be held at low temperatures in order to minimize corrosion.

  (2) *Corrosion.* As indicated earlier, some wastes must be stored in the acid condition, primarily because they would corrode mild steel, even if neutralized. Intense radiation fields produce free radicals in aqueous systems, which can affect corrosion rates. Fortunately, most wastes are oxi-

dizing systems in which peroxide radicals are formed, producing a corrosion-inhibiting oxide layer on metallic surfaces. Reducing systems, however, such as wastes with high chloride content, would tend to induce accelerated corrosion in radiation fields. Where large temperature differences may produce thermal stresses, stress corrosion must be considered.

Cooling coils can be provided in waste tanks for heat removal. Potential coil failure with cooling water contamination, may be regarded as the disadvantage of this method.

Waste can be permitted to boil within the tanks and the evolved vapors condensed and disposed of externally, or returned as liquid to the tank. Unstable conditions can result from this practice, however, for some fission products will concentrate in sludges at the bottom of the tank, and heat generated here can build up as superheat. Eventually the unstable system is disturbed, initiating rapid boiling, and steam is suddenly released at a rate in the order of ten to twenty times the normal rate. Temperatures as high as 176°C. have been observed at tank bottoms (A1).

Contamination control requires that the vapors leaving the storage system be essentially free of activity. Although storage tanks are of massive construction, they have limited resistance to pressures much above or below atmospheric pressure. The tanks, therefore, should be designed to provide adequate relief for both positive and negative pressures.

b. Storage Tanks—Low-Level Waste. If many, very large tanks are to be installed at a site, it is sometimes more economical to build low-level tanks of the same general type of construction as the high-level tanks previously described. Usually, however, cooling provisions will not be necessary. In general, low-level, waste-tank construction should be such as to insure containment of the materials of the chemical nature under consideration, for long periods of time. Provisions should be made for withdrawal of the waste at a later date for possible disposal to the environment. Means should be provided for leak monitoring, and radiation and liquid-level measurement. Usually the tanks will be installed underground, although if the levels of radioactivity are low and the tanks are in a remote location, above-ground storage can be considered. Low-level waste tanks have been constructed of mild steel and of prestressed concrete.

3. *Concentration of Liquid Wastes*

a. Evaporation. Principles. The limits to radioactive decontamination of liquids by evaporation, are determined by the carry-over of activity by entrainment, splashing, and foaming, and the carry-over of volatile radioactivity. Volatiles will be considered in the section on gaseous wastes.

The mechanism of liquid carry-over from boiling liquids in evaporation, has been studied by Vorkauf (V1), O'Connell and Pettyjohn (O1), and by Manowitz et al. (M2). A brief review of the theory follows:

Carry-over can be caused by: (1) entrainment of small liquid droplets by the moving vapor; (2) splashing of the violently boiling liquid directly into the vapor vent line; and (3) foaming of the solution to the extent that foam is carried over into the vapor vent line.

In general, entrainment contributes less to carry-over contamination than does splashing or foaming. Entrainment results when liquid droplets, formed at the surface of the boiling liquid, or formed by bubble breakage, are thrown into the vapor spaces above the boiling liquid and are carried by the vapor stream if their rate of fall is less than the vapor velocity.

O'Connell and Pettyjohn noted that entrainment concentration in the overhead, decreases with increase in heat flux at constant boiling point up to a certain heat flux, but that continued increase in the heat flux caused the solution to boil so violently that liquid is splashed overhead. Both the entrainment rate and the vaporization rate increase as the heat flux increases. However, at low vaporization rates the entrainment rate increases less rapidly than the vaporization rate, so that the concentration of entrainment in the vapor decreases. O'Connell et al., and Manowitz et al., concluded that in a given evaporator there will be an optimum heat flux to give a minimum carry-over. Splashing can be largely eliminated by proper evaporator design; foam carry-over, however, is quite unpredictable. The specific solution to be evaporated must be examined for foaming tendencies in each case. Foaming can be controlled usually by one or more methods, such as dilution, use of a water spray, use of antifoam agents, reduction of the boil-up rate, or installation of a steam-coil foam-breaker in the vapor space. Even moderate foaming, such as is always encountered in the evaporation of dilute radioactive waste, leads to relatively high entrainment carry-over, probably by the mechanism of bubble breakage. Whatever mechanism projects liquid droplets into the vapor space, may also project any suspended solids present. Although suspended solids, such as rust, normally are not considered noxious contaminants, in this case they are likely to have adsorbed considerable amounts of radioactivity, and if carried over into the condensate, will impart radioactive contamination to it by subsequent desorption.

Liquid droplets or foam bubbles containing steam thrown into the vapor space, have an initial size distribution. This distribution may be considerably altered by conditions in the vapor space, e.g., droplets may evaporate until the droplet, surface-tension forces are equalized by the vapor pressure. The size distribution of suspended solids depends upon the size distribution in which they are initially present in the solution.

All particulate matter in the vapor space above the boiling liquid may be considered to belong to one of three groups:

(1) Particles whose rate of fall is so small that they are carried out of the evaporator by the vapor stream.

(2) Particles whose rate of fall is roughly equivalent to the vapor velocity in the evaporator vapor space. It would be expected that particles in this size range would build up to some equilibrium concentration within the vapor space of the evaporator.

(3) Particles whose rate of fall is greater than the vapor velocity. These particles will fall back into the boiling liquid if sufficient vapor space is provided within the evaporator for them to reach the end of their trajectory, or if a suitable baffle is so placed as to deflect the particles back down into the boiling liquid. It is understood that if these particles reach a path of sufficiently high velocity, they may leave the evaporator (splashing). It is readily apparent, then, that even when foaming and splashing conditions are kept under control, the vapor stream leaving any evaporator can and does carry along foam bubbles and entrained liquid and solid particulate matter, and that the concentration, amount, and particle-size distribution of such carry-over, will vary from evaporator to evaporator, and will depend upon the specific liquid being evaporated.

The mechanisms by which entrained particulate matter is removed from a vapor stream have been discussed by Rodebush *et al.* (R3, R2), and by Langmuir and Blodgett (L1). They are direct interception, gravity, inertia, diffusion, and electrostatic forces. In general, the entrainment removal efficiency of devices depending upon impaction and inertial effects, such as cyclones, bubble-cap or Raschig-ring columns, will be good for average particle diameters 100 μ or more, but poor for particles of 10 μ or less in diameter. To remove submicron particles efficiently from vapor streams, filtering media such as Fiberglas or fine-pore filter paper have been used. Here the removal mechanism is predominantly one of diffusion, although for some filter material, electrostatic forces may play an important part (R3).

In general, if particulate matter is in the size and density range where it can be removed by inertial effects, one can expect to find an increase in filtering efficiency with an increase in vapor velocity. On the other hand, a decrease in filtering efficiency with an increase in vapor velocity, usually indicates that diffusion is the primary mechanism of particle removal. Increasing the vapor velocity would have little effect on the filtering efficiency due to direct interception.

Manowitz *et al.*, propose the following equations for a submerged coil evaporator approximately 29 inches in diameter with approximately 2 feet of free board: (D.F. = Decontamination Factor)

$$\text{D.F.} = 4 \times 10^3\, G^{0.75} \text{ (low boil-up rates to splash point)} \tag{1}$$
$$\text{D.F.} = 1 \times 10^{10}\, G^{-3.5} \text{ (above splash point)} \tag{2}$$

where G = boil up rate, lb./(hr.)/(sq. ft.) of disengaging area

For the case of an evaporator with a restriction or baffle, the coefficients in the above equations would be higher, and the exponents would remain about the same.

Foaming would also alter the values of the coefficients and would alter the splash point. If the foam is such that all bubbles break at the surface, Eq. (1) should hold up to the splash point, except that the coefficient should be much lower. If hollow-foam bubbles are readily carried into the vapor space, the splash point should occur at lower boil-up rates than occur in the same equipment without foaming. The splash point should be a function of the height of the vapor space.

For the special case, involving the evaporation of a solution of radiochemicals in the presence of suspended solids, upon which the activities in the solution are readily adsorbed and desorbed (in general, fission products will behave this way), the following equations should be of interest:

$$\text{D.F.} = 42\, G^{.9} \text{ (up to the splash point)} \tag{3}$$
$$\text{D.F.} = 2.9 \times 10^5\, G^{-1.3} \text{ (beyond the splash point)} \tag{4}$$

It should be noted that these decontamination factors are appreciably lower than those for liquid entrainment. The decontamination factors alone are defined as the ratio of fission product activity in the still pot to the fission product activity in the condensed vapor. The above equations apply simultaneously with Eqs. (1) and (2), but only to that portion of the activity which is strongly adsorbed on the suspended solids.

Few data are available on the efficiency of de-entrainment devices. Cyclone separators, bubble-cap columns, or packed columns would not be expected to give additional decontamination to a stream leaving an evaporator by more than a factor of ten, since these devices will remove only particulates down to the 5–10 micron range. Fiberglas filters, however, will give greater decontamination factors. Manowitz et al., propose the following equation to predict the behavior of unbonded 20 μ diameter Fiberglas filters, packing density 5#/ft.³

$$\text{D.F.} = 940 \left(\frac{h}{u}\right)^{0.65} \tag{3}$$

where h = filter height (ft.)
and u = vapor velocity through the filter (ft./sec.)

Metallic wool, steam filters have also demonstrated good performance, but they would not be expected to yield as high a D.F. as a glass wool fiber unless the metallic wool could be obtained in fibers 20 μ in diameter or less, or unless fine glass wool fibers were interwoven in the metallic wool.

Evaporators are used extensively for evaporating liquid wastes at AEC production and research sites.

b. Ion Exchange. (1) *Principles.* When an aqueous solution containing cation M^{+n} is brought into contact with an ion exchange resin, M^{+n} ions in the solution exchange with those cations originally held on the resin by its free acid groups. This may be indicated as a reversible reaction (R_nM being the resin anion and H^+ its original cation)

$$n\,RH + M^{+n} \rightarrow R_nM + n\,H^+$$

The final equilibrium concentrations of M^{+n} and H^+ depend, as in any chemical reaction, principally upon the activities of the two cations and upon the respective affinities of each for the resin. In general, the cation-resin bond strength increases with increasing charge on the cation, and decreases with increasing radius of the hydrated ion. A typical series, from strong to weak, is: Th, La, Ce, Rare Earths, Y, Ba, Cs, Sr, K, NH_4, Na, H.

To free an adsorbed cation from its resin-bound state requires its physical replacement by another cation. If the relative affinity for the resin of the supplanting cation and/or its concentration in the solution phase, is sufficiently greater than that of the ion originally combined with the resin, replacement will be nearly complete. The effective concentration of an ion may be lowered by complex formation. Thus, any cation may be effectively replaced from its resin compound by a relatively dilute solution of a second cation containing a compound which will complex the other cation. In a similar manner, anion exchange resins can be used to remove fission product anions from dilute aqueous solutions. The ion exchange process is, in many cases, strongly pH and temperature dependent.

Colloids, for example, oxides with no ionic charge, are removed by ion exchange processes. Zirconium oxides tend toward sal formation, and in this colloidal form are difficult to remove from solution.

(2) *Applications.* The following important parameters should be considered in deciding whether or not to apply ion exchange processes:

(a) *Solids content.* Since extraneous solids will only tend to blanket the ion exchange surface, ion exchange processes should be applied only to aqueous solutions whose total solids content is less than 1%.

(b) *Chemical nature of waste.* Since ion exchange is a chemical process, the chemical nature of the feed must be controllable and constant. Changes in pH and the presence of complexing agents will markedly affect the efficiency of the process. The presence of trivalent salts of large, hydrated-ion radius (Al^{+3}) as noncontaminants, and the presence of contaminants in colloidal form (Zr) are undesirable.

(c) *D.F. desired*. In general, decontamination factors in the order of 100 can be obtained with mixed ion exchange beds (both anion and cation resins). Under optimum conditions D.F.'s as high as 10^5 can be obtained.

(d) *Engineering characteristics*. Consideration must be given to the need for regeneration and what to do with the regenerant. The use of ion exchange introduces a solids handling problem. Higgins (H3) has devised an ingenious contactor which makes the operation mechanically feasible for remote control operations. In principle, the contactor permits continuous counter-current extraction of the fission products. The only moving parts are the valves and the resin bed itself. Eventually, however, the resin itself must be disposed of. For high-level wastes, radiation effects on the resin must be considered. In general, these will result in lowered resin capacity, although in some cases radiation will cause physical breakdown of the resin.

Walters *et al*. (W1), have proposed an electro-deionization process for radioactive waste treatment. The process consists of partial deionization in a multi-compartment, permselective, membrane cell, and is similar to the production of potable water from sea water. This first step is followed by final deionization of the partially demineralized effluent in a multi-compartment, permselective, membrane cell employing mixed anion exchange and cation exchange granules in the deionization compartment. This process would permit higher volume reductions than for chemically regenerated ion exchangers.

c. Precipitation. (1) *Principles.* Several different mechanisms have been proposed to explain the carrying of trace elements by precipitation. These mechanisms are not mutually exclusive, and one, two, or more may be operative in any particular instance. The simplest ideal type of carrying arises when a trace element is incorporated into the crystal lattice of the carrier precipitate, a solid solution being formed. This will occur, in general, when the precipitated salt of the trace element and of the carrier are isomorphous, and the ions of the two elements are not greatly different in size. For good carrying (high D.F.), it is desirable that most of the trace element should go into the precipitate with only a small proportion remaining in solution. This condition can be realized, in general, if the compound of the carrier which is precipitated is more soluble in the aqueous solution than is the corresponding compound of the trace element. For example, lead sulfate is somewhat more soluble in water than is barium sulfate, and consequently lead sulfate is a good carrier for barium. Similarly, if complex-forming anions are present in the solution and the carrier element forms complexes in solution, but the trace element does not, the carrying will be enhanced.

Many precipitates, especially when freshly formed, have large surface areas and so appreciable amounts of trace elements may be carried by adsorption. It appears, as a rough general rule, that highly charged positive ions are more readily adsorbed than those of lower charge.

Frequently the carrying can be improved by having an excess of the anion present. Because of this effect the order of addition is important and leads to a precipitation technique called "reverse strike." In this technique the solution of carrier and trace element are added to a solution of the precipitating agent, as for example, adding a solution containing rare earths as trace element, and ferric ion as a carrier to a caustic solution. The precipitate is always formed in the presence of hydroxyl ion so that adsorption of the trace element is favored.

If carrying is principally by adsorption, other conditions, such as concentrations, rate of mixing, temperature, and presence of interfering substances, may affect the efficiency of the process.

(2) *Applications.* The following important parameters should be considered in deciding whether or not to apply precipitation processes:

(a) *Solids content.* Selective precipitants may sometimes be found for the removal of radioactive constituents even in the presence of large amounts of inert salts. Therefore, precipitation methods should be considered for decontaminating systems with high solids content.

(b) *Chemical nature of waste.* Precipitation processes will be sensitive to changes in the chemical nature of the feed.

(c) *D.F. desired.* In general, D.F.'s of 10 to 100 can be obtained with single stage precipitation processes. Multiple stage processes will give D.F.'s in excess of 1000 and D.F.'s in the order of 10^5 can be obtained in the precipitation process is combined with an acid adsorption or ion exchange process.

(d) *Engineering characteristics.* In a highly radioactive system, filtration and solids handling are complex operations.

Rupp (R5) mentions the following precipitation methods for mixed and specific fission products:

(1) Ferrous sulfide, ferrous hydroxide, calcium phosphate precipitation of Cs, rare earths, Sr, and Ru.

(2) Precipitation of Cs with silicotungstic acid.

(3) Ferrocyanide precipitation of Cs.

(4) Alkaline carbonate precipitation of alkaline earths and rare earths.

(5) Sulfide precipitation of Hg, Ru, and Te.

(6) MnO_2 precipitation of Ru.

(7) Selective precipitation of Ru with ferric hydroxide at pH 2.5.

(8) Tetraphenyl arsonium nitrate (TPAN) precipitation of technitium.

(9) Rare earth oxalate precipitation.

Ruthenium precipitation techniques are useful in head-end treatments to solvent extraction processes. Scavenge cakes resulting from such operations often constitute major activity fractions of high-level wastes.

Selective Cs and Sr precipitation methods are useful in trying to separate these long-lived elements from the rest of the fission products, in order to permit more rapid disposal of the short-lived residue.

Glueckauf and Healy (G3) have developed a process for Cs and Sr removal which involves precipitation on anionic resins in hydroxide form. This method of precipitation was found to give a good separation of Cs from the rest of the fission products, but the strontium was largely retained, at least in the presence of large amounts of iron contained in the fission product solutions.

4. *Dispersal of Liquid Wastes to the Environment*

The practice of dispersing some liquid waste directly to the ground has been followed at Hanford, Oak Ridge, and at Chalk River. As pointed out by Brown *et al.* (B5), the factors requiring evaluation in considering the feasibility of disposal of radioactive wastes to the ground include:

(1) The chemical and radiochemical content of the waste.

(2) The effectiveness of retention of radioisotopes in the available soil column above the ground water table

(3) The degree of permanence of such retention, as influenced by subsequent diffusion, leaching by natural forces, and additional liquid disposal.

(4) The natural rate and direction of movement of the ground water from the disposal site to public waterways, and possible changes in these characteristics from the overall liquid disposal practices.

(5) Feasibility of control of access to ground water in the affected region.

(6) Additional retention, if any, on sands and gravels in the expected ground water travel pattern.

(7) Dilution of the ground water upon entering public waters.

(8) Maximum permissible concentration (and total quantities) in public waters of the radio elements concerned.

Ground disposal has been practiced in a variety of subsurface sumps ranging from open trenches and ponds to subterranean caverns. The various fission products are selectively adsorbed by soils. In general, the order of retention will be: Pu, Rare Earths, Sr, Cs, Ru, NO_3^-. At Oak Ridge, ruthenium and nitrates were found in sampling wells 85 feet distant from injection points six weeks after the introduction of wastes. Approximately 2×10^8 gallons of active wastes containing several hundred thousand curies of fission products, have been stored in such sites as

Hanford in the first decade of operation (B5). As a result of their pile accident, the Chalk River plant disposed of over a million gallons of liquid waste containing 10^4 curies of long-lived fission products to the ground (M3). At Oak Ridge, thousands of curies of fission products have been disposed of in open waste pits (S2). The specific experiences at these locations, however, cannot be applied to other locations without field and laboratory evaluation of local problems.

At any particular site, surface disposal (up to 1000 feet depth), and deep disposal (10,000–15,000 feet) should be considered. If present on the site, salt domes would be good disposal areas. Cavities can be made in salt either by solution or by mining. In general, no ground water movement occurs through salt, which is also plastic and therefore probably freer from earthquake hazards than most other minerals.

Shale cavities have a wide distribution throughout the country and are low in excavation cost. However, they have many porous zones and little is known of their stability.

Deep disposal may be feasible, but an extensive study should be carried out to insure that conditions in areas under consideration are known to be static; however, the expenses for study, drilling, and monitoring, are likely to be prohibitive.

Although ocean disposal is feasible for small amounts of relatively low-level waste, the transportation expense and the unknown factors involved in dispersing a contaminant to a system from which it may later be extracted by marine life, are such as to make this practice inadvisable.

B. Gaseous Wastes

1. *Stack Dispersal*

The stack is frequently the most effective way of disposing of gaseous wastes. A properly designed stack can make the difference between a safe plant and a plant in which a combination of weather and plant operations may cause costly atmospheric pollution. As has been emphasized earlier, it is imperative that a complete climatological picture be obtained for the site. The meteorologist, from a knowledge of atmospheric diffusion, the terrain, the climatological picture, and the nature and concentration of the radioactive contaminant, determines what the effective stack height should be. The effective stack height is the stack height plus the height above which the effluent plume rises from the stack, owing to the stack draft velocity and/or the buoyancy of the effluent. The design and choice of height for the stack itself, however, is an engineering problem.

A rule of thumb frequently used to estimate effective stack height for low wind speeds, is the result of the investigations of O'Gara and Fleming

(O2). It states that each degree Fahrenheit of smoke temperature above ambient air, is equivalent to 2½ ft. of extra stack height.

The atomic energy industry encounters situations in which the effects of both smoke temperature and velocity must be considered, and for this reason a more exact approach is necessary. When meteorological parameters, such as vertical temperature gradient and wind speed, are taken into consideration as well, the problem becomes quite complicated. The only attempt to take all of these parameters into account is that of Bosanquet et al. (B4).

A simpler approach than that of Bosenquet is afforded by an equation empirically derived by Davidson (D1) from the wind tunnel experiments of Bryant (B6). It does not consider the effect of the vertical gradient of air temperature, but this does not eliminate it in favor of the Bosenquet formula, since the effect of air stability on effective stack height is not definitely known for all conditions. The Bryant-Davidson expression is

$$h = d \left(\frac{V_s}{U}\right)^{1.4} \left(1 + \frac{T}{T_s}\right)$$

where h is the rise of the plume above the stack, d is the stack diameter (ft.), V_s is the stack draft velocity (ft./sec.), U is the mean wind speed (ft./sec.) and T and T_s are the excess and absolute temperature of the stack gas in similar units, respectively.

In engineering terminology, the stack draft often is expressed in pounds per second, since this quantity is invariable for different atmospheric pressures and effluent densities. The effluent velocity (V_2) varies considerably with such values, however, for any fixed discharge rate and stack diameter.

Assuming a specific volume of 12.4 cu. ft./lb. of air at 30 in. Hg and 32°F., it can be shown that

$$V_s = \frac{0.962 D T_s}{d^2 P} \cong \frac{D T_s}{d^2 P}$$

where D is the discharge rate in pounds per second, and P is atmospheric pressure in inches of mercury. This expression is valid when the effluent is mostly air, as in the case of most atomic energy plants.

2. *Equipment for Particulate Removal*

Requirements for air and gas cleaning equipment in the atomic energy field differ from those for conventional equipment in the following respect: (a) cleaning efficiencies are much more severe; (b) equipment life is of prime importance and should be designed for as long as possible; (c) minimum maintenance is necessary; (d) consideration must be made for disposal of the equipment itself or for the components thereof.

The problems to which particulate removal equipment must be applied, include the treatment of reactor cooling gases, the treatment of cell ventilation air, the treatment of dissolver exit gases, and the treatment of incinerator flue gases.

Reactor cooling gases should be treated both at inlets, to decrease particulate matter exposed to neutron fluxes, and at outlets to remove particulates resulting from corrosion, erosion, or fuel element rupture. Operating temperatures sometimes prove a limitation for this application. The treatment of dissolver exit-gases imposes acid fume resistance as a condition in the treatment equipment.

The limit to the effectiveness of a gas cleaning device is its efficiency in the removal of aerosols. Neither electrical nor thermal precipitation has proven practical for the rapid removal of aerosols, and filtration seems to be the best method of removal. Aerosol filters consist of loosely aggregated fibers, and in order to avoid excessive resistance to flow of gas, the mesh of the fiber must be larger than the size of the particle to be removed. There is therefore no screening action; the removal of the particle depends entirely upon a chance collision of the particle with a fiber of the filter. Once having collided, the particle adheres by adsorptive forces.

Very large particles can be precipitated by centrifugal action, as in a cyclone separator. For smaller particles whose diameter is in the order of several microns, the centrifugal action is no longer effective since the inertia of the particle is not sufficient to overcome the resistance of the air. Thus, the air flows around the fibers of a filter in stream lines, and the particles are carried around with these stream lines. There is a range of particle sizes for which a higher velocity will improve the operation of the filter, since the inertial effects will carry the particles across the stream lines into collision with the fibers of the filter. For particles whose diameter is smaller than 1 micron, no inertial effects exist, but the kinetic diffusion becomes of greater importance in such smaller particles. Very small particles ($0.01\ \mu$) are precipitated very rapidly by diffusion. The process is analogous to the condensation of a vapor on a cold surface. The particles most difficult to remove by filtration, are those in the range of 0.1 to 1.0 microns, i.e., smokes. In order to obtain efficient filtration without excessive resistance, the filter must contain fibers of small diameter, approaching that of the particles themselves.

The high efficiency, asbestos-cellulose filter (C.W.S. filter) is one of the best developed for aerosol removal. This filter, which shows a decontamination factor of nearly 5×10^3 for $0.3\ \mu$ particles, improves in performance within a short time after the beginning of service, because of particulate deposition. Usually a glass wool pre-filter is used before the CWS filter, to protect it and to increase its life. One of the major weaknesses of the cellulose-asbestos filter is its inability to withstand more than small amounts of corrosive gases, and its limitation to temperatures below 100°C.

Glass fiber filters have been used for handling dissolver exit gases. This application has involved the use of chemically resistant glass in various diameters from 25 μ to 1 μ placed in layers of varying density and thickness. Studies by Blasewitz and Judson (B3) give performance and life characteristics for pilot studies.

A summary of the operational characteristics of many types of air cleaning equipment has been made by Silverman (S1) and is reproduced in Table II.

3. *Equipment for Vapor Removal*

Radioactive gases created by neutron flux activation, such as A^{41}, and certain radioactive fission product gases (Xe, Kr), are not easily removed by conventional approaches. The noble gases may be condensed and adsorbed on activated charcoal at extremely low temperatures. The cost of such systems per cubic foot of treated air, is so high that the method is feasible only for small volumes. Another approach for such volumes is compression and storage of the gases in chambers, for times of sufficient length to permit these isotopes to decay.

Methods have been developed for the removal of I^{131} from dissolver exit gases. Scrubbing with spray or baffle-plate scrubbers and utilizing caustic adsorbents, yields efficiencies of only 60 to 80%, and removes some oxides of nitrogen simultaneously, through neutralization by part of the caustic. Use of a silver reductor unit, composed of a tower packed with saddles that are coated with silver nitrate and maintained at an elevated temperature, gives the highest removal of iodine compounds (S1). Tests with this reactor have resulted in iodine removal with efficiencies greater than 99.99%.

C. Disposal of Solid Wastes

1. *Collection and Packaging*

Solid wastes can be categorized into three classes—(1) contaminated equipment, (2) contaminated nonburnable trash, and (3) contaminated burnable trash.

Occasionally contaminated equipment, such as stainless steel tanks or pumps, which must be disposed of, can be transferred to the burial ground without secondary packaging. More frequently, however, the equipment is first packaged in a metal-lined, airtight wooden box, or in a plastic bag. This is necessary in order to prevent the leakage of contamination when

TABLE II
OPERATIONAL CHARACTERISTICS OF AIR CLEANING EQUIPMENT

Type of equipment	Particle size range, mass median, microns	Per cent efficiency for size in column 2	Velocity, ft./min.	Pressure loss, inches of water	Approximate Cost/CFM ($)	Current Application in U. S. Atomic Energy Program
Simple settling chambers	50	60 to 80	25 to 75	0.2 to 0.5	0.05	Rarely used
Cyclones, large diameter	5	40 to 85	2000 to 3500 (entry)	0.5 to 2.5	0.10 to 0.25	Precleaners in mining, ore handling, and machining operations
Cyclones, small diameter	5	40 to 95	2500 to 3500 (entry)	2 to 4.5	0.25 to 0.50	Same as above
Mechanical centrifugal collectors	5	20 to 85	2500 to 4000	—	0.20 to 0.35	Same as large cyclone application
Baffle chambers	5	10 to 40	1000 to 1500	0.5 to 1.0	0.05	Incorporated in chip traps for metal turning
Spray washers	5	20 to 40	200 to 500	0.1 to 0.2	0.10 to 0.20	Rarely used; occasionally as cooling for hot gases
Wet filters	gases and 0.1–25 μ mists	90 to 99	100	1 to 5	0.09 to 0.10	Used on laboratory hoods and chemical separation operations
Packed towers	gases and soluble particles	90	200 to 500	1 to 10	0.40 to 0.80	Gas absorption and precleaning for acid mists
Cyclone scrubber	5 μ 5	40 to 80	2000 to 3500 (entry)	1 to 5	0.25 to 0.40	Pyrophoric materials in machining and casting operations, mining and ore handling; roughing for incinerators

Type	(col2)	Efficiency (%)	Capacity (cfm)	Power (HP/1000 cfm)	Pressure drop (in H₂O)	Application
Inertial scrubbers, power driven	8 to 10	90 to 95	—	3 to 5 HP per 1000 cfm	0.15 to 0.25	Pyrophoric materials in machining and casting operations, mining and ore handling
Venturi scrubber	1	99 for H₂SO₄ mist; SiO₂, oil smoke, etc., 60 to 70	12,000 to 24,000 at throat	6 to 30	0.50 to 3.00	Incorporated in air cleaning train of incinerators
Viscous air conditioning filters Dry-spun glass filters	10 to 25 / 5	70 to 85 / 85 to 90	300 to 500 / 30 to 35	0.03 to 0.15 / 0.1 to 0.3	0.004 to 0.006 / 0.02 to 0.04	General ventilation, air General ventilation air; precleaning from chemical and metallurgical hoods
Packed beds of graded fibers 1 to 20 μ, 40 inches deep	1	99.90 to 99.99	20	10 to 30	1.0 to 5.0	Dissolver, exit gas cleaning
High efficiency cellulose-asbestos filters	1	99.95 to 99.99	5 through medial, 250 at face	1.0 to 2.0	0.04 to 0.06	Final cleaning for hoods, glove boxes, reactor air and incinerators
All-glass web filters	1	99.95 to 99.99	5 through medial, 250 at face	1.0 to 2.0	0.07 to 0.10	Final cleaning for hoods, glove boxes, reactor air and incinerators
Conventional fabric filters	1	90 to 99.9	3 to 5	5 to 7	0.30 to 1.00	Dust and fumes in feed materials production
Reverse-jet fabric filters	1	90 to 99.9	15 to 50	2 to 5	0.50 to 1.00	Dust and fumes in feed materials production
Single-stage electrostatic precipitator	1	90 to 99; 90 to 95 on metallurgical fumes	200 to 400	0.25 to 0.75	0.50 to 2.00	Final cleanup for chemical and metallurgical hoods; uranium machining
Two-stage electrostatic precipitator	1 to 5	85 to 99	200 to 400	0.25 to 0.50	0.25 to 0.50	Not widely used for decontamination

the equipment is transferred to the burial pits. Transportation is by means of railroad truck, as will be described later.

Contaminated nonburnable trash, is put into convenient containers that can be sealed. These can be boxes, cartons, or drums. Frequently 50-gallon steel drums are used. Drums are sometimes filled with an annular layer of concrete to provide some shielding for very active trash contained in the void space of the drum.

Burnable trash is sometimes packaged in cardboard containers but more frequently is compressed and baled. It is often more economical to bury baled burnable trash than to incinerate it.

2. *Incineration*

Because of the complexity of solids-handling equipment, incineration will be considered for low-level waste only, and not for high-level, burnable trash.

Because of the hazards of handling radioactivity, special consideration should be given to the following features of incinerator design: (a) packaging, (b) feed device, (c) combustion chamber, (d) fly ash and particulate removal, (e) ash removal.

For the convenience of the operator, waste should be packaged in uniform-sized packages designed for ease of handling. Asphalt-lined bags should be avoided because of the evolution of excessive particulate matter from their incineration. Fiberboard cartons or drums are generally satisfactory.

The feed chute must lead through some system of interlocked, double doors to prevent blowback of radioactive ash or fumes from the furnace.

The primary requisite of the combustion chamber is the insurance of complete combustion. If air is used to support combustion, the air supply should be drawn from somewhere outside the building to eliminate the possibility of radioactive dust being dispersed through the building in the event of a puff resulting from momentarily accelerated combustion. Remote ignition should be provided within the combustion chamber.

Both large and fine particulates of fly ash must be removed from the flue gas. A rough scrubber, such as a Pease-Anthony, will suffice for the large particulates, although the scrub solution, of course, constitutes a liquid waste. Fine particulates can be removed by Fiberglas filters or CWS-type filters.

The ashes must be removed into a system in which dusting will be minimized. Collection in sealed receivers or collection under water may be used.

Decontamination factors of about 10^9 are frequently required for incinerators. This degree of decontamination can be obtained with the equipment described above. However, the necessary storage of filters and scrub solutions is a disadvantage of this method.

Incinerators are not widely used by the AEC.

3. *Burial on Site*

In many cases, the most economical method of disposing of solid wastes is by burial on site. For large atomic energy installations, or possibly for regional burial grounds, the following criteria should be observed.

The estimated life of the burial ground should be at least 100 years. Obviously, it will be advantageous if the area can be located in an empty section of the country. At least one square mile of burial area plus one-half square mile of fringe area, should be sufficient to meet the 100-year limitation.

The geology of the area should be known in as much detail as possible. There should be as few faults as possible, with no history of seismic disturbances. Limestone formations should be avoided. The land should be flat with possibly one hill for air dispersion, and there should be tree and shrub coverage to keep dust down.

Low rainfall in the area is preferable, as is a low ground water table. Ground water and surface water hydrology must be carefully studied. Such a study would probably result in a well-survey system, which could also be used to monitor the burial grounds.

The Oak Ridge National Laboratory has been burying solid wastes on site for many years. Trenches $10 \times 200 \times 15$ feet deep, are dug. Packaged trash and equipment, can be transferred into the trenches from the backs of trucks, by a crane. The trenches are then backfilled with 3 to 4 feet of earth. Occasionally, auger holes 1 foot in diameter and 15 feet deep are dug, for the disposal of especially radioactive pieces of equipment. The use of an auger hole reduces the back-scatter of radiation to the bulldozer operator. The area is gridded and recorded, and this procedure is followed both for permanent marker purposes and for the occasional recovery of valuable equipment after burial.

A description of how the Hanford plant discards "hot" equipment, was published in the May, 1955 issue of *Nucleonics*. The techniques involved are as follows:

A remotely-operated crane in the separations plant lifts the "hot" piece of equipment with attached piping from a cell, places it in a metal-lined wooden box on a railroad flat car, and bolts down a lid. With the hot equipment inside the box, no one can stay within 600 feet of it for more than one-half hour. Ten empty cars separate the locomotive from the 6-ton piece of radioactive "junk" on the trip from the separations plant to the burial ground. Each burial is preceded by a dry run, during which the box is hauled empty from the separations plant to the burial ground, to be sure it will pass all gate posts and power lines safely. If, on the appointed day, the winds blow faster than 10 miles per hour, the operation is postponed. At the burial ground the locomotive stops at a concrete siding at the head of a long trench perpendicular to the rail line. A steel cable attached to the box and looped along the train is then attached to a tractor. The tractor drags the box off the flat car and down a ramp into a trench. The box is then buried by a crane operator who dumps a pile of dirt and rocks into the trench, working from behind a pile of dirt and rocks, and using mirrors at the end of his long boom to see what he is doing.

This description clearly indicates the problems involved in burying high-level solid wastes.

IV. Recovery of Fission Products from Radiochemical Wastes

A. Processes for Recovery of Specific Fission Products

By far the best solution to many of the waste disposal problems discussed, would be waste utilization. The fission products may turn out to be valuable by-products. Quite a bit of research is now going on in the finding of uses for intense radiation fields (M1, B1). Certain of the fission products, particularly Cs^{137} and Sr^{90}, would have advantageous characteristics for such use. Most of the work in developing processes for the recovery of specific isotopes from fission product wastes, has been done by Rupp and his co-workers at the Oak Ridge National Laboratory. The following is a description of the processes that Rupp has selected for a fission product recovery pilot plant to be located at Oak Ridge.

Simplified flow sheets are given in Figs. 2 and 3 for two of the basic types of waste that will be encountered. The Purex type of waste is the simplest of all of the wastes to process, being a nitric acid solution of fission products, corrosion products, and a small amount of other impurities. Advantage is taken of the fact that one can evaporate this waste, and thereby achieve a greatly increased concentration of material, before actual chemical separation of the constituents is started.

The evaporator used for this purpose contains steam coils for heating the liquid and additional heat is injected continuously as steam to supply the additional water required for the distillation of nitric acid, which distills at less than 6 N. The limits to which evaporation can be carried, probably are determined by the amount of corrosion products that accumulate in the solution, and the degree of contamination of the condensate from the evaporator. The concentrated waste solution is stored in a tank equipped with cooling coils, and is passed on to the next main

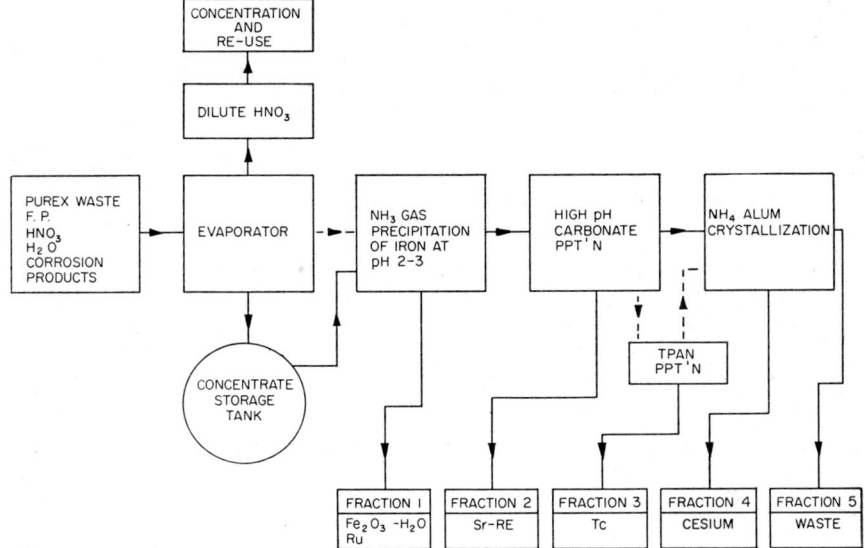

Fig. 2.

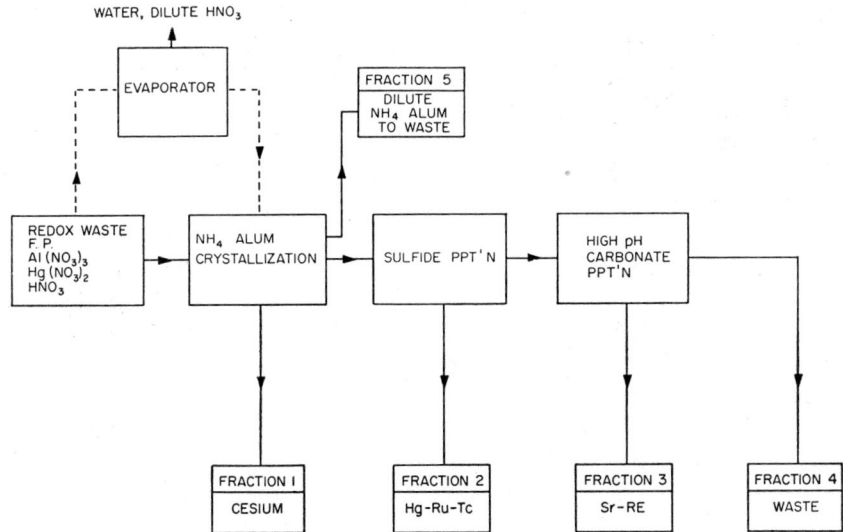

Fig. 3.

step, which is the precipitation of iron at pH 2.5. The precipitation is very similar to the precipitation of iron obtained by the hydrolysis of urea; however, it is more easily accomplished by injecting ammonia gas into the solution, which is vigorously agitated. Most of the iron and ruthenium are brought down at this point, allowing the rare earths, alkaline earths and cesium to pass into the filtrate. A carbonate precipitation is made on the filtrate, bringing down the strontium and rare earths together as fraction no. 2. The solution now contains only cesium, which is taken out in the alum crystallization process as fraction no. 4. If it is desired to remove technetium, a special precipitation with tetraphenyl arsonium nitrate is made and the crude technetium accumulated as fraction no. 3. The slightly contaminated waste that is discharged from the alum crystallizers, is treated before discharge from the plant.

It will be noted from the flow sheets, that the chief difference between the two types of waste processing, is that the Redox waste goes to the alum crystallization first for removal of most of the aluminum in the form of alum. If sulfuric acid is used as the means of sulfate addition in forming the alum crystals, nitric acid is actually generated during the crystallization and it may be feasible to recycle material to the evaporator and remove nitric acid, bringing the concentrated alum back to the crystallizer for another cycle. In this way it may be possible to remove most of the aluminum as the sulfate and the nitrate ion as nitric acid. However, in practice it appears that enough of the aluminum is removed in the first crystallization; so the waste alum solution is passed on to other crystallizers which remove the final traces of cesium. This aluminum waste is then passed on to further treatment to remove traces of strontium and rare earths before discharging from the plant (Fig. 4). The cesium product in this process is accumulated as fraction no. 1 to be concentrated by further fractional crystallization.

Redox-type waste contains considerable mercury, which must be removed. Advantage is taken of the presence of mercury to use it as a carrier for the ruthenium and technetium when this group is precipitated as the sulfides. This involves fairly corrosive chemical solutions, but they can be handled in equipment fabricated with special grades of stainless steel. The filtrate contains only the alkaline and rare earths which are then precipitated as carbonates, the same as the Purex-type procedure. The waste from this step is treated separately.

In the finishing stages of cesium processing, the cesium fraction that is brought to the final crystallization step, represents a large concentration of cesium over that found in the original waste. This is because the cesium can be progressively accumulated in the bed of crystals. By decanting off the supernatant liquor after one crystallization, another batch of raw material containing cesium can be put into the crystallizer and the previous crystals redissolved. Additional ammonium alum is added if necessary, to saturate the solution at 80°C., and a new crop of crystals is grown, this time including all the cesium that was brought down during the previous crystallizations, plus the newly added cesium. As successive crystallizations are made, the crop of crystals becomes progressively richer in cesium.

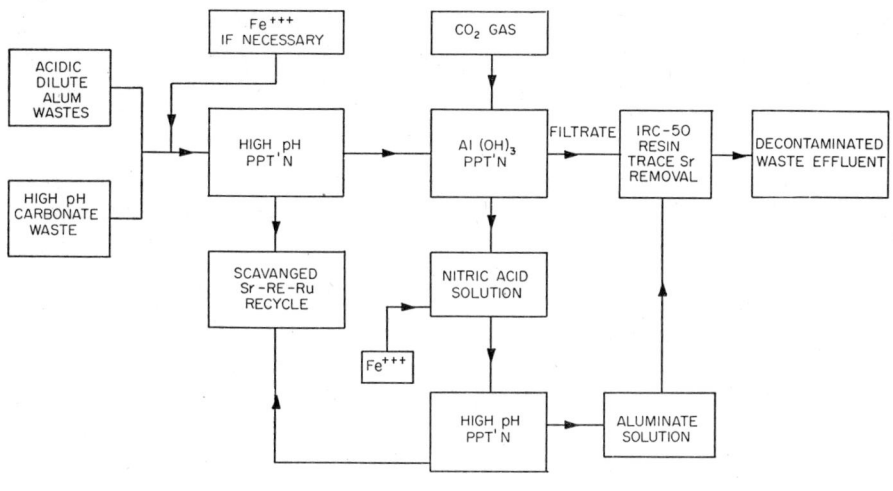

Fig. 4.

This procedure can be followed until the cesium losses in the supernatant liquor become higher than desired. In the semi-works equipment designed for this process, several crystallizers are used, some of which handle the supernatant liquor for recrystallization and recovery of small amounts of cesium that have passed the first crystallization. Others are used to concentrate the cesium by dissolving the alum crystals in water, and then growing smaller batches of crystals to get the cesium into progressively smaller volumes of liquid. If enough cesium is accumulated, pure cesium aluminum alum crystallizes. In this process, Cs^{137} comes down almost radiochemically pure after one crystallization; only insignificant amounts of Ce^{144}, Sr^{90}, and Ru^{106} superficially contaminate the first batches of crystals. However, some inactive fission products, such as rubidium, accompany the cesium and must be removed by selective crystallization near the end of the refining process. At all stages of the process, the various fractions can be recycled to an appropriate previous point in the crystallization cycle, to keep losses at a minimum.

It may seem that crystallization would be a hard procedure to carry out by remote control, but in actual practice with large batches of cesium in 50-gallon crystallizers, this has not been the case. The crystals are of large size, like granulated sugar, and the supernatant liquor can be easily decanted through a screened dip-leg. The remaining crystals are very easily dissolved by adding water and increasing the temperature to 80°C. Speaking in terms of decontamination factors, it is easily possible to obtain a cesium decontamination of 10^2 per crystallization step. Therefore, two successive crystallizations are usually enough to achieve the required decontamination. It has been found that in achieving decontamination factors larger than this, the limiting factor is often not the process itself, but rather the contamination that is picked up from the equipment.

The concentrated cesium alum produced by the crystallization process contains ammonium and aluminum sulfate, which must be removed. The next step is to precipitate the aluminum as the hydroxide, bringing it down in the granular form by the ammonia gas technique. The filtrate contains cesium sulfate and ammonium sulfate and could be transported to a source fabrication plant as a highly concentrated product at this stage.

In the preparation of the sources, the next step is to remove the sulfate by passage of the solution at high flow rates through a bed of anion exchange resin; good control is necessary at this point to avoid too much radiation damage to the highly sensitive anion exchange resin. It has been found possible to do this at the 1000-curie level with an acceptable amount of resin damage. This is the only point in the process where an organic material is used, and work is being continued to find a suitable substitute procedure. The effluent from the column contains cesium hydroxide and ammonium hydroxide; the latter is removed by distillation. The cesium hydroxide is neutralized with hydrochloric acid, and the cesium chloride solution is taken to dryness at 460°C. to remove all traces of water. The dried material is then pressed in a hydraulic press at 20,000 psi to produce pellets which are very compact and hard.

The finishing stages for strontium separation are shown in Fig. 4. The main separation of the rare earths from the alkaline earths is made by ammonia gas precipitation of the rare earths as hydroxides in a carbonate-free medium. The alkaline earths pass into the filtrate and are removed in the next step as the carbonates. Since the separation of rare earth hydroxides and the only moderately soluble alkaline earth hydroxides is not clean, a re-precipitation step is required. The alkaline carbonates are then passed to packaging, either as the dried carbonates, or are first converted to sulfates, oxides, or fluorides for subsequent packaging in multiple-walled, weld-sealed, containers for storage. The filtrate containing some strontium is returned to the plant, waste effluent, decontamination process.

The plant waste streams are kept to a minimum by appropriate recycling, but the waste acidic alum and alkaline carbonate filtrates are further decontaminated, principally from Sr^{90}, by the procedure shown in Fig. 4. This is essentially a high pH scavenge with $Fe(OH)_3$; and is followed by trace Sr^{90} cleanup, by passing the alkaline waste through a bed of IRC-50 resin, which is especially suited for removing alkaline earths from highly alkaline media. Since only trace amounts of radioactivity are present at this stage, the problem of radiation damage to the resin is not serious.

B. Processes for the Recovery of Mixed Fission Products

If a cheap way could be found in which to concentrate mixed fission products and package them in a stable form, these might serve as economical radiation sources. Relatively little work has been done in this area.

One possibility is a process being developed by Hatch and his co-workers (H2). The method involves the adsorption of fission—product ions on montmorillonite clay, and the fixation of the adsorbed ions by heating the clay to a high temperature. The capacity of the clay is found to be about 1 milliequivalent per gram for cation exchange, and a high degree of fixation has been demonstrated. One limitation of the process is that the feed must be free from multivalent inert ions to keep from blanketing the clay with inert material. Investigations are now centering on methods for removing aluminum ions from raw wastes.

Another possibility lies in the use of equipment being developed by Manowitz and co-workers (H2). The equipment, called a continuous calciner, converts aqueous slurries to compact fused salts. One source of fission products of high specific activity would be the MnO_2 scavenge cake used in separations processes, to remove most of the fission products in a head-end treatment prior to a solvent extraction process. MnO_2 may be dissolved in fused caustic in a continuous calciner to form a highly concentrated radiation source.

V. Other Waste Problems

There are occasions when radiochemical wastes are amenable to treatment by biological methods. This may be because the waste contains sewage and must be biologically treated, or because the waste contains organic matter (i.e., laundry wastes) to which a biological treatment may be applicable. The effectiveness of activated sludge in concentrating radioisotopes from liquid wastes has been studied by Kaufman (K1), Christiansen (C1), and others. The use of microorganisms in oxidation ponds has been studied by Gloyna (G1). The removal of radioactivity from liquid streams by trickling filters has been studied by Dobbins (D2) and others.

Occasionally, also, the contaminant of prime concern in a waste stream is the chemical content rather than the radiochemical content. For instance, at uranium (feed) production centers, the fluoride concentration in the waste is the most serious hazard. Liquid wastes of this nature are usually discharged to lagoons or holding ponds rather than to an environmental stream.

VI. Future Problems

Future problems associated with separations and waste treatment processes, will be the result of two major changes in reactor development. These are the use of new materials of construction for fuel cladding or fuels themselves, and the increase in specific activity of processed fuels.

Fuel composition may change from uranium to plutonium, and cladding from aluminum to zirconium to stainless steel. In some cases blankets, moderators, and coolants must be processed, and these will introduce thorium, beryllium, NaK, and bismuth to the chemical process. These changes in materials will present new chemical and corrosion problems in waste treatment processes and waste storage procedures.

An increase in concentration of fission products in the chemical process will introduce many new problems and will magnify many old ones. Much higher decontamination factors will be required in order that process solutions may be safely discharged or re-used. A considerable amount of heat will be generated in the fuel, in the fuel solutions throughout the separations process, and even in the waste streams. The heat generation problem may be complicated by activity adsorbing on walls of containers or on suspended particles, and creating local hot spots within the system.

Accompanying the heat generation problem will be the problem of radiation damage to the fluids in the separation and waste systems, and perhaps to parts of the containers of the systems. Radiation damage to the fluids of the system may include pH changes, oxidation reductions, or gassing. Radiation damage to containers may include embrittling or fracture of gasket and packing material. Corrosion of container walls may be accelerated by radiation decomposition products.

Fission product recovery may alleviate some waste storage problems but will not nullify the need for low-level waste processing. Indeed, any fission product extraction process will, by its inevitable inefficiency, generate low-level wastes. Thus, continued research and development in the field of low-cost waste processing techniques, is probably justifiable.

Another lucrative field for research and development should be that of chemical reagent recovery. The recovery of aluminum nitrate and nitric acid from solvent extraction wastes is now under study. Future separation processes will result in wastes containing bismuth, zirconium, and other critical materials, whose recovery should prove economically feasible.

The development of cheap transportable packages may change the economics of shipping low-level wastes to a regional disposal site.

Several processes intended to reduce the volume of high-level waste and to convert it to a form in which containment is more assured, are now in the development stage. One of these involves the adsorption of fission products on clay, and the subsequent fixing of the activity on the clay by firing (H2). A second involves the conversion of aqueous waste solutions to anhydrous melts by calcining (H2). A third involves the incorporation of wastes into concrete. These processes are presently in the pilot plant stage and it is difficult to assess their final applicability until they have been further developed and tested for a longer period of time.

To summarize, the problems of radiochemical waste disposal are many and varied. Adequate solutions for many of these problems have been applied by the AEC at its own sites, but these solutions may not be adequate for new commercial separation plants. Waste disposal is a field in which ingenuity and imagination can pay off by appreciably improving the feasibility of widespread atomic power.

REFERENCES

A1. Anderson, C. R., and Rohrmann, C. A., Geneva Conference Paper No. 552 (1955).
B1. Ballantine, D. S., *Modern Plastics* **32**, 131 (1954).
B2. Beers, N. R., *Nucleonics* **4**, No. 4, 28–38 (1949).
B3. Blasewitz, A. G., and Judson, B. F., *Air Repair* **4**, 223 (1955).
B4. Bosanquet, C. H., Carey, W. F., and Halton, E. M., *Inst. Mech. Engrs.* (*London*) *Proc.* **162**, 355–367 (1950).
B5. Brown, R. E., Parker, H. M., and Smith, J. M., Geneva Conference Paper No. 565 (1955).
B6. Bryant, L. W., Experiments in a Wind Tunnel, National Physical Laboratory (Great Britain), 1949.
C1. Christiansen, C., *U.S. Atomic Energy Comm.* Wash. **275**, Sanitary Engineering Conf., April, 1954, p. 198.
C2. Church, P. E., and Gosline, C. A., *Bull. Am. Meteorol. Soc.* **29** (2), 68–73 (1948).
D1. Davidson, W. F., Ind. Wastes, *Ind. Hyg. Foundation Amer. Trans. Bull.* 38–55 (1949).
D2. Dobbins, W. E., *U.S. Atomic Energy Comm.* Wash. **275**, San. Eng. Conf., Apr. 1954, p. 38.
D3. Dwyer, O. E., Geneva Conference Paper No. 550 (1955).
F1. Ferguson, D. E., Geneva Conference Paper No. 551 (1955).
G1. Gloyna, E. F., and Steel, E. W., *U.S. Atomic Energy Comm.* Wash. **275**, San. Eng. Conf., Apr. 1954, p. 10.

G2. Glueckauf, E., Geneva Conference Paper No. 398 (1955).
G3. Glueckauf, E., and Healy, T. V., Geneva Conference Paper No. 415 (1955).
H1. Hatch, L. P., *Am. Scientist* **41**, No. 3, 410 (1953).
H2. Hatch, L. P., Regan, W. H., Manowitz, B., and Hittman, F., Geneva Conference Paper No. 553 (1955).
H3. Higgins, I., AECD 3809 Mar., 1955. Dev. of the Continuous Ion Exchange Contactor.
I1. Recommendations of the International Commission on Radiological Protection, ICRP/54/4. *Brit. Inst. Radiol.*, London, 1954.
K1. Kaufman, W. J., Klein, G., Greenberg, A. E., *U.S. Atomic Energy Comm.* Wash. **275**, San. Eng. Conf., Apr. 1954, p. 1.
L1. Langmuir, I., and Blodgett, K., OSRD 3460 (1944). Smokes and Filters.
M1. Manowitz, B., *Chem. Eng. Progr. Symposium Ser.* **50**, 201 (1954).
M2. Manowitz, B., Bretton, R. H., and Horrigan, R. V., *Chem. Eng. Progr.* **51**, 313 (1955).
M3. Mawson, C. A., Geneva Conference Paper No. 12 (1955).
N1. *Natl. Bur. Standards (U.S.) Handbook* **52** (1953).
O1. O'Connell, H. E., and Pettyjohn, E. S., *Trans. Am. Inst. Chem. Engrs.* **42**, 795 (1946).
O2. O'Gara, P. J., and Fleming, E. P., American Smelting and Refining Co. unpublished data.
R1. Reid, D. G., Geneva Conference Paper No. 543 (1955).
R2. Rodebush, W. H., *U.S. Atomic Energy Comm.* AEC Handbook on Aerosols, pp. 60–64.
R3. Rodebush, W. H., Holly, C. E., Lloyd, B. A., OSRD 2050 Filter Penetration by Aerosols (1943).
R4. Rodger, W. A., The Handling of Radioactive Wastes. Talk given before the Natl. Ind. Conf. Board (October 1955).
R5. Rupp, A. F., Oak Ridge National Laboratory, Fission Product Processing. Talk given at Oak Ridge Natl. Lab.
S1. Silverman, L., Geneva Conference Paper No. 571 (1955).
S2. Struckness, E. G., Morton, R. J., and Straub, C. P., Geneva Conference Paper No. 554 (1955).
V1. Vorkauf, H., *Forchungsheft* **341** (1931).
W1. Walters, W. R., Weiser, D. W., Mareck, L. J., *Ind. Eng. Chem.* **47**, 61 (1955).

HIGH VACUUM TECHNOLOGY

George A. Sofer and Harold C. Weingartner

Nuclear Development Corp. of America, White Plains, N.Y.
and Arthur D. Little, Inc., Cambridge, Mass.

I. Introduction.. 117
II. Historical Development of High Vacuum Technology................... 119
   A. Freeze Drying... 120
   B. Reduction of Magnesium... 120
   C. Vacuum Melting.. 121
   D. Isotope Separation... 121
   E. Vacuum-Tube Manufacture... 121
   F. Vacuum Coating.. 122
   G. Vacuum Metallurgy... 122
   H. Vacuum Drying... 123
   I. Future Development.. 124
III. Chemical Engineering and High Vacuum Technology.................... 124
   A. Mass Balance... 124
   B. Fluid Flow.. 125
   C. Heat Transfer.. 129
   D. Mass Transfer.. 131
   E. Reaction Equilibrium... 134
IV. Vacuum Pumps and Gages.. 136
   A. Vacuum Pumps... 137
      1. The Mechanical Vacuum Pump................................. 137
      2. The Mechanical Booster.. 139
      3. Multi-Stage Steam Ejectors..................................... 139
      4. The Oil Ejector.. 140
      5. The Diffusion Pump... 140
   B. Vacuum Gages.. 142
      1. The McLeod Gage... 143
      2. The Alphatron Gage.. 143
      3. The Hot-Filament Ionization Gage............................. 143
      4. The Philips Gage.. 144
      5. The Thermocouple Gage.. 144
V. Conclusion... 145
   Acknowledgment.. 145
   References... 145

I. Introduction

High vacuum is a term generally referred to pressures below 0.1 mm. of mercury. It is understood to be the pressure region where the laws of fluid flow in the turbulent and laminar ranges do not apply, and where molecular flow either partially or totally characterizes the pattern of fluid flow.

The transition from laminar to molecular flow is gradual. This transitional region is called the "slip flow" region. Flow values in this region lie between values predicted by the laws of laminar flow and those predicted by the laws of molecular flow.

A criterion for determining whether high vacuum conditions exist is that the mean free path of gases in a system must be of the same order of magnitude as the dimensions of the system, or larger.

High vacuum is not and should not be regarded as a unit operation in itself. Rather, it should be viewed as a very low pressure range in which conventional unit operations such as heat transfer, fluid flow, distillation, and extraction are put to use in the light of physical and chemical equilibria, reaction rates, and transfer rates characteristic of this pressure range.

The beneficial influences of high vacuum are entirely analogous to those obtained in the use of high pressure for certain processes.

The influence of high vacuum on diffusional unit operations, such as distillation, drying, impregnation, evaporation and desorption, is to reduce the partial pressure of the inert component (usually air) and thereby increase the rate of mass transfer of the diffusing fluids. In addition to affecting the rate of diffusional operations, high vacuum also influences equilibrium conditions. These two factors work together in bringing such operations nearer to completion. Its effect on heat transfer and fluid flow stems from the drastic reduction of intermolecular collisions and the greater significance assumed by gas-to-wall collisions.

Chemical reactions involving gaseous reactants or products can be influenced by high vacuum in the direction of greater molar volume, as in the case of the dissociation of certain metallic oxides into oxygen and free metals, and the reduction of such oxides by carbon with the formation of CO or CO_2.

Quite aside from physical or chemical reactions, an important function served by high vacuum is the provision of collision-free space, such as required in radio and television tubes, and particle accelerators. In these applications, charged particles must travel relatively long distances before reaching their target. Obviously, their path will be unimpeded only when the probability of collision with residual gas molecules is very low. A similar function is served in vacuum coating, where metal vapor is condensed on a suitable substrate some distance from an evaporation source.

The requirement for high vacuum often stems from the need for chemically inert surroundings necessary for handling reactive metals at high temperatures. For metals such as titanium and zirconium, the heat treating atmosphere must be free of oxygen, nitrogen, and hydrogen, as all of these gases tend to enter into reaction and result in a highly embrittled product. Such operations are either conducted under high vacuum, or in an inert gas atmosphere. The inert gas is usually introduced to the heat-treating equipment following the reduction of air pressure to the high vacuum range.

Research is currently being conducted at extremely low pressure, in the range of 10^{-10} to 10^{-14} mm. Hg abs. In commercial processes today the lower range of pressures employed is of the order of 10^{-6} and 10^{-7} mm. Hg abs., and pressures below this range are frequently referred to as ultrahigh vacuums. It should be readily apparent that the only sound basis for designating high vacuums is in terms of absolute pressure.

In the material which follows, the industrial use of high vacuum is discussed. It will be noted that recently gained knowledge has not yet been reduced to systematically applicable data that are useful and available for engineering design purposes. Furthermore, since useful applications of high vacuum are under investigation and development in many different industries, there exist differences in concept and nomenclature which hinder the unification of information.

A commendable action aimed at overcoming this condition is the organization of the Committee on Vacuum Techniques. This organization is devoting a considerable effort to the development of nomenclature and standards in this field.

II. Historical Development of High Vacuum Technology

The attainment of vacuum dates back to the seventeenth century. Torricelli, in 1643, demonstrated that atmospheric pressure can support the weight of a column of mercury 76 centimeters high. The pressure he attained in the evacuated space above the mercury column, must have been well below 10 mm. Hg absolute. Boyle recognized the influence of vacuum by his statement that "Nature abhors vacuum." Otto von Guericke, in 1654, developed a mechanical pump which he used to evacuate the famous Magdeburg spheres.

Not until the latter part of the nineteenth century, however, did vacuum become a very useful tool. In the hands of J. J. Thomson, it played an important part in the discovery of the electron, and in the hands of Thomas Edison, it made possible the invention of the first electric lamp. In later years, vacuum was useful in studying the internal structure of the atom, and in creating an entirely new industry, the electronics industry.

During World War I, the German firm Heraeus Vakuumschmelze found it advantageous to melt various metals, including steel, under pressures in the range from 1 to 50 mm.

The years of World War II witnessed an extensive industrial application of high vacuum in a number of major war efforts. This was made possible by development of the following vacuum pumps: (a) mechanical vacuum pumps with pumping speeds as high as 200 cu. ft./min. and capable of reaching pressures as low as 0.01 mm. Hg; (b) multi-stage, high capacity steam ejectors capable of producing an ultimate pressure of 0.1–0.05 mm. Hg; (c) vapor diffusion pumps with volumetric speeds as high as 25,000 cu. ft./min. and ultimate pressure of the order of 10^{-6} mm. Hg.

The following projects are good examples of wartime high vacuum application.

A. Freeze Drying

Blood plasma was dehydrated from the frozen state under pressures of the order of 0.3 mm. Hg, and temperatures near $-30°C$. The need for dehydration arose because plasma in aqueous solution spoiled, whereas storing as a dry powder drastically reduced deterioration. Dehydration from the solid state permitted easy re-solution in water or saline solution, by virtue of the high surface-to-volume ratio characteristic of the lyophilic structure produced by this technique. The rigidity of the solid state does not permit substantial shrinkage of the drying material. Dehydration from a liquid solution, on the other hand, is accompanied by excessive volumetric shrinkage, and the result is an amorphous coagulum of low surface-to-volume ratio, and hence poor resolution. Penicillin was dehydrated in very much the same manner, then streptomycin and many other pharmaceutical products were similarly processed.

B. Reduction of Magnesium

The Pidgeon process for reduction of magnesium from dolomite contributed a large portion of total output of this critical metal during World War II. This process (B1) consists of heating calcined dolomite and ferrosilicon in horizontal retorts, and collecting the distilled metal on removable condenser linings. Reaction temperature is held at approximately 1150°C., and a pressure of 0.100 mm. Hg absolute is maintained, which serves to protect the magnesium vapor from reoxidation. The reaction can be represented as follows:

$$\text{MgCO}_3 \cdot \text{CaCO}_3 \xrightarrow{\text{heat}} \text{MgO} \cdot \text{CaO} + 2\,\text{CO}_2$$
Dolomite   Calcined dolomite

$$2\,(\text{MgO} \cdot \text{CaO}) + \tfrac{1}{6}\,\text{FeSi}_6 \xrightarrow{1150°\text{C}} 2\,\text{Mg}\uparrow + \tfrac{1}{6}\,\text{Fe} + (\text{CaO})_2 \cdot \text{SiO}_2$$
Ferro-silicon

High vacuum serves further, by reducing the reaction temperature a few hundred degrees below that necessary under atmospheric pressure, for the same reaction yield.

Reduction of calcium from lime by a similar reaction was carried out in retorts similar to those used for magnesium. Here aluminum, rather than silicon, was used as the reducing agent, in order to maintain a sufficiently low reaction temperature.

$$6\,\text{CaO} + 2\,\text{Al} \xrightarrow[0.100 \text{ mm. Hg}]{1150°\text{C}} 3\,\text{Ca}\uparrow + (\text{CaO})_3 \cdot \text{Al}_2\text{O}_3$$

Other low boiling metals, such as lithium, barium, and columbium, have been reduced by similar reactions, but not on the large scale of magnesium or calcium.

C. Vacuum Melting

In order to achieve the maximum degree of homogeneity and purity in fissionable metals, melting under pressures below 1 mm. was utilized in the war years to reduce the content of dissolved and entrapped gases.

Although vacuum melting in this country was limited, Heraeus Vakuumschmelze was producing a large percentage of the nickel-chromium alloys used in Europe by a vacuum melting technique.

D. Isotope Separation

Separation of fissionable uranium-235 from nonfissionable uranium-238, was carried out by diffusion under high vacuum of suitable gaseous compounds of these metals. This application demanded gigantic installations of high vacuum equipment (B3).

E. Vacuum-Tube Manufacture

Although all vacuum tubes require evacuation, the expanded demand for cathode ray tubes imposed the need for pumps with ultimate pressures of the order of 10^{-6} mm. Hg. The distance of travel of electrons in cathode ray tubes is considerably greater than in receiving tubes; the residual gas pressure must therefore be sufficiently lower to prevent collisions with the electron beam.

Following World War II, a number of new applications opened up for high vacuum, and many wartime uses were extended to peacetime needs. Among the post-war applications, which attained significant commercial importance, are the following.

F. Vacuum Coating

This process, which was developed just prior and during World War II, found a nation-wide market in the metallizing of plastics and other materials. During the war, high vacuum was employed in the evaporation of magnesium fluoride onto optical lenses for the purpose of reducing reflection at the air-to-glass interface, thereby improving the light transmission efficiency of multi-lens systems. Since the war, the same technique was found commercially applicable for evaporating aluminum onto plastic novelty articles, and inexpensive aluminum or zinc die castings for use in imitation jewelry.

It was found necessary to pre-coat the substrate with a smooth, glossy, synthetic organic finish prior to vacuum metallizing. This permits the attainment of a highly specular finish for only a fraction of the cost required for polishing. In those applications where the coated article must remain serviceable after appreciable handling, the evaporated aluminum film is protected by a synthetic finish topcoat. At present this finish is still inferior to electroplating in respect to abrasion resistance. Rapid improvements in vacuum coating, however, are likely to make these processes directly competitive.

A similar use of the technique of vacuum evaporation, is put to work in the manufacture of aluminized TV picture tubes. In this case, the picture tube is used as the vacuum chamber and aluminum is flashed from a heated tungsten filament projecting into the neck of the tube. The aluminum deposits on the inside surface of the tube and covers the phosphor screen. When in use, the aluminum film serves to reflect the light emitted from the phosphor screen which would otherwise be lost inside the tube. In this respect, the aluminum film fulfills the same function as the mirror in back of a kerosene lamp.

Although the majority of vacuum coating installations now in use are batch-type units, a few have been made for semicontinuous coating of paper and plastic films such as cellophane, cellulose acetate, and Mylar. Mylar exhibits unusual adhesion characteristics for the vacuum-deposited aluminum film, and is potentially the source of a host of decorative and functional products.

G. Vacuum Metallurgy

Since the end of the war, the use of vacuum melting was extended to include steels, copper, chromium-nickel alloys, titanium, etc. The actual melting process may or may not take place under high vacuum, but adequate time is allowed subsequently for desorption of dissolved gases under high vacuum.

Vacuum-melted steels have been proven to have longer fatigue life, and their freedom from inclusions has spurred their use in the manufacture of miniature ball bearings. Vacuum-melted copper has found applications in large power vacuum tubes, and vacuum-melted chromium-nickel alloys are mainly used for high temperature turbine blades. Because of their high melting point and high chemical reactivity, titanium, zirconium, and molybdenum are now melted under vacuum as a standard technique.

Heat treating, sintering, and brazing of these metals under vacuum has also become an accepted practice. The increasing use of titanium and related titanium alloys, is accompanied by an expansion of the use of these techniques.

H. Vacuum Drying

The use of freeze-drying for pharmaceuticals did not grow appreciably after the end of the war, because it was no longer used for penicillin, and today plasma is no longer being stockpiled. Other vacuum drying processes, however, have come to the fore. In the food industry, the concentration of orange juice under vacuum has revolutionized the marketing of citrus products. Annual sales of vacuum-concentrated citrus juices now exceed $100 million. Concentration of orange juice is carried out at a pressure of 5–25 mm. Hg, maintained by multistage steam ejectors. A four to one concentration is achieved.

Recently, vacuum drying has been extended to hermetically sealed refrigeration and air conditioning compressors. Since the motor windings in such compressors communicate with the passages of the refrigerant, it is of prime importance to reduce the moisture content of these units to an absolute minimum. Failure to reduce this moisture results in ice clogging of the expansion capillary tube. The development of the gas ballast mechanical vacuum pump is a large factor in making this process commercially feasible. In these pumps, fresh air is introduced toward the end of the compression cycle to dilute the water vapor and prevent it from condensing, by maintaining its partial pressure lower than the vapor pressure at the temperature of the pump. Thus by preventing condensation in the vacuum pump, it is possible to pump larger quantities of water vapor without deterioration of the pumping speed and ultimate pressure. In the absence of gas ballast, the condensing water vapor emulsifies with the lubricating oil, finds its way back to the vacuum side of the pump, and prevents the attainment of low ultimate pressure.

I. FUTURE DEVELOPMENT

In the future, vacuum metallurgy, perhaps, will be the fastest growing application of vacuum technology. The increasing use of refractory metals will be a major factor in this growth. Vacuum brazing for high temperature application will assume more significance. Compressor dehydration and continuous coating are also likely to expand. The increasing production of color television will see a second round of expansion in exhaust and aluminizing equipment for the electronics industry. More automatically operated equipment assemblies will be put into use. It is likely that further growth of vacuum applications in the chemical process industries will take place, but further advances of chemical engineering technology in this field will depend upon the commercial incentives which exist in the near future.

III. Chemical Engineering and High Vacuum Technology

In this section the laws that govern high vacuum processes are briefly outlined. Although in many instances development of these laws has not progressed to a point where application to practical situations can be achieved by simple and straightforward calculations, it is useful to recognize the limitations and potentialities revealed by these laws.

A. MASS BALANCE

Mass balance considerations of high vacuum processes might appear strange to the chemical engineer who is accustomed to thinking in terms of pounds or pound-moles. The units of mass commonly used in high vacuum literature are the micron-liter and the micron-cubic-foot, in the United States, and the torr-liter or the micron-cubic meter in Europe. A micron-liter is a unit of mass and not a unit of volume, as it represents the mass of one liter of the gas in question at a pressure of one micron (0.001 mm. Hg abs.), and room temperature (usually meant to be 26°C.). A torr is a pressure of 1 mm. Hg, thus a torr-liter is a thousand times greater than a micron-liter. The units of throughput take the form, micron-liter per second, micron-cubic-foot per minute, or micron-cubic-meter per hour.

The use of such units as measure of throughput permits a simple relationship between throughput, pressure, and the volumetric rate of displacement. Thus if the volumetric rate of displacement is measured in liters per second and the pressure in microns, the mass throughput, measured in micron-liters per second, is simply the product of the volumetric rate of displacement and the pressure. The volumetric rate of displacement is termed "speed," whence the common formula

$$Q = SP \tag{1}$$

where Q is throughput, S is speed, and P is pressure.

When studying the steady state gas flow through a pipe or a series of pipes and fittings, the mass balance principle implies a constant throughput, Q. If the pressure at the mouth of the vacuum pump is known and the speed of the pump at this pressure is given, the product of these two factors gives the constant throughput which must hold at each point upstream of the pump (in the absence of any outgassing from the walls of the flow lines). Since pressure must be higher upstream in order for the gas to flow, it follows that the "speed" will vary along the length of the flow lines inversely as the pressure. "Speed" corresponds to the product VA in the familiar equation

$$W = \rho V A \tag{2}$$

where W = rate of mass flow, ρ = density, V = velocity, and A = area.

The units of speed and throughput in use in the field of high vacuum, although offering a simple short-cut for relating these variables to pressure, lead to great confusion when attempting to write force balance and energy balance equations. The chemical engineer has solved the difficulties arising from simultaneous use of the mass-pound and the force-pound. He has an even greater task in reconciling mass units expressed in terms of pressure and volume to those written in terms of force and true mass.

B. Fluid Flow

Comparatively little experimental work has been done on the flow of gases under vacuum. Although theoretical relationships can be derived for molecular flow (K1) and for viscous flow (P1), few data other than those reported by Cheng (C1) and Brown et al. (B2), throw any light on the behavior of gases in the region of transition from viscous to molecular flow.

Theoretical flow equations were derived for the molecular flow region by Knudsen (K1) as far back as 1909. These equations for molecular flow and Poiseuille's Law for laminar flow, were the basis for vacuum flow computation until the later years of World War II. Normand (N1) was the first to translate these equations into practical forms for engineering applications. In this reference Normand gives useful empirical rules for applying Knudsen's equations to ducts of rectangular cross-section, non-uniform cross-section, baffles, elbows, etc.

Cheng (C1) in collaboration with National Research Corporation was the first to carry out exhaustive experimental tests for correlating pressure drop under high vacuum with such parameters as pipe length, pipe diameter, mean gas pressure, and gas throughput. The range of pressures investigated covered the regions of laminar flow, molecular flow and the region of transition between laminar to molecular flow, labeled as the region of slip flow. Cheng's data were correlated by an equation essentially in the form of Poiseuille's Law, with a correction factor which assumes increasingly greater importance at lower pressures. This equation applies throughout the range of laminar flow, i.e., Reynolds Number < 2100, as well as in the molecular flow region where the mean free path is so large that intermolecular collisions in the gas phase are very infrequent. This equation is

$$Q = \frac{\pi g_c R^4}{16 \mu l} (p_i^2 - p_e^2) \left[1 + 4 \left(\frac{\pi}{2}\right)^{1/2} \left(\frac{2}{f} - 1\right) \frac{\mu}{p_m R (\rho_1 g_c)^{1/2}} \right] \quad (3)$$

where Q is gas throughput, micron-cm.^{3}/sec.

g_c is a constant numerically equal to the acceleration of gravity and having for dimensions

$$\frac{\text{mass-length}}{\text{force-(time)}^2} = 1.33 \frac{\text{g. mass}}{\text{micron Hg} - \text{sec.}^2\text{-cm.}}$$

f is a molecular reflection factor whose theoretical significance lies in the fact that it is the fraction of all incident molecules which are "diffusely" reflected off a wall (the remaining molecules are reflected "specularly," i.e., with angle of incidence equal to angle of reflection)

R is pipe radius, cm.

l is pipe length, cm.

ρ_1 is $\dfrac{\text{mass density}}{\text{pressure}}$, $\dfrac{\text{g.}}{\text{cm.}^3 - \text{micron}}$

p_i is inlet pressure, microns

p_e is exit pressure, microns

p_m is mean pressure, microns

μ is viscosity at atmospheric pressure, poises

Equation (3) reverts to Poiseuille's Law at sufficiently high pressures, where the second term in the large parenthetical factor becomes negligible compared to unity. As the pressure is decreased and intermolecular collisions become less frequent, a flow velocity profile is established where the forward velocity component near the wall becomes a finite value which increases with lower pressures. The reason for this condition is that at these pressures many molecules can stream from the bulk of flow, where the forward velocity is relatively high, to the wall, without suffering collisions with molecules having low forward velocities. Gas flow under such conditions is termed slip flow. Pressures corresponding to this type of flow are such that the second term of the "correction factor" in equation (3) is finite compared to unity. At lower pressures, where the mean free path is considerably larger than the radius of the pipe, the second term in the "correction factor" becomes dominant resulting in the equation for molecular flow

$$Q = \left(\frac{\pi}{2}\right)^{3/2} \left(\frac{2}{f} - 1\right) \frac{R^3 g_c}{l(\rho_1 g_c)^{1/2}} (p_i - p_e) \tag{4}$$

Thus we have one universal equation which applies throughout the entire range of high vacuum. In fitting these equations to experimental data,

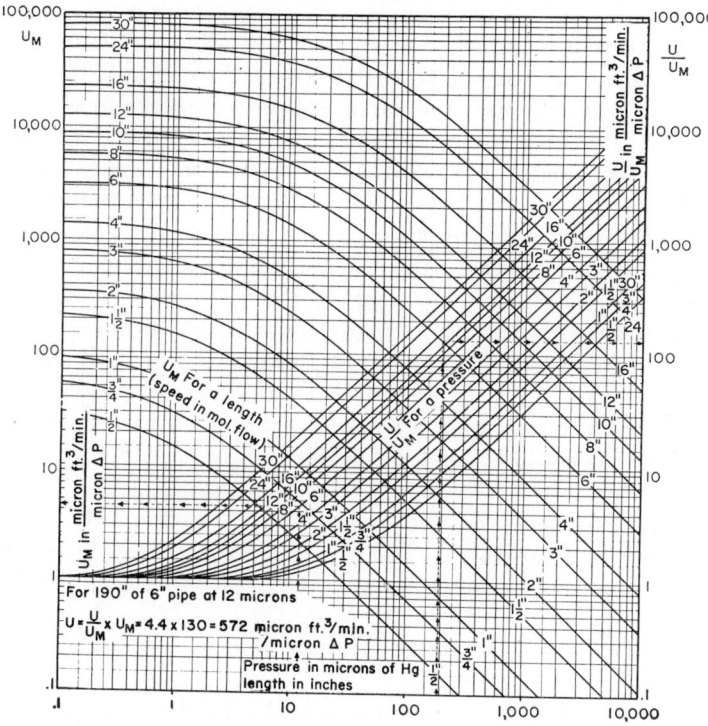

FIG. 1. Brown-DiNardo, and Sherwood curves for pressure drops through pipe.

it was discovered that a molecular reflection factor f, equal to 0.84, best represents the data for glass and copper pipe. A factor f equal to 0.9, gives better agreement with data obtained on iron pipe. The above equations do not apply as such to short pipes or orifices, but are sufficiently accurate for pipes having a length to diameter ratio greater than 10.

A set of curves based on the above equations is given by Brown et al. (B2), which is useful for calculating throughput pipe size, pipe length, and inlet and exit pressures are given. These curves are given in Fig. 1. The molecular conductance U_m is read from the series of curves "U_m for length." The ratio of actual conductance to molecular conductance, U/U_m, is read off the curves "U/U_m for a pressure." The pressure used in this reading should be p_m, the arithmetical mean of the inlet and exit pressures. U can now be calculated from the equation

$$U = \left(\frac{U}{U_m}\right) \cdot U_m \tag{5}$$

and Q is determined from

$$Q = (p_i - p_e)U \tag{6}$$

To solve the common problem where determination of the inlet pressure and the pressure drop is required, given throughput, pipe size, and exit pressure, tedious trial and error solution is necessary when using these curves. The importance of solving such a problem can be appreciated when one considers the condition where a mechanical or diffusion pump of a known size, and hence known throughput, is connected at the end of a pipe of a known size and length. The exit pressure of the pipe is the inlet pressure of the pump, and hence defines a throughput value. In this instance we know the throughput, the exit pressure, and the pipe diameter and we are usually interested in pressure drop across the pipe and the inlet pressure. It is important to have a direct solution for this sort of problem, particularly where more than one pipe size is interposed between a given pump and a vacuum chamber. Lawrance (L2) has rewritten Eq. (3) so that a direct solution can be obtained in each of the following cases:

(1) Given exit pressure, throughput, pipe diameter. Required: pressure drop and inlet pressure

(2) Given exit pressure, inlet pressure, and throughput. Required: pipe diameter

(3) Given exit pressure, inlet pressure, and pipe diameter. Required: throughput.

For the problem where inlet pressure, throughput, and pipe diameter are given, the Lawrance equation can lead to a negative or imaginary value for the exit pressure, if the known variables are not physically compatible. The Lawrance equation is

$$p_i - p_e = \sqrt{p_{ea}^2 + W} - p_{ea} \tag{7}$$

where

$$p_{ea} = p_e + A\left(\frac{2-f}{f}\right)\frac{l}{D} \tag{8}$$

and

$$W = \frac{BK_L l Q}{D^4} \tag{9}$$

Table I gives values for the constants A and B for air and water vapor for a number of commonly used units.

For length-to-diameter ratios greater than 10, K_L can be taken equal to 1. For short pipes of length-to-diameter ratio less than 10, the proper value for K_L can be obtained from Fig. 7 of Lawrance's paper (L2). D in the above equations is the pipe diameter, K_L is a constant. All other symbols have the definitions given in conjunction with Eq. (3).

TABLE I

Units			Air at 26°C.			Water vapor at 26°C.		
p	Q	l and D	A	B	C	A	B	C
micron Hg	micron-liter/sec.	cm.	40.9	11.4	0.033	26.8	5.82	0.0257
micron Hg	micron-liter/sec.	in.	16.1	0.694	0.0051	10.54	0.354	0.00397
micron Hg	micron-ft.³/min.	in.	16.1	0.327	0.0024	10.54	0.167	0.00187

For short pipes, in addition to pressure drop thus calculated, allowance must be made for inlet losses given by the semi-empirical equation

$$\delta p_i = \frac{CQ}{D^2} \tag{10}$$

where C is a constant given by Table I for air and water vapor.

It should be pointed out that the Lawrance equation and the Brown-DiNardo curves are true only at room temperature. Data such as those obtained by Cheng should be extended to gases other than air and water vapor, to non-circular conduits, and to temperatures encountered in vacuum metallurgy, and in freeze drying. Whether or not such work might be undertaken will, of course, depend on its commercial and technological significance.

C. Heat Transfer

The effect of high vacuum on heat transfer is best understood in the light of low molecular density and a low frequency of intermolecular collisions.

Transfer of heat by radiation is not influenced by pressure except in rare cases where absorption of heat in the gas is important. Under reduced pressure, the effect of such absorption would be reduced or become negligible, thus simplifying radiant heat calculations. The relative importance of heat transfer by radiation is increased under vacuum.

Transfer of heat by convection assumes a lower degree of importance in vacua. Since convection is defined as the transfer of heat by the movement of a large number of molecules en masse, this type of heat transfer is non-existent under conditions of molecular flow, where collision between individual gas molecules is a rare occurrence, and where for the large part, gas molecules travel about unaffected by the presence of other molecules.

In the region of slip flow, heat transfer by convection is significant only at a distance from the heat source greater than the mean free path. Within a distance of one mean free path, just as in molecular flow, heat transfer occurs by molecular conduction alone. Kyte et al., describe a method of calculation for heat transfer in slip flow (K2). The technique employed is essentially a trial and error solution whereby a temperature, t_a, is determined at a distance of one mean free path away from the heat source. The rate of heat flow from the hot surface is calculated on the basis of molecular conduction over the distance of one mean free path. The rate of heat flow beyond the distance of one mean free path is calculated on the basis of convection. The correct value of the temperature t_a, is that which results in equal rates of heat flow by conduction and by convection.

Under conditions of laminar flow, the usual natural convection equations can be used. Reference (K2) gives a table of heat transfer equations for spheres and cylinders recommended for use when molecular conduction is a factor, and a second table applicable to natural convection under laminar flow conditions.

The equations for spheres, applicable when the product of the Grashof number and the Prandtl number lies between $10^{1.5}$ and 10^9 are as follows:

(1) Effect of free-molecular conduction negligible:

$$q = 2\pi D k_{w,g} \frac{t_w - t_g}{1 - \frac{1}{1 + 5.01/(GrPr)^{0.26}}} \quad (11)$$

(2) Effect of free-molecular conduction important: (a) for heat transfer by convection,

$$q = 2\pi D' k_{a,g} \frac{t_a - t_g}{1 - \frac{1}{1 + 5.01/(GrPr)^{0.26}}} \quad (12)$$

(b) for heat transfer by molecular conduction:

$$q = \pi D^2 \Lambda_o P \alpha \sqrt{\frac{273.2}{T_a}} \cdot (t_w - t_a) \quad (13)$$

For a correct solution, a temperature t_a must be chosen so as to give equal values for q when calculated by Eqs. (12) and (13).

It will be noted that Eqs. (11) and (12) are identical except that

Eq. (12) is written for the equivalent diameter $D' = D + 2\lambda_a$ at which point the temperature is t_a, rather than the actual diameter D.

q = heat flow, heat units per unit time, Btu/hr.
D = diameter, ft.
$k_{a,g}$ = conductivity of the gas at a temperature $(t_a + t_g)/2$, Btu/ (hr.)(ft.)(°F.)
$k_{w,g}$ = conductivity of the gas at a temperature $(t_w + t_g)/2$
t_a = temperature at one mean free path from the wall, °F.
t_g = temperature of the gas at a large distance from the wall, °F.
t_w = temperature at the wall, °F.
Gr = Grashof number $(D^3 \rho^2 g_o / \mu^2)(\beta \, \Delta t)$
Pr = Prandtl number $(C\mu/k)$
D' = equivalent diameter at one mean free path from the wall, ft.
T_a = absolute temperature, °K.
Λ_0 = free-molecular conductivity of the gas at 0°C., Btu/(hr.)(ft.^{2})(°F.) (micron Hg)
P = pressure of the gas, microns
α = accommodation coefficient, equal to 0.90 for air, 0.50 for helium, and 0.95 for argon
λ_a = mean free path, ft.

Although very useful, the above solution to heat transfer under slip flow and molecular flow conditions, is rather tedious leaving a definite incentive for simplification. Extension of this treatment to other geometrical shapes and elimination of the trial-and-error solution might be the next objective.

D. Mass Transfer

With reduction of pressure the mean free path grows larger, and with it the resistance to mass transfer due to intermolecular collision is progressively diminished. At pressures corresponding to mean free paths larger than the dimensions of the vessel in question, the gas phase resistance to mass transfer is negligible and the only limiting factor on the rate of material movement is the rate of emission from the interface.

Examples of mass transfer under high vacuum are distillation of thermally unstable organic compounds, high-vacuum freeze drying, vacuum concentration of fruit juices, vacuum drying of coffee concentrate, vacuum purification of molten metals, etc. The mechanism of mass transfer in the concentration of fruit juices can be adequately described by the Gilliland-Sherwood equations (G1), since the mean free path is negligible relative to the dimensions of the vessel at the pressures used. These equations show that the mass transfer coefficient is inversely proportional to the pressure. By extrapolation to very low pressures, it should be possible to reach an ever-increasing mass transfer rate. Of course, these equations do not take into account the surface limitation to mass transfer, which at low pressures can be the controlling factor. Under molecular flow conditions the emission from the surface is controlling, and the rate of mass transfer can be fully described by the Knudsen (K1) or Langmuir (L1) equation for surface emission derived from the kinetic theory. Here the rate is independent of the ambient pressure and is proportional to the vapor pressure of the material in question.

$$\frac{dW}{d\theta} = k_1 A \tag{14}$$

$$k_1 = P_v \sqrt{\frac{Mg_c}{2RT}} \tag{15}$$

where $dW/d\theta$ = rate of mass transfer, g./sec.
k_1 = mass transfer coefficient, g./(sec.)(cm.^{2})
A = area of surface in question, cm.^{2}
P_v = vapor pressure of solid or liquid in question, microns Hg
M = molecular weight, g./g. mole, of emitted substance
g_c = constant, numerically equal to the acceleration of gravity whose dimensions are

$$\frac{\text{mass-length}}{\text{force-(time)}^2} = 1.33 \frac{\text{g. mass}}{(\text{micron Hg}) - (\text{sec.}^2)(\text{cm.})}$$

$$R = \text{gas constant} = 62.37 + 10^6 \frac{(\text{microns Hg}) - (\text{cm.}^3)}{(\text{g. mole}) - (°K.)}$$

T = absolute temperature, °K.

The above equations give the maximum rate of evaporation. Although validated by many observations, these equations at times give rates of evaporation as much as 100 times higher than the measured values. This deviation from the Langmuir equation is not limited to a few singular solids or liquids. The same liquid which is noted to obey the Langmuir equation in one test, can be found to deviate widely from it in another. Hickman (H1) has proved that the Langmuir equation will hold true for all liquids he tested, including water, if the interfacial surface is constantly renewed, thereby removing surface impurities which retard the traffic across the interface. A surface once highly emissive becomes "torpid" or inactive upon standing in a glass flask. Upon removal of the impurities collected at the surface by overflowing the liquid into a second flask, the surface loses its torpidity and becomes active again.

The Langmuir equation can be written

$$\frac{dW}{d\theta} = \alpha k_1 A \tag{16}$$

where α is the emission coefficient (also termed the accommodation coefficient). The emission coefficient is an illusive variable lying between zero and unity depending upon the purity of the exposed surface.

Mass transfer data in the region of transition between molecular flow and laminar flow have been recently gathered by Cooke (C2). In a study of sublimation of naphthalene spheres, under vacuum, he discovered that the data collected can be correlated by a concept of "over-all" mass-transfer coefficient equal to the reciprocal of the sum of the resistance in the gas phase and a resistance corresponding to the Langmuir equation for the interface. An emission or accommodation coefficient of 0.10 was found to fit the data.

To illustrate this work, the mass transfer equation for sublimation from spheres at pressures near atmospheric, according to Ranz and Marshall (R1), is

$$\frac{dW}{d\theta} = \frac{2\,D_0 M\,A\,P_v}{R\,T\,Pd_s}[1 + 0.30(Re_s)^{1/2}(Sc)^{1/3}] \tag{17}$$

This can be written as

$$\frac{dW}{d\theta} = \frac{k_2 A}{P} \tag{18}$$

and the principle of additive resistances leads to an equation applicable over the entire range of pressure

$$\frac{dW}{d\theta} = \frac{\alpha k_1 A}{1 + \frac{\alpha k_1 P}{k_2}} \tag{19}$$

where

$$k_2 = \frac{2\,D_0 M\,P_v}{R\,T\,d_s}[1 + 0.30(Re_s)^{1/2}(Sc)^{1/3}] \tag{20}$$

where D_0 = diffusivity of vapor at 0°C. and 1 atmosphere × atmospheric pressure, $\frac{\text{(microns Hg)(cm.}^2)}{\text{sec.}}$

M = molecular weight, g./g. mole
A = area, cm.^{2}
P_v = vapor pressure, microns Hg
R = gas constant = 62.37×10^6 (microns Hg)(cm.^{3})/(g. mole)(°K.)
T = absolute temperature, °K.
d_s = diameter of sphere, cm.
P = system total pressure, microns Hg
Re_s = Reynolds number, $\frac{d_s V \rho}{\mu}$

Sc = Schmidt's number $\frac{\mu}{\rho D}$

k_1 = maximum coefficient of mass transfer for molecular flow, g./sec.-cm.^{2}

k_2 = coefficient of mass transfer, micron Hg-g/cm.^{2}-sec.

α = emission (or accommodation) coefficient

ρ = density, g./cm.^{3}

V = velocity, cm./sec.

μ = viscosity, g./sec.-cm.

D = diffusivity, cm.^{2}/sec.

Inspection of Eq. (19) reveals that at low pressure the second term in the denominator vanishes, and Eq. (19) reduces to Eq. (16). At high pressure the second term in the denominator dominates, and the equation is reduced to Eq. (18).

The biggest obstacle to progress made in establishing mass transfer rates under vacuum will continue to be the variable emission coefficient. Whether or not the accommodation coefficient can become predictable will not be known before a wealth of data is accumulated, and such data will not be accumulated in the near future until there arise commercial incentives sufficient to justify the expense.

E. Reaction Equilibrium

By the same mechanism that high pressure can be used to drive a certain reaction in the direction of smaller molar volume, as for example, in the synthesis of ammonia from hydrogen and nitrogen, high vacuum is employed to shift reaction equilibria in the direction of greater molar volume. Examples of such reactions beneficially carried out under vacuum are the reduction of copper or nickel oxide impurities in copper or nickel melts. Reduction of iron, titanium or aluminum oxides by dissociation under vacuum is impractical because of the extremely low oxygen partial pressures required. For instance, the equilibrium partial pressure of oxygen over titanium oxide at its melting point (1750°C.) is of the order of 10^{-13} mm. Hg, and for aluminum at its melting point, 10^{-50} mm. Hg. The equilibrium partial pressure of carbon monoxide over carbon and oxide mixtures is considerably higher, however. Reduction of metallic oxide impurities in the presence of carbon is, therefore, a common vacuum melting practice for metals like steel where a relatively small concentration of carbon is permissible. The solubility of carbon in molten titanium is very high, so that vacuum melting of titanium in the presence of carbon is generally avoided.

The reaction of titanium with carbon is utilized, however, in the vacuum fusion technique for micro determination of oxygen content in titanium. The conversion of titanium into titanium carbide in the presence of an excess of graphite is permissible so long as all oxygen in the melt can be collected as carbon monoxide.

Vacuum reduction of oxides of the alkali and alkaline earth metals, has proved a very useful technique, of which the Pidgeon ferrosilicon process for reduction of magnesium (B1) has been an outstanding example. Although this process is not currently competitive with electrolytic reduction, the cost differential is not so great that improved vacuum equipment and vacuum techniques could not swing the balance in favor of this process.

TABLE II
Vacuum Thermal Reductions Which Have Been Investigated Recently

Metal	Reaction
Magnesium	$2 \, (MgO \cdot CaO) + \frac{1}{6} \, FeSi_6 \xrightarrow[(0.005 \text{ to } 0.5 \text{ mm. Hg})]{(1100 \text{ to } 1175°)} 2 \, Mg\uparrow + \frac{1}{6} \, Fe + 2 \, CaO \cdot SiO_2$
	$3 \, (MgO \cdot CaO) + 2 \, Al \xrightarrow[(0.005 \text{ to } 0.5 \text{ mm. Hg})]{(1050 \text{ to } 1175°)} 3 \, Mg\uparrow + 3 \, CaO \cdot Al_2O_3$
Calcium	$6 \, CaO + 2 \, Al \xrightarrow[(0.005 \text{ to } 0.1 \text{ mm. Hg})]{(1100 \text{ to } 1175°)} 3 \, Ca\uparrow + 3 \, CaO \cdot Al_2O_3$
Lithium	$6 \, LiOH + 4 \, Al + 6 \, CaO \xrightarrow[(0.005 \text{ to } 0.1 \text{ mm. Hg})]{(1100 \text{ to } 1175°)} 6 \, Li\uparrow + 2(3 \, CaO \cdot Al_2O_3) + 3H_2$
	$\text{Spodumene} \, [LiAl(SiO_3)_2] + Al \xrightarrow[(0.005 \text{ to } 0.1 \text{ mm. Hg})]{(1050 \text{ to } 1150°)} Li\uparrow + \text{complex silicate}$
Barium	$4 \, BaO + 2 \, Al \xrightarrow[(0.1 \text{ mm. Hg})]{(1050 \text{ to } 1200°)} 3 \, Ba\uparrow + BaO \cdot Al_2O_3$
Columbium	$Cb_2O_5 + 5 \, CbC \rightarrow 7 \, Cb + 5 \, CO$

In this process, the free metal, which is more volatile than the other reaction components, is distilled under vacuum and collected in a suitable condenser. A pressure of less than 0.100 mm. Hg is necessary to prevent excessive reoxidation of the metal vapor. The use of vacuum has the advantage of lowering the temperature level of the reaction by several hundred degrees below that needed at atmospheric pressure. Several improvements are foreseen which might enhance the economic feasibility of the Pidgeon vacuum reduction process, e.g., it might be changed from a batch to a semi-continuous process. This might be achieved by introduction of the feed and removal of waste products through air locks. The vapors might be condensed on the walls of a rotary condenser similar in construction to the ice condenser developed for freeze-drying installations. The scraper blades might be rotated slowly enough to permit a substantial build-up of metal and flaking off due to shrinkage away from the wall. The metal flakes might be withdrawn from the condenser through an air lock. Since the operating pressure is about 0.1 mm. Hg, low maintenance mechanical booster pumps might be used. These pumps are much cheaper than ordinary mechanical pumps of the same capacity at this pressure, and are entirely free from the problems associated with the operating fluid of steam jets, oil ejectors, and diffusion pumps. Further description of these pumps is given in Section IV.

Examples of reduction reactions which can be carried out by the above approach are given in Table II (S1).

IV. Vacuum Pumps and Gages

For proper selection of a suitable pumping system, throughput requirements at various pressures should be known. A wide variety of vacuum pumps are available commercially, suitable for widely different ranges of operation. The pump with the lowest ultimate pressure is not universally desirable. A pump which can reduce the pressure in a particle accelerator down to 10^{-7} mm. is usually inadequate for maintaining a vacuum melting furnace at a pressure of 10^{-1} mm. An oil ejector pump capable of handling a high throughput between 1 and 10^{-2} mm. generally will not yield a pressure lower than 10^{-3} mm. Throughput *vs.* pressure requirements can be established on the basis of: (1) pumpdown time to a certain pressure; (2) extrapolation from small-scale operation; and (3) stoichiometric calculation of gas volumes to be handled in a given length of time.

The pumpdown time from atmospheric pressure to 10^{-1} mm. can be used to set the pumping capacity required in that range, if only bulk evacuation is significant. The following equation can be used to determine the pumping speed required, assuming a constant speed between atmospheric pressure and 10^{-1} mm.

$$S = \frac{V}{\theta} \ln \frac{p_1}{p_2} \qquad (21)$$

where, S is the speed of the pump at the chamber to be evacuated (volume/unit time), θ is the time, V is volume, and p is pressure.

Extrapolation from a smaller scale unit is, in many applications, the most satisfactory approach. Scaling upwards may be done in proportion to the weight of the charge, the surface area exposed to vacuum, or the volume of the pumping chamber, depending on whichever of these variables is the controlling factor.

Stoichiometric calculations may be used as a basis for establishing pumping requirements for desorption of gases from metals, or in vacuum freeze-drying, under conditions where the rate of removal of gas is the controlling variable.

A well recommended procedure for determining the pumping requirements is to plot throughput *vs.* pressure over the entire pressure range, preferably on log-log paper, and in units such as micron-cubic feet/min. or micron-liters/sec. The advantage of these units is that a quick mental conversion from throughput to speed is then possible. The throughput curves of selected pumps can then be drawn up on the same plot for comparison with the required curve. Allowance for pressure drop in the lines should be made by using the Lawrance equations or the Brown-DiNardo curves, Section III, A.

A. VACUUM PUMPS

A great variety of pumps are available for producing and maintaining high vacuum. They differ widely in their principle of operation, and the range of pressures for which they are best suited. A brief description of the pumps most commonly used is given below.

1. *The Mechanical Vacuum Pump*

The mechanical vacuum pump consists of an eccentrically mounted rotor driven inside a cylindrical housing. Two types of mechanical pumps are constructed: (a) the rotary piston type, where the rotor comes in close contact with the housing and thus makes the seal between intake and exhaust compartments, and (b) the vane type pump where two vanes, spring-mounted in the rotor, make contact with the walls of the housing and thereby divide the space between the rotor and the housing. Typical construction is shown in Figs. 2 and 3.

The rotors in these pumps are driven at speeds in the neighborhood of 400–500 RPM. Rotational speeds higher than this range result in setting up excessive vibrational stresses. The volumetric displacement or speed of these pumps is, therefore, somewhat limited, though they are capable of handling large gas throughputs in the range approaching atmospheric pressure. Single-stage mechanical pumps are capable of reaching pressures of the order of 0.01 mm. Hg, and two-stage pumps can go down to 0.001 mm. Hg. They can discharge directly to atmosphere and hence can be used as forepumps to back diffusion pumps or oil ejectors. They exhibit a fairly constant volumetric speed from a suction pressure near one atmosphere to about 0.5 mm. Below this pressure their pumping speed decreases, diminishing to zero in the range 0.01 to 0.001 mm. Condensation of water vapor in these pumps materially increases the ultimate pressure that they can produce.

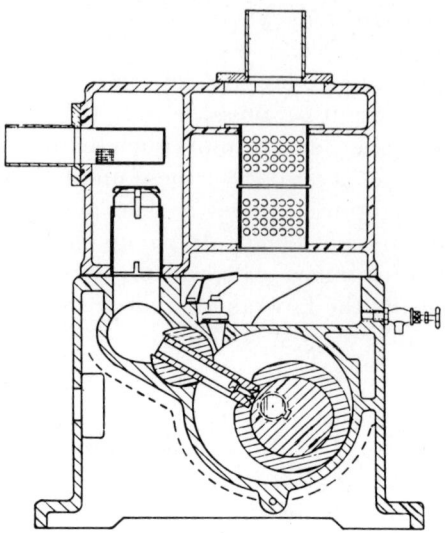

Fig. 2. Rotary—piston mechanical vacuum pump.

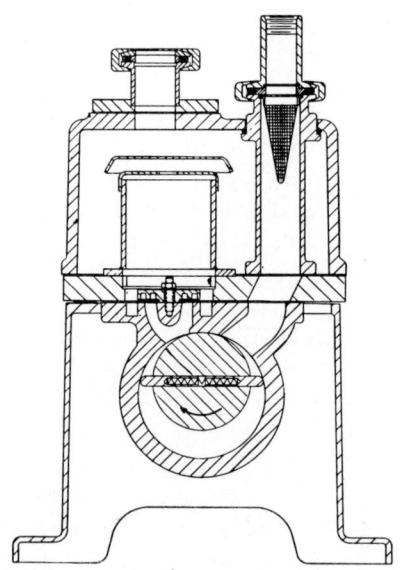

Fig. 3. Vane—type mechanical vacuum pump.

In recent years, the application of gas ballast to these pumps has reduced oil contamination when pumping water vapor.

2. *The Mechanical Booster*

This type of pump generally has two "figure-eight" impellers meshed together as in a gear pump. The impellers are symmetrical relative to their axes of rotation, therefore permitting high rotational speeds, and hence high volumetric displacement. This type of pump has been used both as blower and vacuum pump for several decades; however, its use for producing pressures below 10 mm. Hg dates back only two to three years and is the result of development work aimed at minimizing leakage through the shaft seals. One model of this pump (Kinney KMB-1200) operates at a rotational speed of 1800 RPM and is capable of handling 1000 cu. ft./min. in the range between 10 and 0.001 mm. Hg. These pumps do not discharge directly into the atmosphere; they are backed by forepumps of suitable size. No lubricating or sealing fluids are used, hence there is little chance of contaminating the vacuum system with vapors originating from them. The mechanical booster is not designed for operation with a suction pressure higher than 10–30 mm. Hg. Protective means, such as a pressure switch, are used to open a by-pass line to the forepump, or slow down the rotation of lobe pump as this suction pressure is exceeded. A wide usage is foreseen for these pumps, particularly in vacuum metallurgy, since they require little maintenance and retain their maximum pumping speed over a wide pressure range.

3. *Multi-Stage Steam Ejectors*

When more than three stages of steam jets are used in series, pressures as low as 0.050 or 0.025 mm. Hg can be obtained. A steam ejector system requires no backing pump, the final stage being capable of compression to atmospheric pressure. For many applications, steam ejectors offer great advantages over mechanical vacuum pumps, particularly in processes such as orange juice concentration where large quantities of water vapor are evolved. Steam jets require relatively small initial investment, but consume large quantities of high pressure steam. In general, steam ejectors are not economical below a pressure of 1 mm. Hg. For many applications, steam jets must be avoided as "flash-back" of steam into the vacuum system may cause contamination. If wet steam were to flash back into a titanium melting furnace, the result might be a violent explosion. Steam ejectors are used commonly in installations requiring a continuous demand for pumping large quantities of water vapor between 1 to 50 mm.

4. *The Oil Ejector*

Pumps of this type are exemplified by Consolidated Electrodynamics Corporation's KB-300 and KJ-5000. They consist of a diverging vapor nozzle and a diffusor section similar in construction to a steam ejector. The difference lies in using an organic vapor as the pumping fluid, generated at a boiler pressure of the order of 30–40 mm. Hg. The vapor is condensed and returned to the boiler for re-evaporation. This type of pump can produce pressures as low as 0.001 mm., but has a maximum speed between 0.1 and 1 mm. Hg. Most of the oil ejectors in use are single-stage units. A two-stage oil ejector was developed and is marketed under the commercial designation KJ-5000. Single-stage oil ejectors have highly peaked speed curves, because the diverging nozzle achieves maximum efficiency at a fixed ratio of upstream to downstream pressures. Oil ejectors are more economical than steam jets in their operating pressure range which is between 3 mm. and 0.010 mm. They must be backed by a forepump which can handle their maximum throughput while maintaining a forepressure less than 3 mm. Hg. Accidental exposure of these pumps to atmospheric pressure for more than a few minutes will result in oxidation and thermal breakdown of the oil, and may lead to contamination of the vacuum system. This situation is aggravated if the boiler heaters continue to operate under atmospheric pressure and the temperature of the oil rises above the normal operating level. A typical use of these pumps is for vacuum melting of steel, where large quantities of gas are evolved in the range between 0.30 to 0.05 mm. Hg.

5. *The Diffusion Pump*

A cross-section of a typical diffusion pump is shown in Fig. 4. It consists of an electrically heated boiler where a high boiling organic liquid is vaporized into a vertical stack which feeds two or more annular nozzles. The vapor travels through the nozzles at sonic or higher velocities and discharges into the space between the stack and the wall. The walls are water- or air-cooled and hence act to condense the vapor and return the liquid to the boiler. Pumping takes place as gas molecules, wandering into the vapor jets by their random thermal motion, collide with organic vapor molecules and are given a velocity component in the direction of the vapor stream. Because the pressure in the gas is very low, and hence the mean free path very high, the pumping action is somewhat different from that of an ejector. The difference is that the pressure in the vapor stream can be several orders of magnitude higher than pressure in the gas just above or below the jet. Gas molecules wandering into the vapor stream are trapped through the action of intermolecular collisions inside the jet. As the vapor condenses on the wall, the entrapped gas is diverted downstream of the vapor jet where it can be pumped by another vapor jet.

The pumping action of a given jet is maintained so long as the forepressure, i.e., the downstream pressure, is lower than that required to break through the curtain of vapor. This maximum forepressure is greater the smaller the space between the jet nozzle and the wall of the casing, and the more nearly the direction of vapor flow approaches the axis of the pump. On the other hand, the maximum rate of gas entrainment,

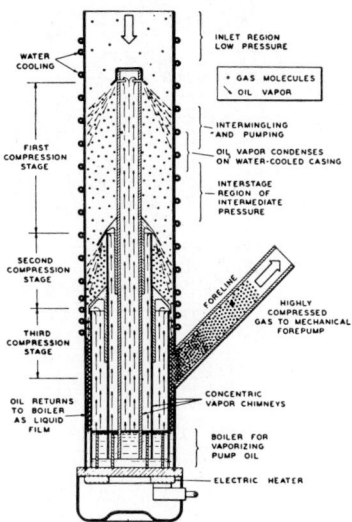

FIG. 4. Cross section of a typical diffusion pump.

since it depends on the rate of collision of gas molecules with the vapor stream, is proportional to the admittance area of the jet and the pressure of the gas. It follows, therefore, that for a given boiler pressure, a high-speed jet is one of low forepressure tolerance, and a high forepressure jet is one of low pumping speed. The use of several jets of progressively lower speeds, but increasingly higher maximum forepressure, yields a pump with the high speed of the first stage and the high forepressure of the final stage.

Although thermodynamically the diffusion pump is highly inefficient, it can achieve a speed as high as 40% to 45% of the ideal pump, which pumps out all molecules streaming to the mouth of the pump by random kinetic motion. The difficulties with breakdown of the pumping fluid noted in connection with the vapor ejector, also apply to the diffusion pump. Silicone oils which are less susceptible to chemical breakdown are now gaining a wider use as diffusion pumping fluids.

There are two types of diffusion pumps: (a) the so called diffusion pump, which operates at a boiler pressure of about 0.5 to 1 mm. Hg and is capable of reaching pressures in the 10^{-6} mm. range; and (b) the booster diffusion pump, which has a boiler pressure of about 3–6 mm. Hg, and has an ultimate pressure of approximately 10^{-4} mm. The difference in performance characteristics of these two types of diffusion pumps is illustrated in Fig. 5.

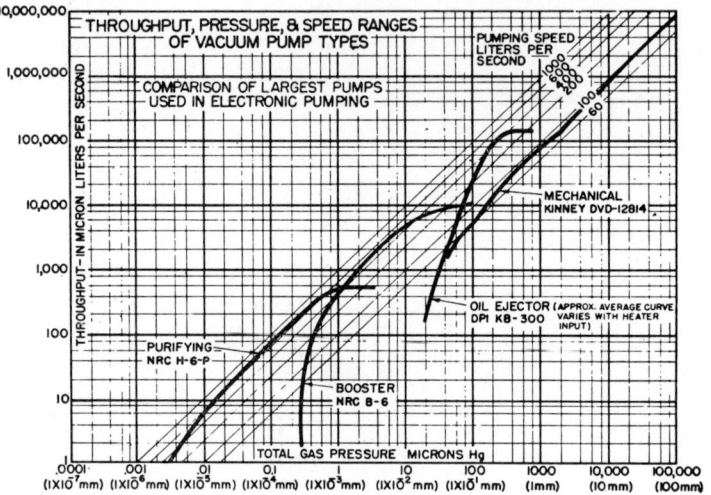

FIG. 5. Throughput *vs.* pressure characteristics of vacuum pump types.

In most applications more than one type of vacuum pump is used. For example, a mechanical vacuum pump might be used in conjunction with an oil ejector and a diffusion pump. In such instances it is necessary to adapt the valving arrangement so that each pump can be opened directly to the vacuum system. By this method it is possible to evacuate the system down to about 2 mm. with the mechanical pump alone, during which time the ejector and diffusion pumps are under no load and are kept under vacuum by a small "holding" mechanical pump. In the range 2 mm. to about 0.050 mm. the ejector pump is opened to the vacuum chamber and the mechanical pump is used for backing the ejector in series. Below 0.050 mm. the diffusion pump might be opened to the system and the ejector and mechanical pumps used to back the diffusion pump in series.

B. Vacuum Gages

Gages which depend on mechanical force for response, such as liquid manometers, Bourdon tubes, and aneroid bellows, are very useful in vacuum measurements, but their range is generally limited to pressures above 1 mm. For lower pressures, several types of gages are available of which the following are the most commonly used.

1. *The McLeod Gage*

This is the only absolute gage available for pressure measurement below 1 mm. Other gages must be calibrated against an absolute standard, at least at one point before their readings can be meaningful.

Its principle of operation is based on Boyle's Law, in that a fixed large volume of the gas at the unknown pressure is compressed to a fixed small volume. The pressure of the gas following compression is measured by a mercury manometer, and the initial pressure is calculated from this pressure and the volume ratio. Pressures as low as 10^{-5} mm. Hg can be accurately measured by this gage. A major cause for error is the presence of condensable vapors in the gas being measured. Under such conditions the pressure obtained with this gage is closer to the partial pressure of noncondensables than the total pressure. It is not exactly equal to the partial pressure of noncondensables, however, because the condensable vapors may exert a significant vapor pressure after condensation.

2. *The Alphatron Gage*

In this gage a radium plaque is the source of alpha particle radiation which ionizes the gas to be measured in proportion to the gas density. The positively charged, ionized gas is collected on a cathode plate. The resulting electric current is directly proportional to the collection rate of ionized molecules and hence to the gas molecular density, and to the pressure. By proper amplification of this current and calibration at one or more points, a microammeter needle can be made to read pressure directly. The latest Alphatron model is capable of measuring pressures accurately from 1000 to 10^{-4} mm. in six linear scales on a single instrument. The pressure obtained is a total pressure reading. However, the reading depends on the composition of the gas, as different gases vary in susceptibility to ionization.

3. *The Hot-Filament Ionization Gage*

The basis of operation of this gage is similar to the Alphatron, except that an incandescent tungsten filament is the source of the ionizing agent. Electrons emitted from the filament, oscillate between a positively charged grid and an ion-collector plate held at a negative potential. The number of ionized molecules is proportional to the molecular density. The same dependence on composition applies for this gage as for the

Alphatron. The hot filament ionization gage is the most accurate gage for pressure measurements between 0.005 and 10^{-7} mm. A modification of this gage, attributed to Bayard and Alpert, has recently enabled accurate pressure readings down to 10^{-13} mm. Hg. Exposure of the gage to atmospheric pressure results in burnout of the tungsten filament. A protective relay circuit is, therefore, employed to turn off the filament current as a pressure rise takes place. An iridium-coated tungsten filament was recently introduced in an attempt to make the gage burnout-proof.

4. *The Philips Gage*

The Philips gage is a cold-cathode ionization gage depending on high voltage rather than high temperature as the source for high velocity electrons. It consists of an anode in the shape of an open-ended pill box and two cathode disks placed a short distance from either end of the anode. A differential of about 2000 volts is maintained between the electrodes, and a strong magnetic field is applied to cause the electrons emitted at the cathodes to move in directly to the anode. They take long spiral paths before reaching the anode, thereby increasing the chances of hitting and ionizing the molecules of the gas whose pressure is to be measured. The main advantage of the Philips gage is that it is burnout-proof and hence permits accidental exposure to atmosphere, but its main drawback is nonlinearity. Furthermore, the action of high voltage discharge in the presence of certain vapors, results in contamination of the electrodes and therefore in erroneous pressure readings.

5. *The Thermocouple Gage*

This gage makes use of the dependence of thermal conductivity on pressure. It is very similar to the hot-wire anemometer, except that the gage is calibrated against gas pressure rather than gas velocity. The thermal conductivity of gases becomes pressure-dependent, when the mean free path is significant in comparison to the dimensions of the apparatus. This gage is insensitive to pressure variations above 1 mm. but is useful in the range between 0.05 and 0.005 mm. Below 0.005 mm., heat loss by conduction through the gas is negligible compared to heat losses through the leads of the hot wire and by radiation. This is a very inexpensive gage and is very widely used for monitoring pressure variation where high accuracy is not of great importance.

Among other gages in use for high vacuum measurements are the Knudsen gage and the Pirani gage. A description of these gages and further detail on the gages discussed above, are given in reference (L3).

V. Conclusion

To say that high vacuum is a useful chemical engineering tool of long standing, is to overlook the absence of a chemical engineering approach in a majority of vacuum applications. High vacuum technology is now making the transition from research laboratories to properly engineered installations. It is the task of the chemical engineer to review the accumulated technology in the light of chemical engineering tools; namely, mass balance, energy balance, rate and equilibrium considerations, and to rewrite this wealth of information in a language familiar and useful to chemical engineering practice.

Acknowledgment

The authors wish to acknowledge their gratitude to National Research Corporation who made this review of high vacuum knowhow possible, by granting company time and technical assistance. Also the authors are indebted to H. A. Steinhertz for reviewing the manuscript and for his many valuable suggestions.

References

B1. Breyer, F. G. *Chem. & Met. Eng.* **49,** 87 (1942).
B2. Brown, G. P., DiNardo, A., Cheng, G. K., and Sherwood, T. K., *J. Appl. Phys.* **17,** 802 (1946).
B3. Benedict, M., and Williams, C., "Engineering Developments in the Gaseous Diffusion Process." McGraw-Hill, New York, 1949.
C1. Cheng, G. K., Sc.D. thesis, Massachusetts Institute of Technology, Cambridge, Mass., 1945.
C2. Cooke, N. E., Sc.D. thesis, Massachusetts Institute of Technology, Cambridge, Mass., 1955.
G1. Gilliland, E. R., and Sherwood, T. K., *Ind. Eng. Chem.* **26,** 516 (1934).
H1. Hickman, K. C. D., *Ind. Eng. Chem.* **46,** (1954).
K1. Knudsen, M., *Ann. Physik.* [4] **28,** 999 (1909).
K2. Kyte, J. R., Madden, A. J., and Piret, E. L., *Chem. Eng. Progr.* **49,** 653 (1953).
L1. Langmuir, I., *Phys. Revs.* **2,** 329 (1913).
L2. Lawrance, R. B., Vacuum Symposium Transactions, p. 55. Committee on Vacuum Techniques Inc., Boston, Mass., 1954.
L3. Lawrance, R. B., *Chem. Eng. Progr.* **50,** 155 (1954).
N1. Normand, C. E., *Ind. Eng. Chem.* **40,** 783 (1948).
P1. Poiseuille, J. L. M., *Société Philomath.*, p. 77 (1838).
R1. Ranz, W. E., and Marshall, W. R., Jr., *Chem. Eng. Progr.* **48,** 173 (1952).
S1. Stauffer, R. A., *Chem. & Ind. (London)*, No. 41, S19 (1948).

SEPARATION BY ADSORPTION METHODS

Theodore Vermeulen

Department of Chemical Engineering, and Radiation Laboratory,
University of California, Berkeley 4, California

I. General Survey... 148
   A. Introduction.. 148
   B. Fluid-Solid Transfer Operations................................. 149
      1. Adsorption... 149
      2. Ion Exchange... 150
      3. Other Operations... 151
II. Physical Factors Affecting Separation Performance................... 153
   A. Equilibrium... 153
      1. Types of Equilibria.. 153
      2. Equations for Highly Favorable or Highly Unfavorable Isotherms... 155
      3. Equations for Moderately Favorable or Moderately Unfavorable Equilibria—The Equilibrium Parameter........................... 158
      4. Isotherms Partly Favorable and Partly Unfavorable............. 160
      5. Ion Exchange Equilibria...................................... 160
   B. Stoichiometric Capacity... 162
   C. Rate Behavior... 163
      1. Rate-Determining Mechanisms.................................. 163
      2. Adsorption Rates... 165
      3. Ion Exchange Rates... 166
III. Binary Fixed-Bed Separations.. 167
   A. Factors Common to All Cases..................................... 167
      1. Variables and Parameters..................................... 167
      2. Material Balance... 172
   B. Limiting Cases of Equilibrium Behavior.......................... 173
      1. Proportionate-Pattern Case (Unfavorable Equilibrium).......... 173
      2. Constant-Pattern Case (Favorable Equilibrium)................ 174
   C. Column Dynamics under Linear Equilibrium........................ 179
      1. General Result... 179
      2. External and Internal Diffusion in Series.................... 182
      3. Longitudinal Dispersion...................................... 182
      4. Analogous Treatment of Heat Transfer......................... 184
      5. Use of Curves to Predict Breakthrough Behavior............... 184
      6. Radial Beds.. 185
   D. Column Dynamics at a Constant Separation Factor................. 185
      1. The Method of Surface Reaction Kinetics...................... 185
      2. External or Pore Diffusion Controlling....................... 189
      3. Solid-Phase (Internal) Diffusion Controlling................. 190
      4. Calculation for Non-Constant N_R or r.................... 190

    5. Combined External- and Internal-Diffusion Resistances............ 191
  E. Note on Multicomponent Saturation............................... 193
IV. Chromatographic Separations.. 194
  A. General Principles... 194
  B. Single Chromatograms: Non-Trace Case........................... 196
  C. "Trace" Chromatograms (Linear Equilibrium)..................... 198
    1. Exact Treatment... 198
    2. Gaussian-Shaped Zones....................................... 198
    3. Design of "Trace" Separations............................... 200
  Nomenclature... 203
  References... 205

I. General Survey

A. Introduction

The emergence of chemical engineering as a professional field of specialized knowledge was catalyzed to a major extent by the systematic classification of apparatus in terms of the Unit Operations. With further progress, the design methods evolved for particular apparatus types have proved equally applicable to other unit operations similar in physical arrangement, material and energy balances, rate behavior, and phase equilibrium. Thus there has been a very extensive development of parallel calculation methods for the separation operations conducted under countercurrent flow conditions—the fluid-fluid operations of distillation, absorption, and extraction.

Among the unit operations, adsorption may be considered a prototype for all fluid-solid separation operations. When it is conducted under countercurrent conditions, the calculation methods required are entirely analogous to those for countercurrent absorption or extraction (H3). Often, however, it is most economical to conduct adsorption in a *semicontinuous arrangement*, in which the solid phase is present as a *fixed bed* of granular particles. The fluid phase passes through the interstices of this bed at a constant flow rate and for an extended period of time. The concentration gradients in the fluid and solid phases display a transient or unsteady-state behavior, and their evolution depends upon the pertinent material balances, rates, and equilibria.

Figure 1 illustrates the two basic methods of fixed-bed operation. In *saturation*, the solute undergoing adsorption is removed continually from the carrier liquid or gas, and accumulates in the solid phase. Such transfer can continue until the concentration on the solid reaches a value corresponding to equilibrium with the concentration in the feed stream, and the column effluent reaches the feed concentration. When the column is saturated to the extent that "breakthrough" of the solute occurs, the flow is interrupted, and the column is then regenerated or eluted by passing through a stream, free of the solute in question, under conditions which favor desorption of this solute.

In elution separation, or *chromatography*, which also is shown in Fig. 1, only a small amount of solution containing the components to be separated is admitted to the column. These solute components are then carried through (eluted) by a carrier phase (the elutant) that initially is free of them. The solutes travel through the column as bands or zones at slightly different velocities. If the column is long enough, the zones will draw apart completely from one another, and may be recovered in the effluent as separate solutions of each individual solute.

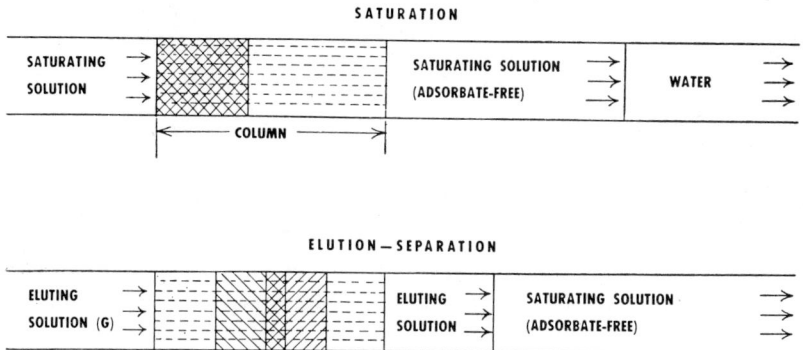

Fig. 1. Schematic diagrams for flow through a fixed-bed adsorption column. Courtesy of *Industrial and Engineering Chemistry*.

The techniques of laboratory- and industrial-scale separations utilizing adsorption and ion exchange have been described comprehensively by Mantell (M3), Cassidy (C2), and Nachod (N1). Treybal (T4) has recently provided a unified and modern chemical engineering approach to fluid-solid separation operations. The present article will treat the problems of data interpretation and apparatus design more extensively than the authors cited, and will give major emphasis to fixed-bed operations.

B. Fluid-Solid Transfer Operations

1. *Adsorption*

The term "adsorption operations," as used in this chapter, is intended to refer to all methods of separation that can be carried out with arrangements of apparatus characteristic of adsorption, or appropriate to it. Any separation operation based upon fluid-solid contact is thus somewhat related to adsorption, and the calculation methods developed for adsorption may prove applicable to it.

Adsorption is analogous to selective condensation of gas molecules or to selective crystallization or fusion from a liquid. It is based upon intermolecular attractive forces of the van der Waals type between a solute and the solid surface. These attractive forces will be most effective and most selective in a region adjacent to the surface and only one solute molecule thick (the monomolecular layer, or monolayer). Although multilayer deposition may develop as the concentration of solute increases toward a condition of normal condensation or precipitation, in general only monolayer adsorption will be involved and the available surface will not even be covered completely.

Since very large surface areas are usually needed, the solids used should be highly porous.[1] The extent of surface involved can be estimated from measurements of adsorptive capacity. With n-butyl amine as adsorptive,[2] for an example, the total volume occupied by one molecule (as liquid) is 1.65×10^{-22} cm.^{3} The area occupied by one molecule is approximately the two-thirds power of the volume, or 3.0×10^{-15} cm.^{2} An efficient adsorbent in contact with a gas phase may hold a volume of adsorbate equal to as much as 1.0% of the particle volume (silica gel has an even larger capacity). One cm.^{3} of such an adsorbent would thus present a surface area of 18 square meters, equivalent to about 5.5×10^6 square feet per cubic foot of particle volume. The precise value of the surface area will vary with the adsorptive species used for the measurement (B12).

The pore structure of catalysts, which is common to adsorbents generally, has been discussed in recent papers by Wheeler (W4).

In the broad sense, adsorption is the selective accumulation of chemical species at all types of phase interfaces. However, only the adsorption of solutes from a fluid phase (gas or liquid) onto a fluid-solid interface will be dealt with here. Such separation methods as froth flotation (D1) and foam fractionation (S3) will not be considered.

2. *Ion Exchange*

Ideally, an exchanger will alter the ionic composition of a solution without changing its over-all concentration of dissolved salts. The exchange material contains either cations or anions that can be displaced by other ionic species of like charge. Ion exchange may be written as a reversible chemical reaction; for example,

---

[1] In this chapter, a *pore* is an element of intraparticle space unfilled by crystalline material, while a *void* is an element of interparticle space in the packed bed.

[2] The *adsorptive* is the solute species undergoing adsorption. Cassidy (C2) reserves the term "adsorbate," often used in the same sense, to denote the combination of adsorbent and adsorptive.

or
$$Ca^{++} \text{ (aq.)} + 2 H^+ \text{ (resin)} \rightleftharpoons Ca^{++} \text{ (resin)} + 2 H^+ \text{ (aq.)}$$
$$Ca^{++} \quad + \quad 2 HR \quad \rightleftharpoons \quad CaR_2 \quad + \quad 2 H^+$$

where R^- or resin represents a stationary anionic site in the polyelectrolyte network of the exchanger phase.

Two distinct kinds of exchange materials have commercial importance. The first comprises claylike materials of high porosity, also called zeolites. The second consists of organic "resins," high-polymer materials which are homogeneous but permeable to water. The diffusional behavior is

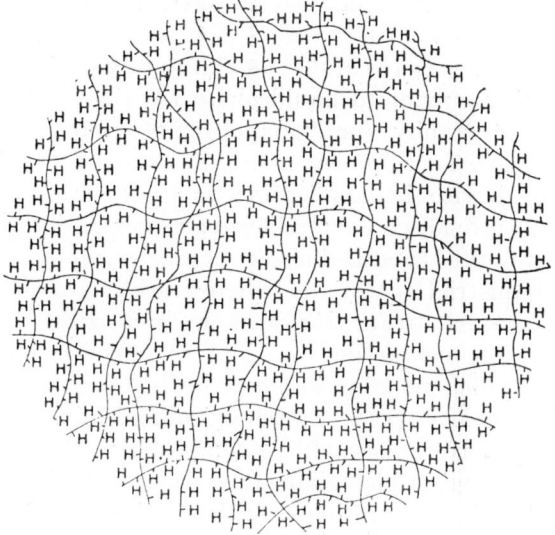

FIG. 2. Structural distribution of exchangeable ions in a synthetic organic resin after W. C. Bauman [p. 50 of reference (N1)].

different for these two kinds of exchanger, the first involving *pore diffusion* and the second homogeneous *intraparticle diffusion*. A schematic structural model of a submicroscopic ion exchange particle (of either kind), as visualized by Bauman and Eichhorn (B4), is given in Fig. 2.

The chemical properties of ion exchange resins are described fully in books by Kunin and Myers (K7) and by Nachod (N1), which give a general technical basis for selecting ion exchange materials for particular applications.

3. *Other Operations*

The theory presented in this article will apply directly to adsorption and to ion exchange operations. A number of related operations can also be conducted under fixed-bed conditions, and their calculations will be closely analogous to those for adsorption. The similarities of these operations to adsorption will be traced qualitatively. The design calculations for these operations will not be treated separately.

a. Fixed-Bed (Partition) Absorption or Extraction. A porous granular solid may have its pores entirely filled with a liquid immiscible with, or nonvolatile in, the fluid to be processed. If the fluid is gaseous, its treatment by the impregnated solid conforms to the definition of absorption; and, if liquid, to extraction. The treating liquid is immobilized by its solid support, and the two in combination are used only in fixed-bed operation. The calculations for such an operation are wholly analogous to those for adsorption.

Partition chromatography (M4) is an elution-separation operation conducted over such a supported liquid-phase treating agent. Vapor-phase partition chromatography (or "gas chromatography") has recently undergone extensive development as an analytical tool. Fredericks and Brooks (F1) deal with its application to hydrocarbon mixtures, and Keulemans (K2) provides a comprehensive report on its uses.

Another instance of fixed-bed absorption or extraction is encountered in small-scale industrial operations, where trace impurities are to be removed from a process stream (for example, hydrogen sulfide from a refinery naphtha). The contacting tower may be filled intermittently with treating liquid, which remains in place during passage of the process stream until it is nearly spent. If no longitudinal mixing of the treating liquid occurs, the performance will be comparable to that of other fixed-bed operations. If longitudinal mixing does occur, a mathematical solution can be derived on the assumption that the mixing is perfect.

b. Leaching. This operation differs from simple dissolution, in having part of the solid phase insoluble in the solvent used for treatment. Its mathematical analysis is analogous to that for adsorption, where the external structure of the solid remains intact during the treatment. Practical examples of leaching are the treatment of such materials as ground sugar beets, ground roasted coffee berries, ores, or freshly formed filter cakes.

c. Drying of Solids. This operation, analogous to leaching, involves the removal of water or another volatile liquid, as vapor in an inert carrier gas. A frequent complication in the drying of vegetable or fruit materials is a decrease in diffusion rates due to shrinkage of pores as the drying proceeds (V1). When this occurs, the mathematical equations for adsorption operations will not apply, although their derivation may provide a model for handling this more complex case.

d. Regenerative Heat Transfer. This operation is commonly conducted by heating a brick checkerwork with hot flue gases until it approaches the entering gas temperature, then cycling a process gas through the checkerwork to raise it to nearly the same temperature. The mathematical treatment of this fixed-bed operation (A3) has served as a model for the linear-equilibrium cases of adsorption (H7) and ion exchange (B5, B8, T2). The cyclic deposition and removal of thermal energy, with its attendant temperature gradients in the fluid and solid phases of the bed, is closely analogous to the deposition and removal of matter in a cyclic or "regenerative" operation.

e. Batch Distillation in a Packed Column. The kettle of a batch still provides the column with vapor feed, and the reflux condenser provides it with liquid feed. If the vapor and liquid feeds were exactly equal, the column would be maintained in a steady state, at total reflux. However, a small net flow of vapor is superposed upon the large refluxing flows. The transient distribution of concentrations in the column under this net flow is comparable to that produced in fixed-bed adsorptions. For example, the separation of two components having constant relative volatility and nearly equal molal volumes, corresponds to binary ion exchange with a moderately nonlinear equilibrium. Because the vapor feed to the column is of changing composition, the problem needs to be treated by the methods of Amundson (A2). The details of the correspondence of variables have not yet been worked out. In the general case, the feed composition will itself depend upon a material balance involving the overhead product composition.

II. Physical Factors Affecting Separation Performance

A. EQUILIBRIUM

1. *Types of Equilibria*

In designing adsorption equipment, the factors to consider are equilibrium, stoichiometry capacity, physical state (size, shape, and manner of packing) of the solid adsorbent, and rates controlling the separation. These subjects will be treated here only from the applicational viewpoint.

First, a number of relations will be considered expressing q^*, the equilibrium concentration in moles of adsorptive per unit weight of solid phase, as a function of c, the concentration in moles per unit volume of fluid phase. For adsorption from a gas, the partial pressure of adsorptive, p, may replace c.

The curve representing this function applies to conditions of constant temperature and is therefore known as an "adsorption isotherm." Three major classes of isotherms can be characterized by their different effects upon column performance. DeVault (D2) first identified the "favorable" equilibrium for which the isotherm is convex upward (Fig. 3a); and the "unfavorable" equilibrium for which the curve is concave upward (Fig. 3b).[3] An intermediate case is that of the linear isotherm.

The slope of a favorable isotherm is thus a decreasing function, while the slope of an unfavorable isotherm is an increasing function, of c. It will be shown later that the equilibrium constants describing the isotherms are numerically larger in "favorable" than in "unfavorable" cases. In fixed-bed adsorption, favorable equilibria lead to relatively sharp concentration gradients in the direction of flow, while unfavorable equilibria lead to more diffuse boundaries.

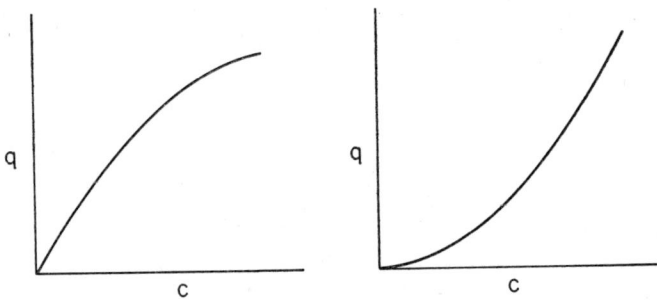

(a) FAVORABLE EQUILIBRIUM   (b) UNFAVORABLE EQUILIBRIUM

FIG. 3. Representative adsorption isotherms.

Isotherms having an inflection point correspond to ranges of both favorable and unfavorable equilibrium. Since separation operations only involve the part of the isotherm which lies below the feed-concentration, the occurrence of an inflection point corresponding to a higher concentration has no effect. If the inflection point occurs at a lower concentration, fixed-bed separation calculations must be carried out by approximate methods to be described below.

For fixed beds, simplified methods of calculation are available when the equilibrium can be considered highly favorable, highly unfavorable, or linear. The general case of nonlinear equilibrium has been solved for only a few types of equilibrium equations (Section II, A, 3). The discussion immediately following (Section II, A, 2) gives a wide selection of equilibrium relationships which can be used empirically in cases allowing the algebraic calculation of column performance; such cases are identified by italicized "type" descriptions.

[3] The terms "favorable" and "unfavorable," not used by DeVault, arise particularly from the consideration of ion-exchange equilibria.

2. *Equations for Highly Favorable or Highly Unfavorable Isotherms*

Research on adsorption as a method of determining surface areas of catalysts has led to the most general equilibrium relations yet available for representing *adsorption from the gas phase*, as developed by Brunauer and co-workers (B13). The widely used Brunauer-Emmett-Teller (BET) equation is

$$\frac{q^*}{q_m} = \frac{K_B p}{[P + (K_B - 1)p][1 - (p/P)]} \qquad (1)$$

Here q_m is the adsorbate concentration when a monomolecular layer has been completely adsorbed; P is the vapor pressure of the pure adsorptive at the temperature of the isotherm, and K_B is an appropriate constant. This equation fits two types of isotherms (the type numbers given are Brunauer's, and the reader is referred to his article for graphic illustrations):

Type 2. Favorable at low, and unfavorable at high, concentrations (if $K_B > 1$).
Type 3. Unfavorable throughout (if $K_B < 1$).

A more general but highly complicated equation of essentially the same nature (B14) serves also to fit two other types of isotherms:

Type 4. Favorable at low and high concentrations, unfavorable at intermediate concentrations.
Type 5. Unfavorable at low, and favorable at high, concentrations.

Empirical equations of intermediate complexity also can be devised to fit isotherms of Types 4 and 5 in any particular instance.

When only monomolecular adsorption is involved, the BET relation (Eq. 1) reduces to the familiar Langmuir isotherm. With $p \ll P$, and $K_B \gg 1$, it becomes (L1)

$$\frac{q^*}{q_m} = \frac{K_L p}{1 + [K_L - (1/P)]p} \approx \frac{K_L p}{1 + K_L p} \qquad (2)$$

where K_L replaces K_B/P. This relation corresponds to

Type 1. Favorable throughout.

Equations 1 and 2 both are based on the assumption that the enthalpy of adsorption does not change as p and q increase. For a particular distribution of enthalpy values, Sips (S5) has developed the following isotherm

$$\frac{q^*}{Q_\infty} = \left[\frac{K_S p}{P + (K_S - 1)p}\right]^M \qquad (3)$$

Here Q_∞ is the concentration reached when the solid is saturated with liquid solute, the constant M characterizes the enthalpy distribution, and K_S is another constant. On strictly empirical grounds, Koble and Corrigan (K5) have proposed a closely related equation which will sometimes be easier to use (N and K_C are empirical constants):

$$\frac{q^*}{Q_\infty} = \frac{K_C p^N}{P^N + (K_C - 1)p^N} \tag{4}$$

The foregoing two equations correspond to the following types of isotherm:

Type 1. Favorable throughout (if $M < 1$ and $M < 2K_S - 1$, or $N < 1$ and $N < K_C/[2 - K_C]$).

Type 2. Favorable at low, and unfavorable at high, concentrations (if $M < 1$ or $N < 1$ only).

Type 3. Unfavorable throughout (if M or N is greater than both criteria).

Type 5. Unfavorable at low, and favorable at high, concentrations (if $M > 1$ or $N > 1$ only).

With $K_S p$ or $K_C p^N \ll 1$, Eqs. (3) and (4) both reduce to the Freundlich relation (n and K_F are arbitrary constants):

$$q^* = K_F p^n \tag{5}$$

This corresponds to Type 1 (if $n < 1$) or Type 3 (if $n > 1$).

The foregoing equations appear to apply whether or not inert carrier gases are present. Regardless of the total pressure (so long as an ideal gas mixture is approximated), only the partial pressure of adsorptive affects the equilibrium. In *adsorption from the liquid phase*, it is generally necessary to assume that the solid is wetted by all the components of the liquid phase, and that all such components are adsorbed to an appreciable extent. For a two-component liquid, designating by N the mole-fraction of the component under consideration, isotherms similar to Eqs. (1–5) can be used on an empirical basis:

$$\frac{q^*}{Q_\infty} = \frac{K_B' N}{[1 + (K_B' - 1)N][K_B'' + (1 - K_B'')N]} \tag{6}$$

$$\frac{q^*}{Q_\infty} = \frac{K_L' N}{1 + (K_L' - 1)N} \quad \text{or} \quad \frac{(1 + K_L'')N}{1 + K_L'' N} \tag{7}$$

$$\frac{q^*}{Q_\infty} = \left[\frac{(1 + K_S')N}{1 + K_S' N}\right]^M \tag{8}$$

$$\frac{q^*}{Q_\infty} = \frac{(1 + K_C')N^m}{1 + K_C' N^m} \tag{9}$$

$$q^*/Q_\infty = N^n \tag{10}$$

Two additional forms will sometimes prove useful. The first is

$$\frac{q^*}{Q_\infty} = \frac{N^l}{N^l + K(1 - N)^l} \tag{11}$$

where l is a suitable constant. Equation 11 can represent the following types of behavior:

Type 1. Favorable throughout ($l = 1$, $K > 1$).
Type 2. Favorable at low, and unfavorable at high, concentrations ($l < 1$).
Type 3. Unfavorable throughout ($l = 1$, $K < 1$).
Type 5. Unfavorable at low, and favorable at high, concentrations ($l > 1$).

Another form is a power-series expression which can be made to conform to any of the five shapes of isotherm that have been discussed here:

$$q^*/Q_\infty = (1 + A + B + C)_N - A_{N^2} - B_{N^3} - C_{N^4} \qquad (12)$$

Any two terms of Eq. (12) can only correspond to either Type 1 or Type 3. Any three terms, potentially, can give Type 2 or Type 5. The sum of four terms may give a satisfactory fit to a Type 4 isotherm (initially favorable, later unfavorable, finally favorable).

An isotherm which is markedly favorable or unfavorable may be approximated by a line or curve intercepting either of the positive axes. Thus, Eagleton and Bliss (E1) used a relation of the type:

$$q^*/Q_\infty = A + (1 - A)_N \qquad (13)$$

If A is positive, the isotherm can be considered favorable throughout; if negative, unfavorable throughout; and if zero, linear.

When a non-ideal binary liquid solution is equilibrated with a solid whose adsorption sites are all identical, a mass-action equilibrium expression can be written which contains the activity coefficients (γ_A and γ_B) of the liquid components:

$$K = \frac{q_A^* N_B \gamma_B}{q_B^* N_A \gamma_A} \qquad (14)$$

The activity coefficients for these species in the solid phase are of course included in the equilibrium constant. This can be shown to reduce to Eq. (7) when $\gamma_A = 1$ and $\gamma_B = 1$ throughout. In general, the activity coefficients are given approximately by

$$\gamma_A = e^{L N_B^2} \qquad (15)$$
$$\gamma_B = e^{L N_A^2}$$

where L is a symmetric or "two-suffix" Margules coefficient. For a homogeneous solution with positive deviations from Raoult's Law, $0 < L < 2$. Combination of Eqs. (14) and (15) yields

$$\frac{q_A^* x_B}{q_B^* x_A} = K e^{L(N_B^2 - N_A^2)} \qquad (16)$$

With positive deviations from Raoult's Law, the partition ratio q_A^*/N_A for adsorption of a component will increase as its concentration decreases.

This corresponds to a Type 2 isotherm. Even if Eq. (16) correctly represents the equilibrium data, it is generally more convenient to fit the isotherm empirically by the use of Eq. (12), (11), or (6).

3. *Equations for Moderately Favorable or Moderately Unfavorable Equilibria —The Equilibrium Parameter*

Of the foregoing equilibrium expressions, only two, namely Eq. (2) (the Langmuir relation) and Eq. (7), have been utilized in a general treatment of adsorption rates.

Although these mass-action expressions are the most frequently used owing to their simplicity, it remains possible that rate treatments based on other equilibrium relations will be derived in the future.

In applying the Langmuir relation to column calculations, the ratio q^*/q_m will be less important than the ratio q^*/q_o^*, where q_o^* is the *maximum* concentration that can be reached in the column, i.e., the solid concentration in equilibrium with the inlet solution concentration p_o. From Eq. (2),

$$\frac{q^*}{q_o^*} = \frac{(1 + K_L p_o) p}{(1 + K_L p) p_o} \qquad (17)$$

Use may be made of a dimensionless *equilibrium parameter*, **r**, defined in this case as

$$\mathbf{r} = \frac{1}{1 + K_L p_o} \qquad (18)$$

In terms of **r**, the isotherm becomes

$$\frac{q^*}{q_o^*} = \frac{p/p_o}{\mathbf{r} + (1 - \mathbf{r})(p/p_o)} \qquad (19)$$

Eq. (7) may be put into the same form by using as the equilibrium parameter

$$\mathbf{r} = 1/K_L' \qquad (20)$$

with the result that

$$\frac{q^*}{Q_\infty} = \frac{N}{\mathbf{r} + (1 - \mathbf{r}) N} \qquad (21)$$

From Eq. (21), the parameter **r** is found to be analogous to the *relative volatility* used in distillation calculations:

$$\mathbf{r} = \frac{N(Q_\infty - q^*)}{q^*(1 - N)} \qquad (22)$$

Thus **r** may also appropriately be called the "separation factor."

The only available general treatment of adsorption rates is based upon

Eqs. (2) and (7) as equilibrium expressions, and thus assumes a *constant separation factor*.

As has already been indicated, any separation involves only the part of the isotherm which lies at or below the feed-concentration level. An empirical fit, based upon the assumption of a constant separation factor, can often be made to this part. For this purpose, Eq. (7) can be rewritten in the form

$$\frac{q^*}{Q} = \frac{K(c_A)_o}{C_o + (K-1)(c_A)_o} \tag{23}$$

where, for empirical use, C_o and Q are regarded as adjustable parameters rather than measured constants for the system. Figure 4 is a logarithmic

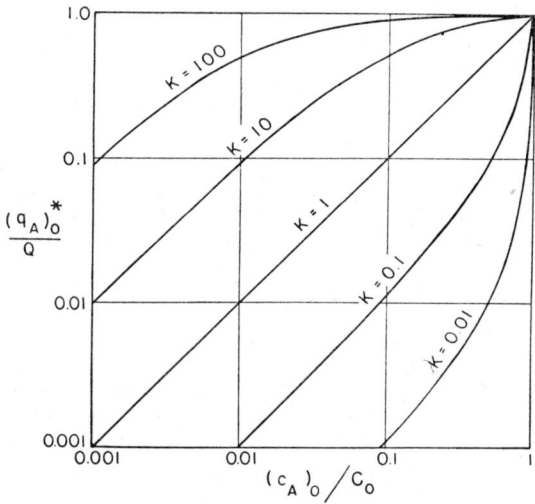

Fig. 4. Dependence of isotherms upon equilibrium constant. The log-log scale is useful for empirical matching with experimental isotherms. Courtesy of *Journal of Chemical Physics* (V5).

plot of Eq. (23). Any isotherm plotted on similar coordinates which shows either a steadily decreasing or a steadily increasing slope can be matched approximately to one curve of this family, and can be described using appropriate values of K, C_o, and Q (V5). An experimental plot of c vs. q is prepared on an identical logarithmic grid. This plot is superposed upon the master plot by sliding it in both coordinate directions until the best fit is obtained to a portion of one of the master-plot curves. The coordinates of the master plot corresponding to $c = C_o$ and $q = Q$ are traced onto the experimental plot, so as to provide the proper numerical values of C_o and Q. The value of K is read directly from the curve which gives the best fit. The equilibrium parameter for this isotherm, as developed in the reference article (V5), is:

$$r = \frac{C_0}{(K-1)(c_A)_0 + C_0} \qquad (24)$$

If an isotherm fits Fig. 4 exactly all the way to $(c_A)_0/C_0 = 1$, C_0 corresponds to the true total solution concentration and Q to q_0.* Equation (24) then becomes applicable to a *binary mixed feed*, to be described (Section III, D, 1b).

It will be seen that $r = 1$ corresponds to the *linear isotherm*. Under certain conditions many of the isotherms considered in the preceding section also become linear.

4. *Isotherms Partly Favorable and Partly Unfavorable*

Algebraic relations for equilibria that are partly favorable and partly unfavorable are included in Section II, A, 2. Such relations can sometimes be used for separation calculations where only the highly favorable part of the isotherm, or only the highly unfavorable part, is involved. In all other such cases the equilibrium data can be utilized numerically, by dividing the concentration range into several sections and determining an average **r** for each interval. The use of non-constant **r** values will be described in Section III, D, 4.

5. *Ion Exchange Equilibria*

The equilibria between *ions of equal valence* (homovalent ions) are usually represented by a simple mass-action relation, analogous to Eq. (7) or (22):

$$K(= K^{\text{II}}) = \frac{q_A c_B}{q_B c_A} \qquad (25)$$

The use of this relation has been justified experimentally by Bauman and Eichhorn (B4) and by Boyd *et al.* (B9). For organic resin exchangers, Gregor (G7) has shown that K varies with changes in resin density that occur progressively during exchange. High concentrations in the solution phase lead to an appreciable uptake of unbound ions by the resin structure, and this uptake also affects the apparent equilibrium. Moreover, the thermodynamic activities of the exchanging ions in solution will change with changing ionic strength; for two ions of equal valence, the changes in activity coefficients frequently will be almost equivalent, and for such cases K will be nearly independent of the total ionic concentration of the solution.

Such factors as these make it difficult to obtain a completely general relation for exchange between *ions of unequal valence* (heterovalent ions). A mass-action expression is again the best approximation. For the exchange

$$A^{(\alpha+)} + \frac{\alpha}{\beta} BR_\alpha \rightleftharpoons AR_\alpha + \frac{\alpha}{\beta} B^{(\beta+)}$$

the equilibrium relation is

$$K = \frac{q_A{}^\beta c_B{}^\alpha}{q_B{}^\alpha c_A{}^\beta} \quad (26)$$

or

$$K\left(\frac{Q}{C_o}\right)^{\alpha-\beta} = \frac{(q_A/Q)^\beta (c_B/C_o)^\alpha}{(q_B/Q)^\alpha (c_A/C_o)^\beta} \quad (27)$$

where Q ($= q_A + q_B$) is the total exchange capacity of the solid, and C_o ($= c_A + c_B$) is the total concentration of ions in solution. For the case

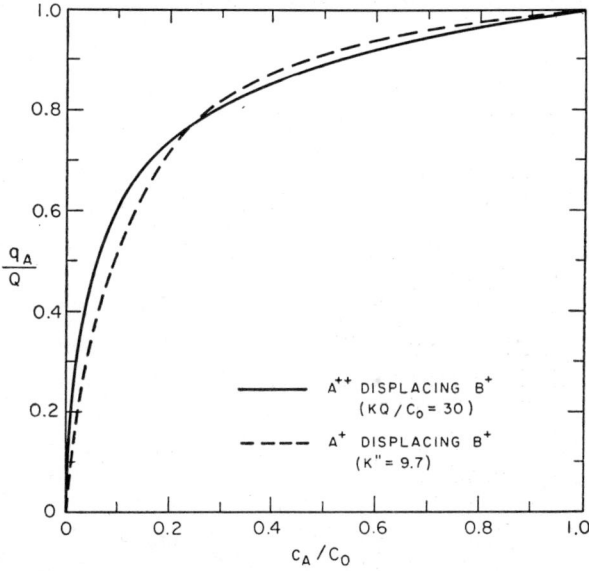

Fig. 5. Typical equilibrium distribution of heterovalent ions between solution and resin, showing its approximation by a second-order equilibrium constant (K^{II}) equivalent to a constant separation factor. Courtesy of *Chemical Engineering Progress* (H6).

of unequal valences, concentrations should be expressed in equivalents rather than in moles.

Data obtained by Gregor (G7) suggest that some complex-ion formation occurs in the resin, of the type

$$A^{(\alpha+)} + X^- + \frac{\alpha-1}{\beta} BR_\beta \rightleftharpoons AXR_{\alpha-1} + \frac{\alpha-1}{\beta} B^{(\beta+)}$$

Walton (W2), and Kunin and Myers (K7), have reviewed the empirical equations that have been proposed to fit ion-exchange equilibria; many of these relate specifically to inorganic zeolites rather than to the synthetic-resin exchangers.

For rate calculations, the equilibrium may need to be represented by Eq. (25), as has been shown by Hiester and Vermeulen (H6). A suitable average value of K^{II} over the entire concentration range is given by

$$K^{II} = [K(Q/C_o)^{\alpha-\beta}]^{2/(\alpha+\beta)} \tag{28}$$

For *trace* concentrations of A, in the vicinity of $c_B = C_o$ and $q_B = Q$, the preferred relation (V4) is

$$K^{II} = [K(Q/C_o)^{\alpha-\beta}]^{1/\beta} \tag{29}$$

The nature of the approximation to Eq. (26) that is given by Eqs. (28) and (25) is shown in Fig. 5.

B. Stoichiometric Capacity

The adsorption or exchange capacity of a column will have a major effect upon the volume of the fluid which can be treated before breakthrough. The equilibrium isotherm for adsorption, as a general rule, will indicate the maximum uptake. For ion exchange (K7) the capacity specified by the manufacturer, or determined experimentally, often can be used without correction.

Occasionally, particularly in cases of high fluid-phase concentration, the capacity must be defined with some care. For material-balance purposes, a suitable boundary must be defined between the solid and the solution. Ordinarily this boundary is the outer surface of the particles. The "solid" concentration at equilibrium, q^*, includes fluid-phase solute in amount $c(1 - \epsilon)\chi$, where $(1 - \epsilon)$ is the non-void fraction and χ is the fractional internal porosity of the particles. The apparent equilibrium constants and particle diffusion rates may therefore vary with c and with χ; the mathematical treatment can be modified accordingly, when necessary.

The rate of uptake of large molecules by an adsorbent or exchanger may lead to apparent variations in capacity. Ion exchange of Dowex 50 in the hydrogen form, with n-butyl amine, indicated to Vermeulen and Huffman (V6) that the normal exchange capacity of the resin could be divided into relatively accessible (60%) and relatively unaccessible (40%) portions. A more likely interpretation, based upon work of Wilson and Lapidus (W8) and Gregor (G7), is that the resin-phase diffusivity declines sharply as the uptake of amine increases. For design calculations, however, it may be necessary to assume a reduced capacity.

The varying uptake due to heterovalence (G7) has been mentioned in Section II, A, 5. For the case of a divalent cation M^{++} and a univalent anion X$^-$, the complex-ion equilibrium might be represented by

$$\frac{q_{MX^+}{}^2}{q_{M^{++}} c_{X^-}{}^2 c_{M^{++}}} = K_{complex} \tag{30}$$

Boyd *et al.* (B8) have reported a variation in the capacity of Amberlite IR-1, a resin containing sulfonic, carboxylic, and phenolic groups.

The uptake of anions, as well as cations, by a cationic exchanger, is predicted by the Donnan theory of membrane equilibria, to increase with the electrolyte concentration in the external solution phase. The cations thus absorbed must be included in the effective resin capacity. Data of Bauman for the HCl uptake in Dowex 50, taken from an experimental plot (N1), are as follows:

Soln. (Cl$^-$), M	0.1	0.2	0.5	1	2	5
Resin (Cl$^-$), M	0.006	0.012	0.03	0.07	0.2	1.2

Resins with a low extent of cross-linking, which have a higher water content, will show a larger uptake of ions from the solution phase. Since the total resin concentration is around 7 M, the effect usually is negligible.

C. Rate Behavior

1. *Rate-Determining Mechanisms*

The sequence of molecular-scale processes involved in an exchange adsorption, in which species B initially adsorbed on the solid is displaced by species A initially in solution, can be grouped into the following four steps, as outlined by Klotz (K4):

a. Fluid-phase external diffusion, sometimes called film diffusion (or F mechanism). Counter-diffusion of A from the bulk fluid to the outer surface of the solid particle and of B from the particle to the bulk fluid.

The rate of material transfer of A into the particle may be expressed:

$$\frac{dq_A}{d\tau} = k_f a_p \frac{\epsilon}{\rho_b} (c_A - c_A^*) \tag{31}$$

where k_f is the fluid-phase mass-transfer coefficient, a_p is the external area of particles per unit bulk volume of packed column, ϵ is the fraction of external voids, τ is time, and ρ_b is bulk density of the packing.

b. Fluid-phase pore diffusion, in porous bodies whose pores are freely accessible to the bulk fluid outside. Counter-diffusion of A through the pores of the particle to the point where exchange occurs and of B from the point of exchange in the pore surface back to the outer surface of the particle.

The pore-diffusion rate for a sphere, as adapted from Barrer (B2), is

$$D_{\text{pore}} \left(\frac{\partial^2 c_i}{\partial r^2} + \frac{2}{r} \frac{\partial c_i}{\partial r} \right) = \chi \frac{\partial c_i}{\partial \tau} + \rho_p \frac{\partial q_i'}{\partial \tau} \qquad (32)$$

where D_{pore} is the diffusivity, ρ_p is the density of the adsorbent particle, c_i is the fluid-phase concentration of component within the particle at radius r, and q_i' ($= q_i - \chi c_i$) will generally be in equilibrium with c_i. The mean concentration of the entire particle, of total radius r_p, is

$$q = \frac{3}{r_p^3} \int_0^{r_p} q_i r^2 dr \qquad (33)$$

Here r_p is the outer radius of the particle. These equations will normally be written in terms of the component being adsorbed. If exchange adsorption is involved, with consequent counter-diffusion, this will enter into the evaluation of D_{pore}.

c. Reaction, or Phase Change. Desorption of B from the solid phase at a pore surface or at the outer surface, and adsorption of A in its place.

For one-component adsorption, the rate is

$$\frac{dq_i}{d\tau} = k_i' \left[p(q_m - q_i) - \frac{1}{K_L} q_i \right] \qquad (34)$$

For exchange between two components, usually from the liquid phase,

$$\frac{d(q_A)_i}{d\tau} = k_i \{ c_A[(q_A)_o^* - (q_A)_i] - \mathbf{r}(q_A)_i(C_o - c_A) \} \qquad (35)$$

Here q_i or $(q_A)_i$ is the solid-phase concentration at the surface, and k_i' or k_i is the rate of surface reaction. When the surface-reaction equation is used empirically for the entire rate behavior, k_i will be replaced by k_{kin}.

Equation 34 may be put into the form of Eq. (35). Use of Eq. (2) in the Langmuir form replaces q_m by q_o^*, and also introduces p_o; correspondence is obtained if $k_i = k_i'/(1 - \mathbf{r})$ or $k_i' q_m/q_o^*$.

Surface reaction is usually very fast compared to the rates of material transfer, so that experimental values of k_i are not known.

d. Solid-phase internal diffusion, sometimes called particle diffusion or P mechanism. This includes diffusion through a homogeneous, permeable (i.e. absorbing), non-porous solid; diffusion in a mobile, adsorbed phase covering the pore surfaces of a porous solid whose crystalline portion is impermeable; or diffusion in an absorbing liquid held in the pore spaces of a solid.

The rate of internal diffusion is expressed by

$$D_p \left(\frac{\partial^2 q_i}{\partial r^2} + \frac{2}{r} \frac{\partial q_i}{\partial r} \right) = \frac{\partial q_i}{\partial \tau} \qquad (36)$$

Here D_p is the diffusivity, and q_i is the solid-phase concentration at radius r. This equation has been solved only for the irreversible- and linear-equilibrium cases of fixed-bed operation (Sections III, B, 2 and III, C, 2). It is usually approximated by the linear-driving-force relation (G5)

$$\frac{dq_A}{d\tau} = k_p a_p (q_A{}^* - q_A) \tag{37}$$

where $k_p a_p$ ($= 60 D_p/d_p{}^2$) is the mass-transfer coefficient, q_A is the concentration of A averaged over the entire particle, and $q_A{}^*$ is the concentration the particle would have if it were in equilibrium with the instantaneous, fluid-phase concentration at the outer surface of the particle.

The sequence of steps will be (a)-(c)-(d) in ion exchange with synthetic-resin materials, and in partition absorption or extraction. For simple adsorption (or ion exchange on inorganic zeolites) the sequence will usually be (a)-(b)-(c), occasionally (a)-(c)-(d), and rarely (a)-(b)-(c)-(d).

2. Adsorption Rates

a. Fluid-phase external diffusion rates appear to conform to the general mass-transfer correlations developed by (among others) Wilke and Hougen for gases (W6), and McCune and Wilhelm (M1) and Gaffney and Drew (G1) for liquids. Evidence of this general agreement has been provided by Dryden (D6), who found an additional resistance attributable to pore diffusion; Dodge and Hougen (D3); Eagleton and Bliss (E1); and others. The correlation of Wilke and Hougen, for example, is expressed in the present notation as

$$k_f = \frac{U}{H_f a_p} = 1.82 U \left(\frac{d_p U \epsilon \rho}{\mu}\right)^{-0.51} \left(\frac{\mu}{\rho D_f}\right)^{-0.67} \tag{38}$$

where U is the mean linear velocity, $U\epsilon$ is the superficial velocity, H_f the height of a transfer unit (or HTU), μ the viscosity of the fluid, ρ the density of the fluid, D_f the bulk diffusivity of the solute in the fluid, and d_p is the effective particle diameter.

b. Fluid-phase pore diffusion rates have been discussed extensively by Wheeler (W4) for the related problem of solid-catalyzed reactions. For diffusion in liquids, and in gases under high pressure, the pore diffusivity is given approximately by

$$D_{\text{pore}} = \frac{D_f \chi}{2} \tag{39}$$

For gases at ordinary pressures in fine pores, Knudsen-flow diffusion is encountered if the molecular mean free path is larger than the pore radius. Wheeler's work gives a relation applicable to all combinations of bulk diffusion and Knudsen flow:

$$D_{\text{pore}} = \frac{D_f \chi}{2} (1 - e^{-2\bar{r}\bar{u}/3D_f}) \tag{40}$$

where the average pore radius $\bar{r}$ is given approximately by the ratio of particle volume to internal pore surface; and $\bar{u}$ is the average molecular velocity:

$$\bar{u} = (8RT/\pi M)^{1/2} \tag{41}$$

with R the gas constant per mole and M the molecular weight of the solute. At large ratios of mean free path to pore radius, the pore diffusivity becomes $2\chi\bar{r}\bar{u}/3$.

Calculations based solely upon a constant D_f do not include the effects of mass flow or of changing composition, but these will generally be negligible.

3. *Ion Exchange Rates*

For exchange in inorganic zeolites, the same rate relations should apply as for adsorption from a liquid phase.

With the synthetic-resin exchangers, the external mass-transfer rate again corresponds to Eq. (38). Data of a number of investigators (B8, G2, M7, N2, S4) have been reviewed by Hiester (H4), and have led to the construction of Fig. 6. The functions plotted, which will be discussed fully in Section III, D, 5, correspond to the relation

$$\frac{H_R D_p}{d_p^2 U} = \alpha_1 \left(\frac{D_f}{d_p U}\right)^{1/2} \frac{D_p}{D_f} + \alpha_2 \tag{42}$$

where H_R is an HTU based upon the reaction-kinetic mechanism, and α_1 and α_2 are constants. The horizontal section of the curve corresponds to internal diffusion and the sloping section to external diffusion. An earlier correlation was given by Vermeulen and Hiester [Fig. 6 of reference (V4)] based upon some of the same data; however, the channeling correction factor $(d_w/d_p)^{1/3}$ introduced there has proved generally to be unnecessary. This reference indicates that *longitudinal dispersion* will need to be considered at $(d_p U)$ values below 10^{-3} cm.^{2}/sec.

The D_p values for univalent cations in Amberlite IR-1, a sulfonated phenol-formaldehyde resin, are in the range of 2 to 3×10^6 cm.^{2}/sec., or about 0.15 times the diffusivity in solution (B8). For divalent cations values of 7 to 10×10^{-7} cm.^{2}/sec., and for trivalent ions values of 3 to 4×10^{-7}, are indicated by chromatographic runs on Amberlite

IR-1 (V4). The D_p values for Dowex 50, a typical sulfonated polystyrene resin, have been shown by Boyd and Soldano (B7) to depend greatly upon the extent of cross-linking as determined by the divinylbenzene (DVB) content of the resin. For 8% DVB, the D_p values are similar to Amberlite IR-1. For 12% DVB, the usual commercial grade, self-diffusion

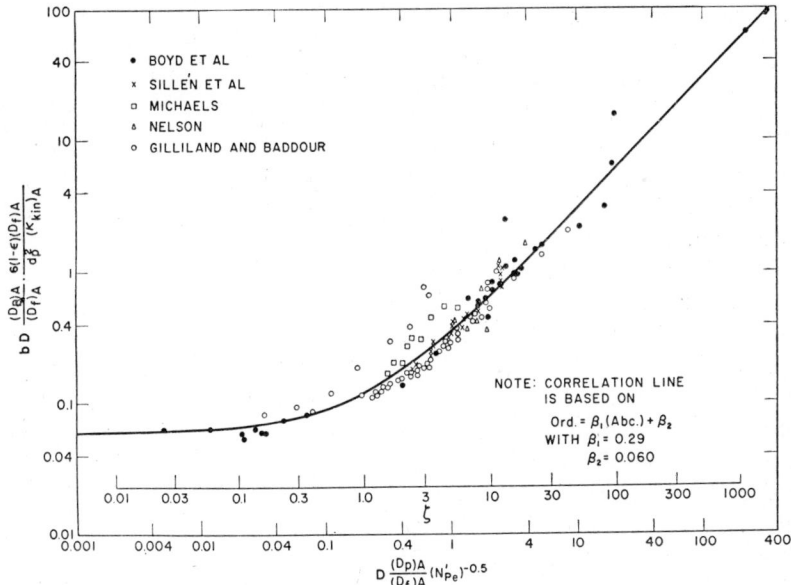

Fig. 6. Logarithmic correlation of ion exchange rates in fixed-bed columns. Courtesy of *A. I. Ch. E. Journal* (H4).

values at 25°C. are: Na^+, 2.9×10^{-7}; Zn^{++}, 2.9×10^{-8}; Y^{+++}, 3.2×10^{-9}; Th^{++++}, 2.2×10^{-10}. Data are also reported for a typical anionic exchanger (B7).

III. Binary Fixed-Bed Separations

A. Factors Common to All Cases

1. *Variables and Parameters*

Column performance studies for fixed-bed columns are concerned with the concentration history of the column effluent—that is, with the variation in concentration as a function of time or of volume of effluent. Concentration-history calculations for fixed-bed adsorption and ion-exchange columns must make use of one or another of a group of specialized results which take the place of a general solution to the problem. The specialized results can be identified on the basis of controlling mechanism (external diffusion, internal pore diffusion, internal solid-phase diffusion, or longitudinal diffusion) and of equilibrium (unfavorable, linear, favorable, or completely irreversible).

The same principles are involved in two different types of operation: saturation, in which one or more solute components are separated from the solvent, and chromatography, in which two or more solute components are separated from one another. Saturation operations will be discussed in this section, and chromatography in the next. The terminology will be that adopted in previous papers by Vermeulen, Hiester, and co-workers (H1-6, V3-6).

The relative concentrations in the column and in the effluent depend upon several physical variables. Once the variables are defined and interrelated, the algebraic and numerical results can be given in relatively simple form. The successful development and design of adsorption operations involves the optimization of all the variables, through the use of equations or curves that really correspond to the physical situation. Unless the entire range of possible behavior is understood, the designer will risk using an inappropriate method. The principal variables are:

Column volume, v. For a point within (or at the downstream end of a column, v will be measured from the upstream end. If h is the height and S is the superficial cross-sectional area, $v = hS$. If ϵ is the void fraction for the column as packed, $v\epsilon$ is the effective volume of fluid in the column.

Volumetric flowrate, F. The residence time for fluid in the column is $v\epsilon/F$. The linear flowrate is given by $U = F/S\epsilon$, and the superficial flowrate is $U\epsilon = F/S$.

Fluid volume passing through the column. The fluid is assumed to enter the column with a sharp front, at time $\tau = 0$, and to flow at constant and uniform velocity. V is the volume that has entered the column up to a particular time τ. At that instant, $V - v\epsilon$ is the volume that has passed out of the column (of volume v) and $\tau - (v\epsilon/F)$ is the elapsed time as measured from the time of the initial arrival of the fluid front at the column exit.

Stoichiometric capacity of the column, $q_o^* \rho_b v$ for adsorption, or $Q\rho_b v$ for simple binary ion exchange. This variable has been discussed in Section II, B.

Material-transfer rates, as reviewed in Section II, C.

Particle diameter, d_p, and pressure drop. For a given column performance, the pressure drop is lowest when the adsorbent particles are spherical and of closely uniform size. The external mass-transfer rate increases inversely as $d_p^{3/2}$, and the internal rate increases inversely as d_p^2. The pressure drop variation will depend upon the Reynolds number, but is about proportional to $U^{3/2}$ and also inversely to $d_p^{3/2}$. Pressure-drop relations are reviewed by Drew and Genereaux (D4), Brownell (B11), and Ergun (E3). In many instances increasing the pressure drop, by using a longer column of smaller diameter, will substantially improve the resin utilization.

The analytical or graphical solutions for concentration are usually obtained in a dimensionless form, which provides the greatest generality. This requires that the column and solution variables be assembled into the following dimensionless parameters:

a. Solution Concentration Ratio, **x**. The column calculations to be considered here usually involve a constant solute concentration in the feed, which can be taken as unity on a relative scale of concentration. The parameter **x**, based on this value, then expresses the ratio of solute concentration in the solution at any downstream point to the feed value; that is,

$$\mathbf{x}_A = c_A/(c_A)_o \tag{43}$$

If the entire system contains only one solute, the subscript is dropped

$$\mathbf{x} = c/C_o \tag{44}$$

If a two-component mixture is fed to a column containing one or both of the components, **x** will be replaced by λ as the corresponding variable (see Section III, D, 1b).

b. Solid Concentration Ratio, **y**. The solute concentration in the solid phase, that is reached after continued contact with successive portions of entering feed, is taken as unity on a relative scale. The parameter **y** then expresses the ratio of solute concentration on the solid to the maximum attainable concentration.

For adsorption of a single solute,

$$\mathbf{y} = q/q_o^* \tag{45}$$

In ion exchange,

$$\mathbf{y}_A = q_A/Q \tag{46}$$

or, alternatively,

$$\mathbf{y}_A = q_A/(q_A)_o^* \tag{47}$$

For a mixed binary feed, or for a column with uniform partial presaturation, **y** will be replaced by ω (Section III, D, 1b).

c. Equilibrium Parameter, or Separation Factor, **r**. For exchange adsorptions (Eqs. 7, 20, 22, 25), **r** is given by $(1/K_L')$ or $(1/K^{II})$. Simple relations express **r** in certain other instances of adsorption (Eqs. 18, 24).

d. Distribution Ratio, **D**. This is the limiting value reached as saturation is approached. For simple adsorption,

$$\mathbf{D} = q_o^* \rho_b / C_o \epsilon \tag{48}$$

For simple binary ion exchange,

$$\mathbf{D} = Q\rho_b/C_o\epsilon \tag{49}$$

This dimensionless ratio may be compared with K_d used by Boyd (B9), Mayer and Tompkins (M6), and others in the chemical literature on ion exchange. In their work, $K_d = Q/C_o$.

e. Number of Transfer Units, $\mathbf{N}$, or Σ. In fixed-bed operations, it is desirable to obtain a concentration history curve that is steep by comparison to the total volume of solution handled in any one cycle. For saturation, a steep breakthrough curve makes it possible to utilize a large fraction of the theoretical capacity of the resin. For chromatography, steep-sided zones provide bettter recoveries and purities of the individual components.

The relative steepness or "sharpness" improves as the volume of the resin bed increases and also as the exchange rate increases. In order to use generalized mathematical results, these two factors must be combined into a dimensionless column capacity parameter Σ, which is entirely analogous to the number of transfer units, NTU or $\mathbf{N}$, as defined by Chilton and Colburn for differentially continuous separations (C3).

For external diffusion controlling,

$$\mathbf{N}_f \ (= \Sigma_f) = k_f a_p v \epsilon / F \tag{50}$$

If pore diffusion contributes or controls, k_f is replaced by k_f'. For solid-phase internal diffusion controlling,

$$\mathbf{N}_p \ (= \Sigma_p) = k_p a_p \mathbf{D} v \epsilon / F \tag{51}$$

with $k_p a_p = 60 D_p / d_p^2$ as given by Glueckauf (G5). For both resistances appreciable, the over-all NTU's are

$$\mathbf{N}_{Of} = \frac{b_f v \epsilon}{\left(\dfrac{1}{k_f a_p} + \dfrac{1}{k_p a_p \mathbf{D}}\right) F} \tag{52}$$

$$\mathbf{N}_{Op} = \frac{b_p \mathbf{D} v \epsilon}{\left(\dfrac{\mathbf{D}}{k_f a_p} + \dfrac{1}{k_p a_p}\right) F} \tag{53}$$

where b_f and b_p are coefficients with values usually near unity (Section III, D, 5). These will depend upon $\mathbf{x}$, but frequently are evaluated at $\mathbf{x} = 0.5$.

f. Number of Reaction Units, Column-Capacity Parameter, or Bed-Thickness Modulus, $\mathbf{N}_R$, or s. The general treatment of ion-exchange rates is based upon a surface reaction-kinetic driving force which approximates either an external or an internal material-transfer driving force. By analogy to Hurt's definition of the number of reactor units (H8), a "reaction unit" will be used here as the counterpart of the transfer unit. The number of reaction units (NRU) is given by

$$\mathbf{N}_R (= s) = k_{\text{kin}} Q v \rho_b / F \tag{54}$$

where, for adsorption from the gas phase, Q represents q_m. This dimensionless quantity has also been called the column-capacity parameter by Hiester and Vermeulen (H6); it corresponds also to the thickness modulus of Hougen and Marshall (H7). It is related to the "number of theoretical plates," N_e, used by Mayer and Tompkins (M6) and Martin and Synge (M4); usually $\mathbf{N}_R \approx 2N_e$.

Mathematical analysis is needed before $\mathbf{N}_R$ or $\mathbf{N}$ can be related directly to the concentration behavior. The foregoing relations serve primarily to define the height of a transfer (or reaction) unit, H. With $\mathbf{N} = 1$, Eqs. (50) and (51) give

$$H_f = U/k_f a_p \tag{55}$$

$$H_p = \frac{U}{k_p a_p \mathbf{D}} = \frac{d_p^2 U}{60 D_p \mathbf{D}} \tag{56}$$

With $\mathbf{N}_R = 1$, Eq. (54) gives

$$H_R = U\epsilon/k_{\text{kin}} \rho_b Q \tag{57}$$

The over-all values H_{Of} and H_{Op} can be similarly defined. At $\mathbf{r} = 1$, $H_{Of} = H_{Op} = H_R$. Elsewhere they may differ by a moderate percentage, as will be shown later.

g. Throughput Ratio, Z. This parameter reaches unity when the volume of feed which has passed through the column becomes stoichiometrically equivalent to the adsorption (or exchange) capacity of the column. The stoichiometric volume, V_{stoic}, can be defined by the relation

$$C_o V_{\text{stoic}} = q_o^* \rho_b v \tag{58}$$

The throughput parameter is then

$$\mathbf{Z} = \frac{V - v\epsilon}{V_{\text{stoic}}} = \frac{C_o(V - v\epsilon)}{q_o^* \rho_b v} \tag{59}$$

The number of column void-volumes of fluid that have passed through the column, divided by the distribution ratio, also gives the throughput parameter:

$$\mathbf{Z} = (V - v\epsilon)/\mathbf{D} v\epsilon \tag{60}$$

The midpoint of the concentration-history curve ($\mathbf{x} = 0.50$) will usually correspond to a value of $\mathbf{Z}$ near unity.

h. Solution-Capacity Parameter, or Solution Volume (or Time) Modulus,

Θ *or* t. This variable enters into the differential equations as a dimensionless time. However, it can be expressed as the product of $\mathbf{N}$ (or $\mathbf{N}_R$) and $\mathbf{Z}$:

$$\Theta_f = \mathbf{N}_f \mathbf{Z} \tag{61}$$
$$\Theta_p = \mathbf{N}_p \mathbf{Z} \tag{62}$$
$$t = \mathbf{N}_R \mathbf{Z} \tag{63}$$

Thus the ratio t/s or Θ/Σ used by Hiester and Vermeulen is expressed here by $\mathbf{Z}$.

2. *Material Balance*

The conservation equation, for an infinitesimal thickness of bed at any given cross section v, expresses the fact that any loss of component A from the solution flowing through the section must equal the gain of component A on the solid and in the solution at that section:

$$-\left(\frac{\partial c_A}{\partial v}\right)_V = \rho_b \left(\frac{\partial q_A}{\partial V}\right)_v + \epsilon \left(\frac{\partial c_A}{\partial V}\right)_v \tag{64}$$

It is convenient to consider the volume of saturating solution that has flowed through this cross section, $V - v\epsilon$, as a variable replacing the feed volume, V. By a fundamental property of partial derivatives,

$$-\left(\frac{\partial c_A}{\partial v}\right)_V = \epsilon \left(\frac{\partial c_A}{\partial (V - v\epsilon)}\right)_v - \left(\frac{\partial c_A}{\partial v}\right)_{V-v\epsilon}$$
$$= \epsilon \left(\frac{\partial c_A}{\partial V}\right)_v - \left(\frac{\partial c_A}{\partial v}\right)_{V-v\epsilon} \tag{65}$$

Hence, Eq. (64) simplifies to

$$-\left(\frac{\partial c_A}{\partial v}\right)_{V-v\epsilon} = \rho_b \left(\frac{\partial q_A}{\partial V}\right)_v \tag{66}$$

Introduction of Eqs. (44), (45), and (59) leads to

$$-\left(\frac{\partial \mathbf{x}}{\partial v}\right)_{Zv} = \left(\frac{\partial \mathbf{y}}{\partial Zv}\right)_v \tag{67}$$

The NTU or NRU can also be introduced, to give the entirely dimensionless form

$$-\left(\frac{\partial \mathbf{x}}{\partial \mathbf{N}}\right)_{ZN} = \left(\frac{\partial \mathbf{y}}{\partial Z\mathbf{N}}\right)_\mathbf{N} \tag{68}$$

It is this *equation of continuity*, rather than the rate equations used with it, that reflects the special behavior of the fixed-bed systems.

B. Limiting Cases of Equilibrium Behavior

1. *Proportionate-Pattern Case (Unfavorable Equilibrium)*

The proportionate-pattern case is a classical one in the theory of chromatography, and was treated by DeVault (D2), Walter (W1), Wilson (W7), and Weiss (W3). It is *assumed* that equilibrium is maintained everywhere in the column, that is, that **N** approaches infinity, due to high mass-transfer rates or to long residence times.

The conservation relation, Eq. (67), is rewritten in the form

$$-\left(\frac{\partial \mathbf{x}}{\partial v}\right)_{Zv} = \left(\frac{\partial \mathbf{x}}{\partial Zv}\right)_{v} \frac{d\mathbf{y}}{d\mathbf{x}} \tag{69}$$

Rearrangement gives

$$\frac{d\mathbf{y}}{d\mathbf{x}} = -\frac{(\partial \mathbf{x}/\partial v)_{Zv}}{(\partial \mathbf{x}/\partial Zv)_{v}} = \left(\frac{\partial Zv}{\partial v}\right)_{\mathbf{x}}$$

$$= v\left(\frac{\partial Z}{\partial v}\right)_{\mathbf{x}} + Z \tag{70}$$

Following DeVault, this relation can be integrated at constant **x** and constant $d\mathbf{y}/d\mathbf{x}$, to give

$$\frac{d\mathbf{y}}{d\mathbf{x}} = Z + \frac{a}{v} \tag{71}$$

where a is a constant of integration. Equation (71) will be valid only in the range of positive **Z** values that gives $0 < \mathbf{x} < 1$. The constant a may be evaluated with the aid of the material-balance integral

$$\int_0^\infty (C_o - c) d(V - v\epsilon) = q_o^* \rho_b v \tag{72}$$

or

$$\int_{Z=0}^{\mathbf{x}=1} (1 - \mathbf{x}) dZ = 1 \tag{73}$$

This relation is comparable to Eq. (66) or (67), but is written for the entire column rather than for a differential section.

For the case of a *constant separation factor*, first treated by Walter (W1), Eq. (23) leads to

$$\frac{d\mathbf{y}}{d\mathbf{x}} = \frac{r}{[(1 - r)\mathbf{x} + r]^2} \tag{74}$$

and to $a = 0$. If Eqs. (71) and (74) are combined, there results

$$\mathbf{x} = \frac{(r/Z)^{1/2} - r}{1 - r} \tag{75}$$

The limits of validity are: $x = 0$ at $Z = 1/r$; $x = 1$ at $Z = r$. This is the desired concentration-history equation. It provides a *proportionate pattern*, because x depends upon Z only and not upon N or v. In this case the relative sharpness of the breakthrough curve cannot be increased by lengthening the column.

Where other isotherms apply, the same procedure may be applied; that is, the derivative can be introduced into Eq. (71), and the result introduced into Eq. (73) in order to evaluate a. It is noted that the Freundlich isotherm with $n = 2$, in the form $y = x^2$, leads to a breakthrough line of constant slope.

If the equilibrium ceases to be unfavorable, with $r < 1$, dx/dZ will take on a negative slope. As this is prohibited by the material-balance relation, the breakthrough curve for "equilibrium" must be drawn continuously vertical in the concentration region for which $r \leq 1$.

At $r \leq 2$, the equilibrium-limit breakthrough given by Eq. (75), is approached only at very high N values (>500). For $r \geq 10$, however, this limit will apply at nearly all N values (≥ 10). The range of validity can only be estimated by comparison with the general result (Section III, D, 1).

2. *Constant-Pattern Case (Favorable Equilibrium)*

It will next be *assumed* that a region of r values exists where the effluent concentration pattern is independent of column length (instead of proportional to it). This is equivalent to $(\partial Zv/\partial v)_x$ constant; the difference of V values for two different values of x (e.g., $x = 0.05$ and $x = 0.95$) remains constant regardless of the length of the column.

With this assumption, Eq. (70) again applies, and its integration gives

$$y = (\text{const.}) \, x \qquad (76)$$

Since y and x have the same limits,

$$y = x \qquad (77)$$

This relation becomes the continuity condition for the constant-pattern case.

In order to obtain realistic concentration-history curves, the rate can no longer be assumed infinite. Instead, the breakthrough becomes dependent upon rate. Since $V_{0.95} - V_{0.05}$ is independent of v, it follows that $Z_{0.95} - Z_{0.05}$ is inversely proportional to v. Thus, in agreement with the results of the previous section, the breakthrough curves are relatively steeper at higher v or higher N.

The constant-pattern case was first identified and discussed by Bohart and Adams (B6), Wicke (W5), and Sillén (S4). Michaels termed it the "constant exchange zone" case (M7), while Dryden (D6) has called it the "constant pattern."

a. External Diffusion. For irreversible adsorption ($r = 0$), Eqs. (31) and (48) lead to

$$d\mathbf{x}/d\tau = k_f a_p \mathbf{D} \mathbf{x} \tag{78}$$

Integration for a column of given v, with evaluation of the constant by Eq. (73), and substitution of Eqs. (50) and (59), gives the result of Drew, Spooner, and Douglas (D5), which has been applied also by Selke and Bliss (S2),

$$1 + \ln \mathbf{x} = k_f a_p \mathbf{D}(V - v\epsilon - V_{\text{stoic}})/F$$

or

$$\ln \mathbf{x} = \mathbf{N}_f(\mathbf{Z} - 1) - 1 \tag{79}$$

Equation (79) applies between $\mathbf{Z} = 1/\mathbf{N}_f$ (below which $\ln \mathbf{x} = -\mathbf{N}_f$) and $\mathbf{Z} = 1 + 1/\mathbf{N}_f$, at which $\mathbf{x}$ becomes unity. Figure 7 shows a breakthrough curve calculated from Eq. (79) for $\mathbf{N}_f = 4$.

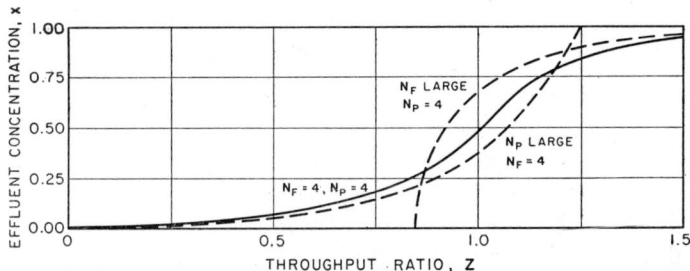

FIG. 7. Breakthrough curves under irreversible conditions, for fluid-phase, particle-phase, and combined resistances controlling. Courtesy of *Industrial and Engineering Chemistry.*

This method is applicable to partially reversible adsorption ($0 < r < 1$), for which $\mathbf{x}^*$ must be replaced by a function of $\mathbf{y}$ (hence, $\mathbf{x}$) that is evaluated from the appropriate isotherm. A graphical method has been developed by Eagleton and Bliss (E1, T4) which can be used when the analytical relations become unmanageable.

Michaels (M7) has solved Eq. (31) for the case of a *constant separation factor.* His result is

$$\frac{1}{1 - r} \ln \frac{\mathbf{x}_2(1 - \mathbf{x}_1)}{\mathbf{x}_1(1 - \mathbf{x}_2)} + \ln \frac{1 - \mathbf{x}_2}{1 - \mathbf{x}_1} = \mathbf{N}_f(\mathbf{Z}_2 - \mathbf{Z}_1) \tag{80}$$

If this relation were extended to a value of $r = 1$, the rate expressed as $dy/d\tau$ would fall to zero for all finite values of $\mathbf{N}_f$. Again, the range of validity must be established by comparison with a general solution (Section III, D, 1). The following limits are found:

r	0	0.2	0.5	0.8
Minimum N	(4)	10	25	75

b. Internal Solid-Phase Diffusion. An exact solution for the *irreversible* constant-pattern breakthrough has been provided by Wicke (W5), Boyd et al. (B8), and others:

$$\mathbf{x} = 1 - \frac{6}{\pi^2} \sum_{n=1}^{\infty} \frac{1}{n^2} e^{-n^2[\psi N_p(Z-1)+0.97]} \tag{81}$$

As in Eq. (51), $\mathbf{N}_p = 60 D_p \mathbf{D} v \epsilon / F d_p^2$. For this case ($\mathbf{r} = 0$), $\psi = 4\pi^2/60$. Glueckauf and Coates (G5) have provided the linear-driving-force approximation of Eq. (37), which in the irreversible case becomes

$$\frac{d\mathbf{x}}{d\tau} = \frac{60 D_p}{d_p^2} \mathbf{D}(1 - \mathbf{x}) \tag{82}$$

which integrates to

$$\mathbf{x} = 1 - e^{-[N_p(Z-1)+1]} \tag{83}$$

If used in this form, Eq. (83) gives a good fit to Eq. (81) in the lower part of the curve, when $0 < \mathbf{x} < 0.6$. However, the slopes in the asymptotic, high-concentration region will be high by a factor of $15/\pi^2$. If conversely a match is made in the high-concentration region, a moderately large error results in the lower part of the curve.

For a partially irreversible adsorption, i.e. one with highly favorable equilibrium, and with a *constant separation factor*, Glueckauf and Coates's result becomes:

$$\frac{\mathbf{r}}{1-\mathbf{r}} \ln \frac{\mathbf{x}_2(1-\mathbf{x}_1)}{\mathbf{x}_1(1-\mathbf{x}_2)} + \ln \frac{1-\mathbf{x}_1}{1-\mathbf{x}_2} = \mathbf{N}_p(\mathbf{Z}_2 - \mathbf{Z}_1) \tag{84}$$

The limits of validity here are the same as listed above for Eq. (80).

A quadratic-driving-force approximation has been developed by Vermeulen (V3). For $\mathbf{r} = 0$ this fits Eq. (81) closely over the entire range of $\mathbf{x}$. This can be extended to the general form:

$$\frac{d\mathbf{y}}{d\tau} = \frac{60 \psi D_p}{d_p^2} \cdot \frac{(\mathbf{y}^*)^2 - \mathbf{y}^2}{2(\mathbf{y} - \mathbf{y}_o)} \tag{85}$$

From Glueckauf's recent work (G3), the author finds

$$\psi = \frac{\pi^2}{\pi^2 \mathbf{r} + 15(1 - \mathbf{r})} \tag{86}$$

For the completely irreversible case, Eq. (85) integrates to

$$x = [1 - e^{-\psi[N_p(Z-1)+0.93]}]^{1/2} \tag{87}$$

This result has been used for extrapolation at $r = 0$ into the region of N_p values below 4, where y obeys Eq. (81) or (87) with an exponential term $e^{-N_p(Z-Z_b)}$ for which $N_p Z_b$ ($= \Theta_b$) is a known function of N_p (V3). In this region x no longer exhibits a constant pattern. The resulting curves for x give a much improved fit to data of Vermeulen and Huffman (V6) and Dryden (D6). The relative simplicity of Eq. (87), combined with good accuracy, should make it a useful replacement for Eq. (81) in many areas of application.

Fujita (F2) has solved the linear-driving-force case for adsorption of two solutes in the "irreversible" region. His treatment assumes that the two solutes have different rate coefficients, and that the equilibrium constant between them is not unity. The results generally are not obtained in closed form, but Fujita's article gives plots of the results for several typical cases.

c. *Pore Diffusion.* This problem has been solved for the irreversible case ($r = 0$), in unpublished work by Acrivos and Vermeulen. The fluid-phase concentration in the pores is assumed negligible compared to the solid-phase value. Equation (32) then becomes

$$\frac{D_{\text{pore}}}{r^2} \frac{\partial}{\partial r}\left(r^2 \frac{\partial c_i}{\partial r}\right) = \rho_b \frac{\partial q_i}{\partial \tau} \tag{88}$$

Differentiation of Eq. (33) gives

$$\frac{\partial}{\partial r}\left(\frac{dq}{d\tau}\right) = \frac{3r^2}{r_p^3} \frac{\partial q_i}{\partial \tau} \tag{89}$$

Combination of Eqs. (88) and (89) gives

$$\frac{dq}{d\tau} = \frac{3 D_{\text{pore}} r^2}{\rho_b r_p^3} \frac{\partial c_i}{\partial r} \tag{90}$$

In the irreversible case, the concentration wave entering the particle will saturate each spherical layer before it penetrates further. That is, c_i cannot exceed zero at a particular radius $r = r_i$ until q_i at that r has become equal to the saturation value q_m. With $c/C_o = q/q_m = 1 - (r_i/r_p)^3$ by material balance over one particle, Eq. (90) leads to

$$\frac{dr_i}{d\tau} = -\frac{k_{\text{pore}}}{15} \frac{r_p^2 + r_p r_i + r_i^2}{r_i} \tag{91}$$

where

$$k_{\text{pore}} = \frac{60 D_{\text{pore}} C_o}{\rho_b q_m d_p^2} \tag{92}$$

Integration of Eq. (91) can be followed by numerical evaluation of x. The result fits the empirical relation

$$x = 0.557[N_{pore}(Z - 1) + 1.15] - 0.0774[N_{pore}(Z - 1) + 1.15]^2 \quad (93)$$

with $N_{pore} = 60 D_{pore} v / F d_p^2$. Whereas (at $r = 0$ only) the external-diffusion curve has a finite limit at $x = 1$, and the solid-phase internal-diffusion curve has a finite limit at $x = 0$, the pore-diffusion curve has two such limits; at $x = 0$, $Z = 1 - (1.15/N_{pore})$, and at $x = 1$, $Z = 1 + (2.43/N_{pore})$.

d. Combined External- and Internal-Diffusion Resistances. Where these two mechanisms have similar rates, external diffusion will tend to predominate at low extents of breakthrough and internal diffusion will have more of a retarding effect as full saturation is approached. The rates given by Eqs. (31) and (85) can be set equal, in terms of interface concentrations that are in mutual equilibrium:

$$k_f(x - x_i) = k_p D \psi \frac{y_i^2 - x^2}{2x} \quad (94)$$

In the initial stages of breakthrough, under *irreversible* conditions, $x_i = 0$ and $y_i < 1$, so that breakthrough follows the external-diffusion curve (Eq. 79). Eventually y_i reaches unity, and abruptly x_i becomes finite; Eq. (87) then applies. Equation (94) can be solved for x at this transition point. The integration constants in Eqs. (79) and (87) must be modified by introducing Eq. (73), as described elsewhere by the author (V3). The curve resulting from $N_f = 4$ and $N_p = 4$ is shown in Fig. 7.

For partially reversible conditions, evaluation of x_i and y_i is more difficult. The method of Section III, D, 6 is applicable.

e. Surface Reaction-Kinetics Controlling. This empirical case is useful, at r values between zero and 0.5, for treatment of the combined-mechanism region and of pore diffusion, or for preliminary interpretation of data when the mechanism is not known. Equations (35) and (77) give

$$\frac{dx}{dZ} = N_R(1 - r)x(1 - x) \quad (95)$$

Integration gives

$$\frac{1}{1 - r} \ln \frac{x}{1 - x} = N_R(Z - 1) \quad (96)$$

This result was derived by Walter (W1) and Sillén (S4), and much earlier in the case of $r = 0$ by Bohart and Adams (B6). Like the combined-mechanism curve of Fig. 7, Eq. (96) gives breakthrough curves which are asymptotic at x values of both zero and unity.

f. Application to Mixed-Bed Deionization. The removal of a salt M^+X^- from water in a single bed utilizes the three reactions:

$$HCat + M^+ = MCat + H^+$$
$$AnOH + X^- = AnX + OH^-$$
$$H^+ + OH^- = H_2O$$

where Cat represents cationic resin, and An, anionic resin. As in simple adsorption, but unlike ordinary ion exchange, the total ionic concentration of the effluent has a very small value. In this region of concentrations, external diffusion is likely to control.

The adsorption is irreversible ($r = 0$) because of the neutralization reaction. Sharp breakthrough curves are obtained, and thus high flow rates are possible (K7, C1).

An idealized limiting situation can be considered, where the two resins are present in stoichiometric proportions. Then

$$(Q\rho_b v)_{HCat} = (Q\rho_b v)_{AnOH} = C_o(V - v_T\epsilon)/Z \tag{97}$$

with the total bed volume $v_T = v_{HCat} + v_{AnOH}$. Equation (79) can then be applied to calculate the breakthrough curve.

C. Column Dynamics under Linear Equilibrium

1. *General Result*

Linear equilibrium involves constant-separation-factor conditions, with the value of the factor r equal to unity. In binary ion exchange involving only one component in the feed, $r = 1$ corresponds to $K = 1$, a condition which may be approached but seldom is realized exactly. In ion exchange with a mixed feed, r tends toward unity as the feed molefraction of the exchanging component $[(c_A)_o/C_o]$ diminishes (Section III, D, 1b). For simple adsorption, $r = 1/(1 + K_L p_o)$ by Eq. (18), and the case of $r = 1$ corresponds to a wide range of weak adsorption where $K_L p_o \ll 1$ as has been noted by Wicke (W5), Thomas (T1), and Hougen and Marshall (H7).

When $r = 1$, the same form of equation is obtained from several different rate-determining mechanisms. For external diffusion, Eq. (31) becomes

$$\left(\frac{\partial y}{\partial \tau}\right)_N = \frac{k_f a_p(\mathbf{x} - \mathbf{y})}{D} \tag{98}$$

For internal diffusion, Eq. (37) or (85) gives approximately

$$\left(\frac{\partial y}{\partial \tau}\right)_N = k_p a_p(\mathbf{x} - \mathbf{y}) \tag{99}$$

By analogy to Eqs. (98) and (99), the pore-diffusion relation is approximately

$$\left(\frac{\partial y}{\partial \tau}\right)_N = k_{\text{pore}}(x - y) \tag{100}$$

The surface-reaction expression, from Eq. (35), is

$$\left(\frac{\partial y}{\partial \tau}\right)_N = k_i C_0 (x - y) \tag{101}$$

In each case, then,

$$\left(\frac{\partial y}{\partial ZN}\right)_N = x - y \tag{102}$$

Integration of Eqs. (102) and (68), jointly, has been carried out by Anzelius (A3) and Schumann (S1). The results can be expressed as

$$x = J(N, ZN) \tag{103}$$
$$y = 1 - J(ZN, N) \tag{104}$$

where the function J of two variables s and t is given by

$$J(s,t) = 1 - \int_0^s e^{-t-\xi} I_0(2\sqrt{t\xi}) d\xi \tag{105}$$

and I_0 is a modified Bessel function of the first kind. $J(s,t)$ is related to Thomas's function $\phi(s,t)$ in the following manner (T1)

$$J(s,t) = 1 - e^{-s-t} \phi(s,t) \tag{106}$$

It is also related to Brinkley's function g, of which a punched-card table is available (B10):

$$J(s,t) = 1 - g(\sqrt{s}, \sqrt{t}) \tag{107}$$

Another relation useful in the evaluation of J is

$$1 - J(s,t) = J(t,s) - e^{-t-s} I_0(2\sqrt{st}) \tag{108}$$

Since $I_0(0) = 1$ it is apparent that the lower boundaries of $J(s,t)$ are

$$J(0,t) = 1; \quad J(s,0) = e^{-s}$$

It can also be shown that the upper limits of J are:

$$\lim_{s \to \infty} J(s,t) = 0; \quad \lim_{t \to \infty} J(s,t) = 1$$

In the region where the variables of the argument are both greater than 10, use has been made of an asymptotic expansion due to Onsager and given by Thomas (T2) which reduces to:

$$J(s,t) = \frac{1}{2}\left\{1 + \text{erf}(\sqrt{t} - \sqrt{s}) + \frac{e^{-(\sqrt{s}-\sqrt{t})^2}}{\pi[\sqrt{t} + \sqrt[4]{st}]}\right\} \tag{109}$$

accurate to within 1% when $\sqrt{st} \geq 6$, where (for any number z)

$$\text{erf}(z) = \frac{2}{\sqrt{\pi}} \int_0^z e^{-\eta^2} d\eta \qquad (110)$$

as given in standard tables. At $\sqrt{st} \geq 60$, the last term of Eq. (109) can be dropped.

Figure 8, on logarithmic-probability coordinates, shows the behavior of the **J** function. The concentration histories, as plotted against time on

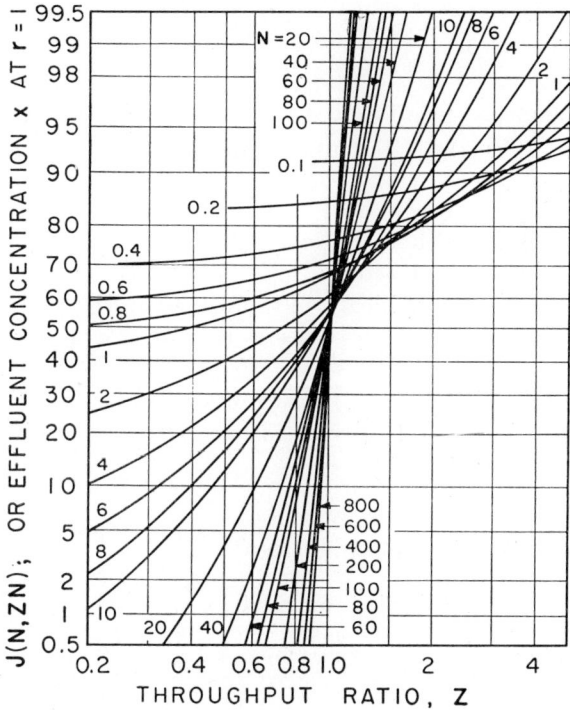

Fig. 8. Dependence of the **J** function upon NTU and throughput ratio. Courtesy of *Chemical Engineering Progress* (H6).

linear scales, normally are S-shaped. The probability scale for **x** largely eliminates the curvature of such plots and also makes it possible to plot accurately those values that are either very small or very near to unity. The logarithmic scale for **Z** makes it possible to compare experimental **x**-time plots directly with the theoretical curves; this curve-fitting technique was utilized in analogous heat-transfer calculations by Furnas (F3), and in ion exchange work by Beaton and Furnas (B5).

The **J** function is also given by Hougen and Marshall (H7) on logarithmic coordinates, by Furnas (F3) on linear coordinates, and by Klinkenberg (K3) in nomographic form. Goldstein (G6) and Klinkenberg have analyzed the behavior of the **J** function under various limiting conditions.

2. *External and Internal Diffusion in Series*

At $r = 1$, for combinations of Eqs. (98) and (99), or (98) and (100), the transfer resistances can be added; for instance,

$$\frac{1}{k_{\text{kin}} C_o} = \frac{D}{k_f a_p} + \frac{1}{k_p a_p} \tag{111}$$

and the mathematical results are still expressed by Eqs. (103) and (104). However, Eqs. (99) and (100) do not give precise results at **N** values below 50.

The exact integration of Eq. (32) or (36) under linear-equilibrium conditions, in conjunction with Eqs. (1) and (68), has been carried out by Rosen (R1) and by Kasten *et al.* (K1). Rosen's computed results have been presented in graphical and tabular form. His variables have the following correspondence to the ones used here: $x = \mathbf{N}_p/5$; $y/x = 2\mathbf{Z}/3$; $x/\nu = \mathbf{N}_f$.

Rosen, and also Wicke (W5), have given an asymptotic relation for solid-phase diffusion or for pore diffusion, at $r = 1$, which can be expressed as

$$\mathbf{x} = \tfrac{1}{2}[1 + \operatorname{erf} \tfrac{1}{2} \sqrt{\mathbf{N}}\, (\mathbf{Z} - 1)] \tag{112}$$

Equation (109) gives for the same cases

$$\mathbf{x} = \tfrac{1}{2}[1 + \operatorname{erf} \sqrt{\mathbf{N}}\, (\sqrt{\mathbf{Z}} - 1)] \tag{112a}$$

If **Z** is near unity, these two expressions are numerically equivalent.

3. *Longitudinal Dispersion*

The effect of longitudinal dispersion in a column, as a factor which reduces the sharpness of breakthrough curves, has been considered by Barrow and co-workers (B3), Ledoux (L4), and others. Wicke (W5) has discussed the combined effect of longitudinal dispersion with internal diffusion, while Lapidus and Amundson (L3) have reviewed the combined effect with external diffusion. Van Deemter *et al.* (V2) have recently studied this problem with respect to gas chromatography.

For the case where fluid-particle equilibrium is closely approached, and the breakthrough slope is determined entirely by longitudinal dispersion, the foregoing authors (L3, W5) give a result equivalent to:

$$x = \frac{1}{2}\left[1 + \mathrm{erf}\left\{\frac{1}{2}\sqrt{\frac{Uv^2\epsilon}{ESV}}\frac{V - V_{\mathrm{stoic}} - v\epsilon}{v\epsilon}\right\}\right] \quad (113)$$

where E is the effective dispersivity, cm.^{2}/sec. In order to avoid an indeterminate error-function argument when V_{stoic} vanishes, it is convenient to define a modified throughput ratio, z:

$$z = \frac{V}{V_{\mathrm{stoic}} + v\epsilon} \quad (113\mathrm{a})$$

This function satisfies the relation $\mathbf{D}(Z - 1) = (\mathbf{D} + 1)(z - 1)$; hence, for large $\mathbf{D}$, $z \approx Z$. An effective number of transfer units for longitudinal dispersion is defined by

$$\mathbf{n} = (\mathbf{D} + 1)Uv/SE \quad (113\mathrm{b})$$

Equation (113) then becomes

$$\mathbf{x} = \tfrac{1}{2}[1 + \mathrm{erf}\ \tfrac{1}{2}\sqrt{\mathbf{n}}\ (\mathbf{z} - 1)] \quad (113\mathrm{c})$$

which is seen to have the same form as Eq. (112). In the absence of channeling, and with the eddy dispersivity large relative to the molecular diffusivity of the fluid, the value of eddy dispersivity determined for gas flow in catalyst beds can be used to estimate $\mathbf{n}$. Since the eddy dispersivity is near $2d_p\epsilon U$, for laminar flow (J1), often $\mathbf{n} \approx (\mathbf{D} + 1)v/2d_p\epsilon S$. The height of one effective transfer unit is then $H_\mathrm{D} \approx 2d_p\epsilon/(\mathbf{D} + 1)$.

Calculations of this type are particularly applicable to the flushing or rinsing of ion exchange columns with distilled water. In this case, $\mathbf{D} = 0$. For beds with $\mathbf{D} = 0$ and $\mathbf{n} < 20$, Jacques and Vermeulen (J1) have found that Einstein's random-walk approach (E2) gives a result which is intermediate to the diffusion equation (A4, M8) and the Poisson distribution (K6), and is thus most apt to apply to actual packed-bed conditions:[4]

$$\mathbf{x} = 1 - \mathbf{J}(\mathbf{nz}, \mathbf{n}) \quad (113\mathrm{d})$$

This reduces to Eq. (113c) at large values of $\mathbf{n}$.

---

[4] For use in interpreting infinitesimal pulse-function distributions, or chromatograms (cf. Section IV, C, 2), with longitudinal dispersion controlling, differentiation of Eq. (113d) yields

$$\mathbf{x} \propto e^{-\mathbf{n}-\mathbf{nz}}I_0(2\mathbf{n}\sqrt{\mathbf{z}}) \quad (113\mathrm{e})$$

At large $\mathbf{n}$, this becomes

$$\ln \mathbf{x} = \ln \mathbf{x}_{\max} - \tfrac{1}{4}\mathbf{n}(\mathbf{z} - 1)^2 \quad (113\mathrm{f})$$

which has the same form as Eq. (166).

4. Analogous Treatment of Heat Transfer

Under conditions of constant heat capacities for the solid checkerwork of a heat regenerator, and for the fluid phases releasing or withdrawing thermal energy, this problem can be solved as a linear-equilibrium case. The relative temperatures are

$$\mathbf{x} = \frac{T_f - T_o}{T_\infty - T_o}; \qquad \mathbf{y} = \frac{T_p - T_o}{T_\infty - T_o} \tag{114}$$

where T_o is the initial temperature of the bed, T_∞ the temperature it will reach at saturation, T_f the instantaneous temperature of the fluid, and T_p the instantaneous average temperature of the solid.

The controlling resistance may be taken as external to the particle:

$$\frac{\partial \mathbf{y}}{\partial \tau} = \frac{k_H a_p \epsilon}{c_p} (\mathbf{x} - \mathbf{y}) \tag{115}$$

where k_H is the heat-transfer coefficient, and c_p is the heat capacity of the solid per unit packed volume. Also, c_f is the heat capacity per unit volume of the fluid, in the following relations:

$$\mathbf{D}_H = c_p / c_f \epsilon \tag{116}$$

$$\mathbf{N}_{Hf} = \frac{k_H a_p}{c_f} \frac{v\epsilon}{F} \tag{117}$$

$$\mathbf{Z}_H = \frac{c_f}{c_p} \frac{V - v\epsilon}{v} \tag{118}$$

$\mathbf{N}_H$ represents the number of heat-transfer units.

If, instead, the conductance of the particle is limiting, it is true approximately that

$$\frac{d\mathbf{y}}{dt} = \frac{60 D_H}{d_p^2} (\mathbf{x} - \mathbf{y}) \tag{119}$$

where D_H is the thermal diffusivity of the solid. Here

$$\mathbf{N}_{Hp} = \frac{60 D_H}{d_p^2} \frac{v\epsilon}{F}$$

and $\mathbf{Z}_H$ and $\mathbf{D}_H$ are unchanged.

In either of the above cases, the relative temperatures are given by Eq. (103) or (113) (or Fig. 8) for the effluent, and by Eq. (104) for the checkerwork. Rosen's numerical solutions (R1) can also be applied directly, if solid-phase conduction accounts for part (or all) of the transfer resistance.

5. Use of Curves to Predict Breakthrough Behavior

For a column of given height or volume, the smaller of the calculated $\mathbf{N}$ values will control unless the values are of the same magnitude. If the column dimensions, the flow rate, and the cycle period are all specified, the dimensionless parameters N and Z can be calculated, and (if $r = 1$) x can be read from Fig. 8. For example, assume that $k_f a_p$ is known to be 20 min.$^{-1}$; $\epsilon = 0.4$; $\rho_b = 0.5$ gm./ml.; and $q_o{}^* = 50$ mg./gm. In addition, design values are specified for $v = 75$ ml., $C_o = 0.5$ mg./ml., and $F = 30$ ml./min. It follows that $D = 125$ by Eq. (48), and $N_f = 20$ by Eq. (50). The effluent time $(\tau - v\epsilon/F)$, corresponding to $Z = 1$, is indicated by Eq. (59) to be $Dv\epsilon/F$ or 125 min. If it is desired to calculate x at a time $(\tau - v\epsilon/F) = 75$ min., the coordinates $Z = 0.60$ and $N = 20$ in Fig. 8, show that $x = 0.10$.

If the breakthrough value is specified along with N, the chart can be used to find Z. If x and τ are specified and the height of the column is to be determined, a trial-and-error solution must be made on Fig. 8; or the value of N can be read off a plot of x against NZ, since here $NZ \approx (k_f a_p/D)$.

6. *Radial Beds*

The use of annular cylindrical beds, fed through a central channel and drained at the periphery, is of potential interest in cases where high throughput rates and wide but shallow beds are desired. Radial-flow geometry is also characteristic of one operating method used in paper chromatography.

Lapidus and Amundson (L2) have solved this problem for a constant volumetric flow rate. For $r = 1$, the result is again given by Eq. (103).[5] The dimensionless parameters have their usual values, with v calculated as

$$v = \pi l(R_e{}^2 - R_i{}^2) \tag{120}$$

where l is the axial depth of the bed; R_e is the exterior radius; and R_i is the interior radius. In this case the external mass-transfer coefficient k_f, being a function of the linear velocity, will vary with the radius.

The derivations for radial beds include the general reaction-kinetic case with constant separation factor, analogous to the treatment of axial-flow beds in the next section. An equation identical in form with Eq. (121) is obtained by Lapidus and Amundson.

D. Column Dynamics at a Constant Separation Factor

1. *The Method of Surface Reaction Kinetics*

a. Thomas's Solution. The most general relation that has been developed for breakthrough behavior is that of H. C. Thomas (T1), which

---

[5] The two terms in the argument of J are inverted in the reference (L2) and in the dissertation of Hiester (H1), relative to Eq. (103).

includes the equilibrium parameter **r** as an independent variable along with the number of transfer units **N** and the throughput ratio **Z**. Equations (35) and (68) have been solved, to give

$$x = \frac{J(rN_R, ZN_R)}{J(rN_R, ZN_R) + e^{(r-1)N_R(Z-1)}[1 - J(N_R, rZN_R)]} \quad (121)$$

and a similar relation (cf. Eq. 156) for **y**. An extensive graphical representation of **x** has been given by Hiester and Vermeulen (H6), and numerical values have been computed and tabulated by Opler and Hiester (O1).

It is apparent that Eq. (121) contains the **J** function as a limiting case, at **r** = 1. This equation also has been shown to reduce to the constant-pattern result (Eq. 96) with $r \ll 1$, and to the proportionate-pattern result (Eq. 75) with $r \gg 1$, in work by Hiester and Vermeulen (H6) and Gilliland and Baddour (G2). Goldstein (G6) has reviewed this result from a mathematical viewpoint, and has presented limiting forms which give an accurate approximation in certain regions; his variables u, s, and y correspond respectively to the present **x**, **N**, and **ZN**.

In the writer's view, the Thomas equation is the most important single result in the entire theory of adsorption-column performance. By reference to this result, all other solutions are found to be classifiable in terms of their **r** and **N** values.

b. Generalized Behavior of Binary Systems. Thomas's result has been applied mainly to two-component exchange-adsorption systems in which a fluid containing one component interacts with a solid phase containing initially only the other component; or to one-component adsorption involving solid that initially is entirely free of adsorbed material. However, Thomas (T2) has also shown that elution from a completely saturated bed and saturation of a completely eluted bed are complementary processes, each of which has an equilibrium constant that is the reciprocal of the constant for the other. Amundson (A2) has extended Thomas's methods to obtain an algebraic framework for numerical integration of the binary case in which the initial solid-phase composition varies with position in the column, and the feed composition varies with time (but also under the restriction of constant total flow rate).[6]

The problem to be considered now is one of exchange adsorption with *uniform partial presaturation* (V5). The initial solid-phase and entering fluid-phase concentrations are each to be uniform (independent of position or time, respectively), but may each involve any desired proportion of each of the two components A and B.

In a column that is uniformly presaturated with component A to any

---

[6] On p. 819 of reference A2, the definitions of parameters $F'(x)$ and $G'(y)$ have been interchanged.

specified level (for example, by mixing A-resin and B-resin; or by saturating completely with a solution containing both A and B ions), the initial concentration $(q_A)_o$ corresponds to an equilibrium concentration $(c_A)_o^*$ in the solution phase at a given total solution concentration C_o. With a feed at $(c_A)_o$, only the amount $[(c_A)_o - (c_A)_o^*]$ will be removed from each unit volume of solution under the most favorable conditions. Likewise, if $(q_A)_o^*$ is the resin concentration that would be in equilibrium with feed solution at $(c_A)_o$, the resin at most will accumulate only $[(q_A)_o^* - (q_A)_o]$ moles of A per unit weight. The equilibrium is given by Eq. (21), with **x** replacing **N**.

In order to utilize the saturation functions derived for pure feed and pure resin, a solution saturation-fraction must be defined,

$$\lambda = \frac{c_A - (c_A)_o^*}{(c_A)_o - (c_A)_o^*} = \frac{\mathbf{x} - \mathbf{x}_o^*}{\mathbf{x}_o - \mathbf{x}_o^*} \qquad (122)$$

and a resin saturation-fraction,

$$\omega = \frac{q_A - (q_A)_o}{(q_A)_o^* - (q_A)_o} = \frac{\mathbf{y} - \mathbf{y}_o}{\mathbf{y}_o^* - \mathbf{y}_o} \qquad (123)$$

where

$$\mathbf{x}_o = (c_A)_o/C_o, \qquad \mathbf{x}_o^* = (c_A)_o^*/C_o$$
$$\mathbf{y}_o = (q_A)_o/Q, \qquad \mathbf{y}_o^* = (q_A)_o^*/Q$$

The numerical solutions for $\lambda = \lambda(\mathbf{r},\mathbf{N},\mathbf{Z})$ and $\omega = \omega(\mathbf{r},\mathbf{N},\mathbf{Z})$ will correspond exactly to those derived for **x** and **y**, if the parameters can be redefined so that λ and ω, respectively, will replace **x** and **y** in the rate and material-balance relations (Eqs. 68 and 35) to give:

$$\left(\frac{\partial \omega}{\partial Z \mathbf{N}_R}\right)_{\mathbf{N}_R} = \lambda(1 - \omega) - r\omega(1 - \lambda)$$

$$= -\left(\frac{\partial \lambda}{\partial \mathbf{N}_R}\right)_{Z\mathbf{N}} \qquad (124)$$

This leads to the relations:

$$r = \frac{(K^{II} - 1)\mathbf{x}_o^* + 1}{(K^{II} - 1)\mathbf{x}_o + 1} \qquad (125)$$

$$\mathbf{N}_R = \frac{kQ\rho_b}{(K^{II} - 1)\mathbf{x}_o^* + 1} \frac{v}{F} \qquad (126)$$

and

$$Z = \frac{[(K^{II} - 1)\mathbf{x}_o + 1][(K^{II} - 1)\mathbf{x}_o^* + 1]}{DK^{II}} \frac{V - v\epsilon}{v\epsilon} \qquad (127)$$

To use this method for simple adsorption (instead of exchange adsorption), the equations become

$$r = \frac{1 + K_L p_o^*}{1 + K_L p_o} \tag{128}$$

$$\mathbf{N}_R = \frac{kQ\rho_b}{1 + K_L p_o^*} \frac{v}{F} \tag{129}$$

and

$$\mathbf{Z} = \frac{(1 + K_L p_o)(1 + K_L p_o^*)}{\mathbf{D} K_L p_o} \frac{V - v\epsilon}{v} \tag{130}$$

The theory for partially presaturated columns serves to unify the calculation methods for exchange adsorption and Langmuir adsorption, by including within its framework several situations which previously have had to be solved as isolated problems. These situations are those of elution of a completely saturated column, and of saturation and chromatography of trace components (the trace linear-equilibrium situation).

c. *Elution from a Completely Saturated Column.* In saturation, $\mathbf{x}_o = 1$ and $\mathbf{x}_o^* = 0$. In elution of the same component, by means of desorption or of the reverse exchange, $\mathbf{x}_o^\dagger = 0$ and $(\mathbf{x}_o^\dagger)^* = 1$, where the dagger (†) designates the parameters for the elution cycle. From Eqs. (122–130), with the unlabeled parameters indicating the saturation cycle, it can be shown (V5) that

$$\lambda^\dagger = \mathbf{x}^\dagger = 1 - \mathbf{x} \tag{131}$$
$$\omega^\dagger = \mathbf{y}^\dagger = 1 - \mathbf{y} \tag{132}$$
$$\mathbf{r}^\dagger = 1/\mathbf{r} \tag{133}$$
$$\mathbf{N}_R^\dagger = \mathbf{r}\mathbf{N}_R \tag{134}$$
$$\mathbf{Z}^\dagger = \mathbf{Z} \tag{135}$$

These results are well known from Thomas's work (T2), but serve to confirm the correctness of the uniform-partial-presaturation theory.

d. *Behavior of Trace Components.* The binary trace situation represents a linear equilibrium, because of the small extent of saturation of the adsorbent by a trace component, A. For ion exchange, with $c_A \ll C_o$, it follows that $C_o = c_A + c_G \approx c_G$, where c_G is the concentration of the gross or carrier ion. For adsorption, $K_L p_o \ll 1$. In both processes, generally $\mathbf{x}_o^* = 0$, although this limitation is not a necessary one. From Eq. (125) or (128),

$$\mathbf{r} = \frac{C_o}{(K^{II} - 1)(c_A)_o + C_o} \text{ or } \frac{1}{1 + K_L p_o}$$
$$\approx 1 \tag{136}$$

A trace component can be defined, for practical purposes, as one for which $\mathbf{r}$ approaches unity within acceptably close units. It is generally necessary that $\mathbf{r}$ lie between 0.90 and 1.10, and hence that

$$[(K^{II} - 1)(c_A)_o/C_o] < 0.10 \tag{137}$$

Thus, for a value of K^{II} large compared to unity, the peak value of c_A will need to be very small relative to C_o. For chromatographic separations under "trace" conditions, additional restrictions are sometimes necessary (V4).

For the trace-adsorption or trace-exchange case, $\mathbf{N}_R$ is unchanged from the "gross" value (Eq. 54), as given by Eq. (126). In this case, from Eq. (127),

$$\mathbf{Z} = \frac{1}{\mathbf{D}K^{II}} \frac{V - v\epsilon}{v\epsilon} \approx \frac{1}{\mathbf{D}_{\text{trace}}} \frac{V - v\epsilon}{v\epsilon}$$

$$\approx \frac{(c_A)_o(V - v\epsilon)}{(q_A)_o{}^* \rho_b v} \tag{138}$$

with $\mathbf{D}_{\text{trace}} = (q_A)_o{}^* \rho_b / (c_A)_o \epsilon$.

2. *External or Pore Diffusion Controlling*

From Eq. (19), (21), or (23), the equilibrium can be written:

$$\mathbf{x} = \frac{\mathbf{ry}}{1 + (\mathbf{r} - 1)\mathbf{y}} \tag{139}$$

For simple adsorption, or for exchange adsorption, Eq. (31) or (98) transforms to

$$\left(\frac{\partial \mathbf{y}}{\partial \mathbf{ZN}}\right)_{\mathbf{N}} = \mathbf{x} - \mathbf{x}^* \tag{140}$$

Introducing Eq. (139),

$$\left(-\frac{\partial \mathbf{y}}{\partial \mathbf{ZN}}\right)_{\mathbf{N}} = \frac{\mathbf{x}(1 - \mathbf{y}) - \mathbf{ry}(1 - \mathbf{x})}{1 + (\mathbf{r} - 1)\mathbf{y}} \tag{141}$$

which corresponds to a result of Adamson and Grossman (A1, G8). This has the form of Eq. (35), and integrates to give Eq. (121), if

$$\mathbf{N}_R = \frac{\mathbf{N}}{1 + (\mathbf{r} - 1)\mathbf{y}} \tag{142}$$

where $\mathbf{N}$ is $\mathbf{N}_f$ or $\mathbf{N}_{\text{pore}}$. The average value of $\mathbf{y}$ may be taken as 0.5, if $\mathbf{r} < 1$; or as $1/(\mathbf{r} + 1)$, if $\mathbf{r} > 1$. Hence

$$\mathbf{N}_R = \frac{2\mathbf{N}_f}{\mathbf{r} + 1} \text{ or } \frac{2\mathbf{N}_{\text{pore}}}{\mathbf{r} + 1}, \text{ if } \mathbf{r} < 1 \tag{143}$$

$$\mathbf{N}_R = \frac{\mathbf{N}_f(\mathbf{r} + 1)}{2\mathbf{r}} \text{ or } \frac{\mathbf{N}_{\text{pore}}(\mathbf{r} + 1)}{2\mathbf{r}}, \text{ if } \mathbf{r} > 1 \tag{144}$$

With the use of these substitutions, the results tabulated for Eq. (121) can be used widely for diffusion-controlled rates. As will be discussed in Section III, D, 4, Eq. (142) can be used without averaging y, if greater accuracy is needed.

3. *Solid-Phase (Internal) Diffusion Controlling*

The equilibrium can be written

$$y = \frac{x}{r + (1 - r)x} \tag{145}$$

Eqs. (37) and (145) give

$$\left(\frac{\partial y}{\partial ZN_p}\right)_{N_p} = y^* - y$$

$$= \frac{x(1 - y) - ry(1 - x)}{r + (1 - r)x} \tag{146}$$

This also integrates to give Eq. (121), if

$$N_R = \frac{N_p}{r + (1 - r)x} \tag{147}$$

For an average value of $x = 0.5$,

$$N_R = \frac{2N_p}{r + 1} \tag{148}$$

4. *Calculation for Non-Constant N_R or r*

If either of these parameters varies, the general reaction-kinetic solution can still be used to develop a breakthrough curve for any given pattern of behavior. An experimental curve may be divided into several regions, each of which is small enough for the correction factor to be essentially constant. In terms of curve-matching, this corresponds to matching different segments of a breakthrough curve to different master curves.

Based upon the equivalence of slopes at related points or within related small increments of the kinetic and the diffusional curves, the following procedure has been used in constructing the numerical plots of solutions of the diffusional equations (H6). The over-all concentration range is divided into four segments, as follows:

x, range, %	x_{avg}, %
1–10	3
10–50	30
50–90	70
90–99	97

Within each segment, the values of N_R and r are evaluated for the average x. From a reaction-kinetic plot of x vs. Z corresponding to an appropriate **r**, as calculated from Eq. (121), **Z** is read at the correct $\mathbf{N_R}$ for the concentrations at each end of the range, and a $\Delta \mathbf{Z}$ is computed. The latter values are plotted consecutively on linear-coordinate paper, and are integrated graphically, as by Eq. (73), to locate the $\mathbf{Z} = 1$ point and thus to establish the **Z** coordinates of the entire curve. Families of breakthrough curves for external and internal diffusion controlling, based on such constructions, are available (H6).

5. *Combined External-and Internal-Diffusion Resistances*

Under linear-equilibrium conditions, the combined effect of two diffusional resistances in series is treated by simple addition of the resistances. In the irreversible case, either the external or the internal resistance alone controls, with the latter over-taking the former part-way along the breakthrough curve (Section III, B, 2d). However, when the resistances are in relatively equal balance, with $3\mathbf{N}_p > \mathbf{N}_f > 0.3\mathbf{N}_p$, it is possible throughout the range of **r** values to calculate the breakthrough behavior by assuming an equivalent reaction-kinetic resistance.

It has been shown (H4) that the diffusional resistances can be combined by the relation:

$$\frac{b\epsilon}{k_{\text{kin}} Q \rho_b} = \frac{1}{k_f a_p} + \frac{1}{k_p a_p \mathbf{D}} \tag{149}$$

Equation (38) provides the form of relation for expressing k_f. The exponent given there for N_{Sc} was an assumed value, and experimental values are frequently nearer to -0.50. A simplified relation may therefore be written,

$$k_f = \alpha U (N_{Pe}')^{-0.50} \tag{150}$$

where $N_{Pe}' = d_p U \epsilon / 6(1 - \epsilon) D_f$ is the Peclet number for material transfer, and α is found empirically to be 0.38. With κ_{kin} replacing $k_{\text{kin}} Q \rho_b / \epsilon$, Eq. (149) becomes

$$b\mathbf{D} \frac{D_p 6(1-\epsilon)}{d_p{}^2 \kappa_{\text{kin}}} = \beta_1 \mathbf{D} \frac{D_p}{D_f}(N_{Pe}')^{-0.50} + \beta_2 \tag{151}$$

where $\beta_1 = \epsilon/6(1-\epsilon)\alpha$, and $\beta_2 = (1-\epsilon)/10$. This is the equation plotted in Fig. 6; and it is equivalent to Eq. (42) for a system of constant **b**, **D** and ϵ.

The correction term **b** in Eqs. (149) and (151) is given by

$$\mathbf{b} = \frac{q_o{}^*(c - c_i) + C_o(q_i - q)}{c(q_o{}^* - q) - q(C_o - c)/K} \tag{152}$$

Since the interface concentrations c_i and q_i cannot be measured directly, a mechanism parameter which depends upon their behavior is also defined:

$$\zeta = \frac{q_o^*(c - c_i)}{C_o(q_i - q)} \qquad (153)$$

The relation between **b** and ζ is evaluated numerically by using successive assumed values of c_i (and q_i, in equilibrium with it) to determine both **b** and ζ. Usually c/C_o is taken as 0.5. In view of the low sensitivity of **b**

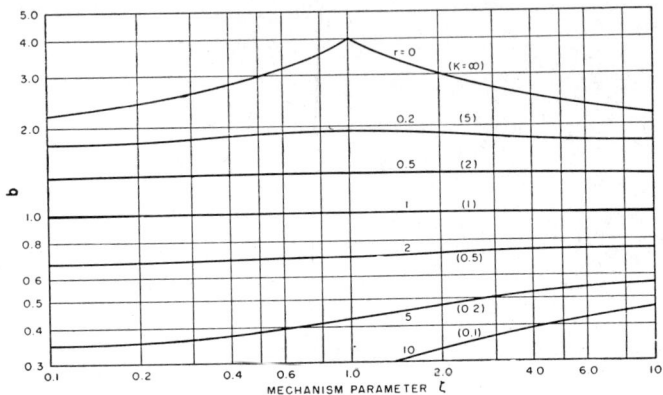

Fig. 9. Correction term for combining diffusional resistances. Courtesy of *A. I. Ch. E. Journal* (H4).

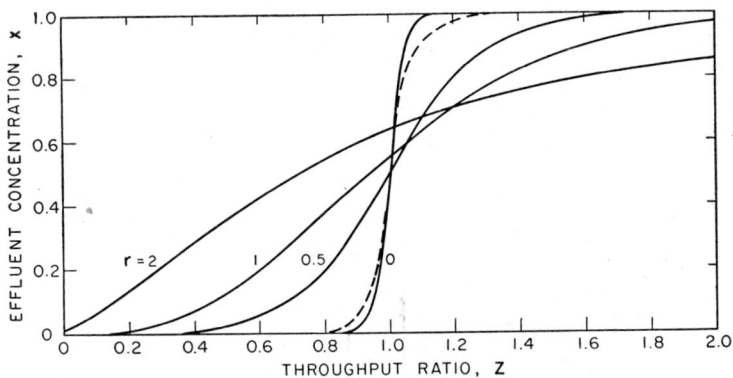

Fig. 10. Breakthrough curves at constant N_f and N_p, showing effect of changing equilibrium parameter **r**. At **r** = 0, the solid curve is given by the reaction-kinetic treatment (Eq. 96), and the dashed curve by the diffusional relations (Eqs. 79, 87, 94).

to changes in ζ, constant-pattern behavior is assumed for $K > 1$, with $c^*/C_o = 1/(1 + K)$; and proportionate-pattern behavior for $K < 1$, with $c^* \approx c$. Values of **b** calculated in this way are shown in Fig. 9.

Both in data interpretation and in equipment design, ζ must be evaluated from known or estimated mass-transfer coefficients or diffusivities, by the relations

$$\zeta = \mathbf{D}\frac{k_p}{k_f} = 4.8\mathbf{D}(\mathrm{N}_{\mathrm{Pe}}')^{-0.5}D_p/D_f \tag{154}$$

Thus ζ is proportional to the abscissa of Fig. 6, and is shown in that figure on an auxiliary scale.

For calculating the over-all NTU's by Eqs. (52) and (53), the external-diffusion value $\mathbf{b}_f$ is given by $\mathbf{b}$ at the appropriate ζ, divided by $\mathbf{b}$ at $\zeta = \infty$. The internal diffusion value $\mathbf{b}_p$ is given by $\mathbf{b}$ at the appropriate ζ, divided by $\mathbf{b}$ at $\zeta = 0$.

From Eq. (149) it follows that

$$\begin{aligned}\frac{\mathbf{b}}{\mathbf{N}_R} &= \frac{1}{\mathbf{N}_f} + \frac{1}{\mathbf{N}_p} \\ &= \frac{1}{\mathbf{N}_f}\frac{1+\zeta}{\zeta} = \frac{1+\zeta}{\mathbf{N}_p}\end{aligned} \tag{155}$$

At constant $\mathbf{N}_R$ and increasing $\mathbf{r}$ (decreasing K), the breakthrough curves become more shallow. At constant $\mathbf{N}_p$ and $\mathbf{N}_f$ as $\mathbf{r}$ increases, $\mathbf{N}_R$ will also decrease because of the behavior of $\mathbf{b}$; this intensifies the diminution in slope. Figure 10 gives a family of breakthrough plots for $\mathbf{N}_p = 20$ and $\mathbf{N}_f = 20$ at several $\mathbf{r}$ values, as calculated from the equivalent reaction-kinetic curves, and indicates a very great effect of equilibrium upon the concentration history.

E. Note on Multicomponent Saturation

The attention of many workers has been given to the equilibrium-limited case (proportionate-pattern) of multiple adsorption (D2, W1, W3, W7, among others). In the constant-pattern case, Fujita's work has already been discussed (Section III, B, 2b). Also, using the theoretical-plate approximation to a packed column, plate-by-plate calculations can be carried out in the constant-pattern case exactly as for continuous (countercurrent) distillation; this treatment is suggested from work on chromatographic and "displacement" problems by Mair (M2), Spedding (S6), and Glueckauf (G3). Moreover the linear-equilibrium result can be extended, in a nearly trivial fashion, to any number of components.

For the general multicomponent case of non-linear equilibrium, no calculation method has yet been developed for determining the rate-dependent, breakthrough behavior. An approximate calculation can be made by considering the feed mixture as a single component with an effective average $\mathbf{r}$, then estimating how the total adsorption wave (the terminology is from K4) will be divided into zones containing the individual components. With a feed containing two solutes, the one less strongly adsorbed will often over-ride the other and appear first in the effluent; this behavior is utilized in the laboratory technique of *frontal analysis*, described by Tiselius (T3).

It is likely that progress in this field will come only after many additional data from column operations have been assembled systematically according to their respective rate and equilibrium values.

IV. Chromatographic Separations

A. General Principles

The essentials of chromatography have been discussed in Section I, A, and are illustrated in Fig. 1. Although saturation operations have been investigated at length by rate-theory approaches, the case of chromatographic elution from an incompletely saturated column has received little attention from the rate standpoint.

In the development of a chromatogram, or pattern of discrete solute zones, movement of each zone through the adsorbent occurs by the following mechanism. Fluid on the upstream (trailing) side of a zone is undersaturated with respect to the adsorbed component and continually takes it into solution. Passing beyond the peak of the zone to the downstream (leading) side, the same fluid is supersaturated relative to coexisting adsorbent and hence gradually redeposits the solute component.

A zone obeying a linear equilibrium would retain a constant rectangular shape, in the event that complete equilibrium could be maintained between solution and resin at each point in the column at all times during its elution. Actually, equilibrium cannot be maintained, and hence the zone undergoes a continual spreading; an initial rectangular shape changes to a flattened-top bell shape, and this in turn develops into a fully peaked bell shape.

A relative sharpening of each zone occurs as the length of column traversed increases, because the spreading of the zone is approximately proportional to the square root of the column length;[7] whereas the total volume of elutant[8] required, for complete elution of the zone, is almost directly proportional to the column length.

Vermeulen and Hiester have shown that the maximum useful charge to a column, in the linear-equilibrium case, will increase with the square root of column length.[9] Thus, at a constant flow rate, the volume of

---

[7] This is precisely so in the linear-equilibrium case ($r = 1$). All other r values give somewhat more spreading.

[8] "Elutant" is preferred by the author and his co-workers for the agent effecting an elution; however, "eluent" and "elutriant" are also used by many writers.

[9] For ion exchange, in the non-trace case, the saturation step is assumed to be conducted at the same total fluid concentration C_0' as is used during the elution period. Under trace conditions, calculations can easily be made for $C_0 \neq C_0'$.

charge that can be processed per unit volume of adsorbent will be inversely proportional to the square root of column length. The systematic selection of optimum values of elutant concentration, pressure drop through the column, flow rate, total cycle period, charge period, recycle feed rate, multiple cycling, concentration of complexing agent, and number of stages of operation, has been analyzed for representative systems (V4, H2).

The feed may be introduced to the column as a narrow band of high (non-trace) concentration, as is customary in many laboratory separations. In elution, such a zone will first assume a flattened-top asymmetric

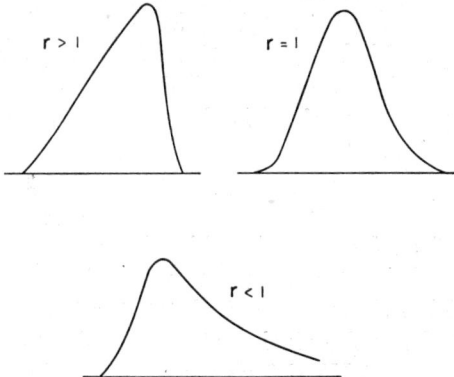

Fig. 11. Variation in symmetry of chromatograms. Ordinates represent concentration; abscissas, time. Courtesy of *Industrial and Engineering Chemistry*.

shape, then a peaked asymmetric shape, and ultimately, because of the reduction of the peak concentration level, a peaked symmetric shape consistent with trace conditions. It is observed that the optimum quantity of material that can be separated at maximum efficiency by a given column, when fed in this manner, is nearly identical with the optimum quantity of material fed under trace conditions (Section III, D, 1d) as an initially broader band of smaller amplitude. Further, the column behavior for a charge fed under non-trace conditions becomes nearly identical with that predicted by the comparatively simple calculation methods for fully trace conditions.

If $r = 1$, the concentration history for a chromatographic zone takes on a mirror-image symmetry with reference to its peak concentration, as the zone travels through a column of considerable length. Many asymmetric curves can therefore be attributed to deviations of r from unity—i.e., to non-trace conditions. It has been shown that the r values for a saturation and a corresponding elution are reciprocals of each other

(Section III, D, 1c). Also, for a column of given length, the smaller the value of **r** for saturation or for elution, the steeper is the saturation or elution curve.

If $r < 1$ for saturation, the leading edge of the zone—the first to emerge in the effluent—will be steeper than the trailing edge; if $r > 1$ for saturation, the leading edge will be more sloping than the trailing edge. Representative zone shapes for $r < 1$, $r = 1$, and $r > 1$ are given in Fig. 11. Because **r** depends upon concentration, the symmetry or asymmetry of an elution curve for any component depends upon the maximum concentration levels of that component in the solution and on the resin, relative to the concentration levels of the carrier component.

In the trace case, the interactions between various solutes present can be neglected, provided not only that $c_A \ll C_o$ (or c_o^*), but also that $q_A \ll Q$. The non-trace case has been derived only for a single solute, in simple adsorption; or for a mixed feed of solute and eluting component, in exchange adsorption.

B. Single Chromatograms: Non-Trace Case

The method of surface-reaction kinetics at a constant separation factor, as developed by Thomas for saturation conditions, is again applied. Values of $\mathbf{N}_R$ should be calculated for the real or hypothetical saturation step, and for the elution step, by using the methods of Section III, D. An apparent NTU for saturation can be obtained from the calculated value for elution ($\mathbf{N}_R^\dagger$), by the relation $(\mathbf{N}_R)_{app} = \mathbf{N}_R^\dagger/r$. Where part or all of the resistance is in the fluid phase, $(\mathbf{N}_R)_{app}$ will not be equal to $\mathbf{N}_R$ as determined for saturation; then, the combined saturation and elution should be described by the geometric mean of these two $\mathbf{N}_R$ values.

The equations derived in this section apply to either external or internal diffusion, or to both together, but probably not to longitudinal dispersion or diffusion as the controlling factor (as may occur particularly in gas chromatography). However, the equations of the next section, for "trace" chromatograms, should apply equally well regardless of which mechanism controls the shape to the curve.

A fluid containing the saturating component A, at concentration $(c_A)_o$, is considered to have passed through the column during a time interval τ_{sat}, which corresponds to a volume V_{sat} and a throughput ratio Z_{sat}. At this instant, the eluting fluid begins to pass through the bed at an unchanged flow rate. This elutant stream differs from the previous fluid only in the omission of saturating component (or in replacement of saturant by additional eluting component to keep C_o constant).

Equations (35) and (68) have been solved in this case by Hiester and

Vermeulen (H1, H5) and by Goldstein (G6). The boundary conditions are:

At $v = 0$, $\lambda = c_A/(c_A)_o = 0$
At $Z' = 0$, as a function of Z_{sat} and N_R,

$$\omega = \frac{q_A}{(q_A)_o^*} = \frac{1 - J(Z_{sat}N_R, rN_R)}{J(rN_R, Z_{sat}N_R) + e^{(r-1)N_R(Z_{sat}-1)}[1 - J(N_R, rZ_{sat}N_R)]} \quad (156)$$

The resulting equations are

$$\lambda' \text{ (denom.)} = J(rN_R, ZN_R) - J(rN_R, Z'N_R) \quad (157)$$
$$\omega' \text{ (denom.)} = J(Z'N_R, rN_R) - J(ZN_R, rN_R) \quad (158)$$

where

$$\text{(denom.)} = J(rN_R, ZN_R) + e^{(r-1)N_R(Z-1)}[1 - J(N_R, rZN_R)]$$
$$+ e^{(r-1)N_R(Z'-1)}[1 - J(rZ'N_R, N_R)] + J(Z'N_R, rN_R) - 1 \quad (159)$$

where the unprimed values are measured from the start of the saturation period, and the primed values (') from the start of the elution period. It can be shown that, for large values of Z_{sat}, these equations will reduce to the form for elution from a completely saturated bed (Section III, D, 1). This limit will occur near $Z_{sat} = 2\sqrt{\pi/N_R}$.

A discontinuity in fluid concentration will occur at the assumedly sharp elutant front. As the front travels down the bed, this discontinuity decreases and rapidly becomes negligible.

The function Z_{sat} can be described as a "memory" term. Equations for multiple saturation and elution involving more than one memory term are also given by Hiester (H1).

Goldstein (G6) has provided approximation functions to aid in evaluating Eqs. (157–159). Baddour and co-workers (B1) have demonstrated their utility for matching experimental curves. For example, with $r < 1$ and $Z_{sat} < 1 - r$, the following rules apply: Below $Z = Z_{sat} + r$, λ' is given by x from Eq. (96); and above $Z = 1/r$, from Eq. (75). Between these two limits, Goldstein finds

$$\frac{1}{\lambda'} = \frac{1-r}{(r/Z)^{1/2} - r} - (4\pi N_R)^{1/2}(rZ)^{1/4}[1 - (Z'/r)^{1/2}]e^{N_R[(\sqrt{rZ'}-1)^2 - (1-r)Z]} \quad (160)$$

The elution of two non-trace solutes (H$^+$, K$^+$) by a third component (Na$^+$) has been studied by Baddour and Hawthorn, under conditions where the equilibrium constant for exchange was less than unity between one component (H$^+$) and the elutant, and this component was eluted first; while the equilibrium constant for the other component was greater than one. Equations (157–160) were found to apply, provided the leading band (H$^+$) was calculated as if its inlet concentration were that of both solute components (H$^+$ plus K$^+$); with V_{sat} in the calculations, reduced in value to correspond to the total amount of H$^+$ actually introduced. For the trailing band, the concentration curve could be calculated just as if the leading component had not been present. The extension of this method to related situations awaits further experimental study.

C. "Trace" Chromatograms (Linear Equilibrium)

1. *Exact Treatment*

The value of trace conditions for obtaining sharp separations has been discussed in Section IV, A, and the criteria in terms of K^{II} and $(c_A)_o/C_o$ are given in Eqs. (136), (137), and (29). As already stated, "trace" conditions are characterized especially by a value of the equilibrium parameter **r**, equal to unity.

From two different approaches, Stene (S7), and Rosen and Winsche (R2) have shown that the sequence of saturation and elution can be described by the difference between the concentration curve for saturation starting at $Z = 0$ and continuing through the elution period, and a second curve representing non-existent saturant which starts at $Z' = 0$. Vermeulen and Hiester (V4) have demonstrated that, with **r** = 1, Eq. (157) shows just this property:

$$\lambda' = c_A/(c_A')_o = \mathbf{J}(\mathbf{N}_R, Z\mathbf{N}_R) - \mathbf{J}(\mathbf{N}_R, Z'\mathbf{N}_R) \qquad (161)$$

Figure 12 shows the transition from complete saturation to incomplete saturation that occurs, at **r** = 1 and $\mathbf{N}_R = 80$, as $\mathbf{Z}_{sat}$ is progressively reduced. Boyd and co-workers (B8) and Glueckauf (G4) have also considered the chromatogram as a composite of separate saturation and elution curves.

For maximum utilization of adsorbent, it is desirable to select a charge period which will give λ_{max} between 0.50 and 0.85; thus, as will be shown, $\mathbf{Z}_{sat}$ should be between $\sqrt{\pi/\mathbf{N}_R}$ and $\sqrt{3\pi/\mathbf{N}_R}$. At λ_{max} values below 0.50, the same quantity of adsorbent serves to recover smaller amounts of solute. At still higher $\mathbf{Z}_{sat}$ values, on the other hand, the overlap between successive solute components increases so that the utilization of adsorbent levels off and gradually declines.

2. *Gaussian-Shaped Zones*

The lower two curves of Fig. 12 are seen to be symmetrical, and similar in shape to each other. Such curves can be represented by a Gaussian-distribution equation, provided that (a) $\mathbf{N}_R > 50$, and (b) $Z < \sqrt{\pi/\mathbf{N}_R}$.

Equation (109) can be used with the last term neglected, at high $\mathbf{N}_R$. With the introduction of Eq. (110), Eq. (161) becomes:

$$c_A'/(c_A')_0 = \tfrac{1}{2}\,\text{erf}\,\sqrt{N_R}\,(\sqrt{Z}-1) - \text{erf}\,\sqrt{N_R}\,(\sqrt{Z'}-1)$$

$$= \int_{\sqrt{N_R}(\sqrt{Z'}-1)}^{\sqrt{N_R}(\sqrt{Z}-1)} e^{-\eta^2}\,d\eta \tag{162}$$

If the upper limit is close to the lower limit, the theorem of the mean allows the integral to be approximated by the product of an intermediate value of its integrand, multiplied by the interval of the independent

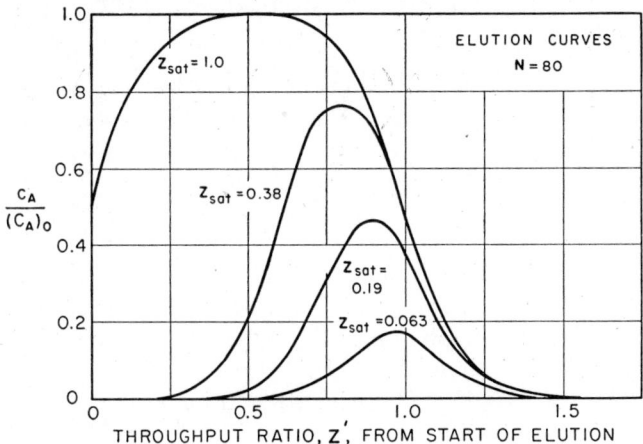

Fig. 12. Elution curves for varying charge periods.

variable. A suitable intermediate value of the variable, η, is taken at the midpoint of the interval. Thus

$$\frac{c_A'}{(c_A')_0} = \frac{\sqrt{N_R}\,(\sqrt{Z}-\sqrt{Z'})}{\sqrt{\pi}}\, e^{-N_R(0.5\sqrt{Z}+0.5\sqrt{Z'}-1)^2} \tag{163}$$

Further transformation (V4) gives

$$\frac{c_A'}{(c_A')_0} = \frac{Z_{\text{sat}}\sqrt{N_R}}{2\sqrt{\pi}}\, e^{-(N_R/4)(Z'+0.5Z_{\text{sat}}-1)^2} \tag{164}$$

Glueckauf (G4) has obtained a similar result.

The peak of the zone is given by

$$(c_A')_{\max} = (c_A')_0 Z_{\text{sat}} \sqrt{N_R}/2\sqrt{\pi}$$

$$= \frac{\sqrt{N_R}\,[(c_A)_0 V_{\text{sat}}]}{2\pi[V'_{\max} + 0.5(C_0/C_0')V_{\text{sat}} - v\epsilon]} \tag{165}$$

where $V'_{\max}$ is the volume of elutant that has entered the column when the peak is reached; $V'_{\max} = (D'_{\text{trace}}+1)v\epsilon$, different for each solute.

The shape of the zone is given by

$$\ln \frac{(c_A')_{max}}{c_A'} = \frac{\mathbf{N}_R}{4} (Z' + 0.5 Z_{sat} - 1)^2$$
$$= \frac{\mathbf{N}_R}{4} \left(\frac{V' - V'_{max}}{V'_{max} + 0.5(C_o/C_o')V_{sat} - v\epsilon} \right)^2 \quad (166)$$

This relation is used to evaluate $\mathbf{N}_R$ from the half-width of the zone, measured at the half-height or the $1/e$ height.

Mayer and Tompkins (M6), and Matheson (M5) have used the theoretical-plate approach to derive

$$\ln \frac{(c_A')_{max}}{c_A'} = \frac{N_c}{2} \left(\frac{\mathbf{D}'_{trace} + 1}{\mathbf{D}'_{trace}} \right) \left(\frac{V' - V'_{max}}{V'_{max}} \right)^2 \quad (167)$$

where N_c is the number of equilibrium contacts, in accordance with Mayer and Tompkins's definition of a theoretical plate. At large $\mathbf{D}'_{trace}$,

$$\mathbf{N}_R = 2N_c \quad (168)$$

Thus the number of equilibrium stages is directly proportional to column length, if the linear flow rate remains unchanged. The role of contact time, which is obscured in the plate theory, now becomes evident. Wherever mass-transfer is independent of flow rate, a diminution of the flow velocity through a bed of constant length will increase the effective number of theoretical plates in inverse proportion. The number of plates in any one column may vary for the different components, just as $\mathbf{N}_R$ may, although usually the variation is not great.

The correspondence between $\mathbf{N}_R$ and N_c makes it possible to transform all the relations developed for reaction units or transfer units into calculations for theoretical plates. Nevertheless, the theoretical-plate approach is based upon a much less precise model, and is applicable to fixed beds only in special cases (such as linear equilibrium, or constant pattern). Like many approximations, the theoretical-plate method is generally derived and applied without exact knowledge of its limitations.

3. Design of "Trace" Separations

To separate two or more solutes by chromatography, in a pilot-plant unit designed for maximum economy, the operating conditions will generally be established in the following order (V4):

a. The adsorbent composition will be selected, along with the levels of concentration (or partial pressure), that give the greatest difference in distribution ratio ($\mathbf{D}'_{trace}$) values for the different components.

b. The average charge rate of feed mixture is considered a fixed condition. This may be based on a trace component which is present in largest quantity, or whose equilibrium parameter, r, is farthest from unity. It will be expressed as moles fed per unit time; or, when divided by the concentration in the charge, as an average volumetric charge rate.

c. The recovery requirements and D_{trace} values, together, determine the N_R values for each component. For each component whose concentration follows a Gaussian curve, the J-function curve has a new use: it represents the integrated recovery of material at the appropriate N_R and $(Z)_{apparent}$ (or $Z' + 0.5Z_{sat}$).

When the ratios of D'_{trace} values (for short, D) and of N_R values (for short, N) for two components A and B are known, the J curves can be used to relate N_R values to the degree of separation in the following way:

i. Let A represent the component with lower D, which is thus eluted first. The maximum allowable loss of A into fractions containing B is specified; the remaining A will all be recovered ahead of this point.

ii. At an ordinate on Fig. 8 numerically equal to the recovery of A, read for any assumed N_A the corresponding Z_A.

This value must be multiplied by the ratio D_A/D_B to determine Z_B. Also N_A must be converted to the corresponding N_B.

iii. At the values of N_B and Z_B thus calculated, the ordinate on Fig. 8 is read. This gives the loss of B into fractions containing A.

iv. If the recovery of B is greater than required, assume a smaller N_A and repeat. Conversely, if it is insufficient, assume a larger N_A.

v. *Use of recycle.* A moderate reduction in N_R, and hence also in resin volume, may be obtained by taking a low recovery of A per pass at the requisite purity. A separate cross-contaminated fraction containing both A and B can then be collected for recycling in a subsequent portion of feed, before taking the B fraction.

vi. *Multiple cycling.* Whenever the separation factors are close to unity, the components undergoing separation will occupy only a small part of the column at any one time. If, in a single cycle of adsorption and elution, the portions of effluent containing trace components comprise less than half the total volume of effluent, as occurs in many binary separations, the utilization of the resin can be improved by adsorbing a second feed charge from a solution of the same carrier composition as the elutant, before elution of the first charge has been completed. In this way, the effluent can become an almost continuous succession of trace zones.[10] Figure 13 shows a schedule of multiple cycling proposed for a radium-barium separation (V4).

d. The volumetric flow rate is then determined. $(Z_A)_{sat}$ is taken as $\sqrt{\pi/N_A}$. The total Z_A in one cycle is calculated as D_Z/D_A times the Z_Z

---

[10] Multiple cycling is not feasible in cases of *graded elution*, which involve a progressive change in elutant composition; for such a case, however, this entire calculation procedure must be modified substantially.

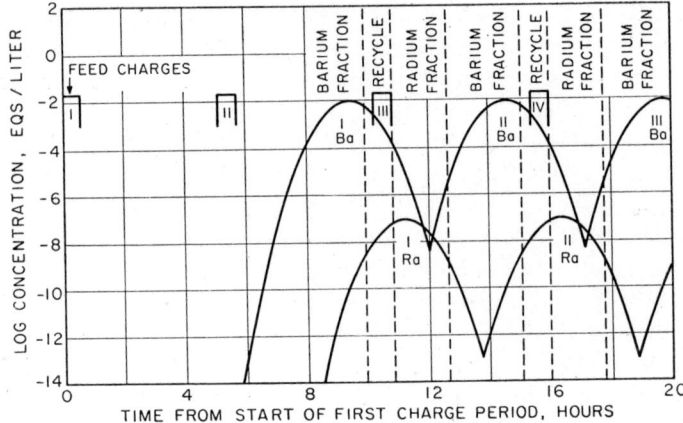

Fig. 13. Effluent composition predicted for first column of a radium-barium separation process.

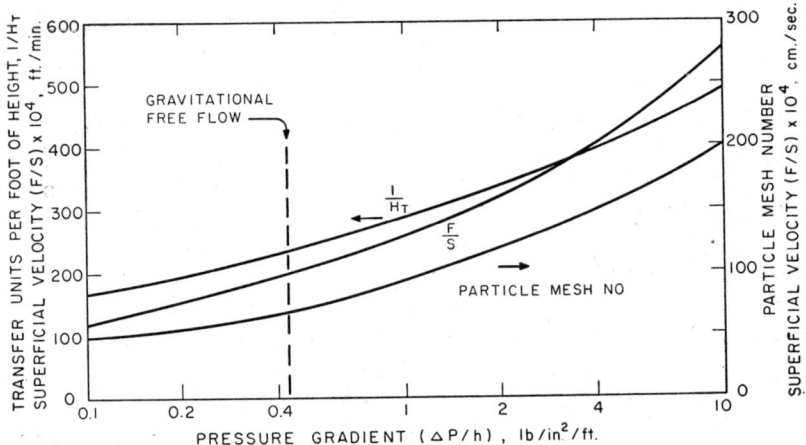

Fig. 14. Typical relation of operating variables for ion exchange or liquid-phase adsorption. Courtesy of *Industrial and Engineering Chemistry*.

required to elute completely the last component (Z) in the column. The total flow rate F is then given by

$$F = \frac{(Z_A)_{total}}{(Z_A)_{sat}} \times \text{(volumetric charge rate)} \qquad (169)$$

e. The column volume and the cycle period remain to be specified. It will be desirable, for liquid-phase adsorptions at least, to use N_{Pe}' as near as is practicable to 20, but no lower because of the onset of longitudinal diffusion. This variation in N_{Pe}' is accomplished by decreasing the particle volume, with an attendant increase in pressure drop per unit height. Figure 14 shows the relations between d_p, $\mathbf{N}_R$, $\Delta P/h$, and F/S calculated for a representative instance of ion exchange.

The linear flow rate U assumed here, together with the known volumetric flow F, fixes the cross-sectional area S.

Once the range of economic N_{Pe}' values has been selected, Fig. 6 can be used to determine κ_{kin} for the component on which calculations are based. By Eq. 54, with $\kappa_{kin} = k_{kin} Q \rho_b / \epsilon$, the column volume is

$$v = \mathbf{N}_R F / \kappa_{kin} \epsilon \qquad (170)$$

From S and v, the length of the column is now established. Several trial calculations will usually be needed to achieve the most economical bed geometry.

Nomenclature

a constant of integration, in Eq. 71

a_p effective mass-transfer area between fluid and particles, ft.^{2}/ft.^{3} (or cm.^{2}/cm.^{3})

A, B, G, Z components

A, B, C coefficients in power-series isotherm (Eqs. 12–13)

c concentration of solute in the fluid phase, lb.-mole or lb.-equiv./ft.^{3} (or gm.-mole or gm.-equiv./cm.^{3})

c heat capacity

C_o total concentration of solutes in the fluid phase

d_p diameter of sphere equal in volume to actual particle, ft. (or cm.)

D diffusivity, ft.^{2}/hr. (or cm.^{2}/sec.)

E effective longitudinal dispersivity, ft.^{2}/hr. (or cm.^{2}/sec.)

F volumetric flowrate, ft.^{3}/hr. (or cm.^{3}/sec.)

g Brinkley's function; see Eq. (107)

H height of contact volume equivalent to one transfer unit or reactor unit, ft. (or cm.)

I_o Bessel function of zero order with imaginary argument; see Eq. (105)

k rate coefficient for kinetics (ft.^{3}/hr. lb.-mole, or cm.^{3}/sec. gm.-mole)

K equilibrium constant for exchange or adsorption, in appropriate concentration or pressure units

l axial depth of a radial bed

l, m, n exponents

L, M exponents

L symmetric Margules factor for activity-coefficient expression (Eq. 15)

M molecular weight

n integer in infinite series (Eq. 81)

N mole fraction

N_c number of equivalent contacts (theoretical plates)

N_{Pe}' Peclet number for mass transfer, $N_{Re} \times N_{Sc}$, dimensionless

N_{Re} Reynolds number for flow past particles, $d_p U \epsilon \rho / 6(1 - \epsilon) \mu$, dimensionless

N_{Sc} Schmidt number, $\mu/D\rho$, dimensionless
p partial pressure of solute
P vapor pressure, lb.-force/ft.^{2} (or gm.-force/cm.^{2})
q concentration of solute in the particle phase, lb.-mole or lb.-equiv./lb. (or gm.-mole or gm.-equiv./gm.) of air dried particles
Q ultimate exchange capacity of the particle phase; equivalent to q_m, for adsorption
r radial distance within spherical particle, ft. (or cm.)
r_p radius of exterior surface of particle, ft. (or cm.)
$\bar{r}$ average pore radius, ft. (or cm.)
R radial distance within adsorbent bed
S cross-sectional area of contactor, ft.^{2} (or cm.^{2})
s, t variables in the argument of J analogous to $\mathbf{s}, \mathbf{t}$
$\bar{u}$ average molecular velocity, ft./hr. (or cm./sec.)
U actual mean linear flow rate of fluid phase relative to solid, ft./hr. (or cm./sec.)
v bulk-packed volume for contact, ft.^{3} (or cm.^{3})
V volume of saturating feed entering column, ft.^{3} (or cm.^{3})

J mathematical solution for column-saturation function; gives $\mathbf{x}$ at $\mathbf{r} = 1$
$\mathbf{n}$ number of effective transfer units with longitudinal dispersion; see Eq. (113b)
$\mathbf{N}$ number of transfer units; see Eqs. (50–51)
$\mathbf{N}_H$ number of heat-transfer units; see Eq. (117)
$\mathbf{N}_R$ number of reaction units; see Eq. (54)
$\mathbf{r}$ equilibrium parameter
$\mathbf{s}$ column-capacity parameter for fixed-bed operation, based on reaction-kinetics calculation
$\mathbf{t}$ solution-capacity parameter for fixed-bed operation, based on reaction-kinetics calculation
$\mathbf{x}$ ratio of fluid-phase concentration of a component to that of all components, c/C_o ($\mathbf{x}_A$, etc.)
$\mathbf{y}$ ratio of particle-phase concentration of a component to total for solid at saturation with the feed, q/Q
$\mathbf{z}$ modified throughput ratio for longitudinal dispersion; see Eq. (113a)
$\mathbf{Z}$ throughput parameter, or ratio of actual volume of effluent to the stoichiometric volume; $\mathbf{t}/\mathbf{s}$ or Θ/Σ

DIMENSIONLESS PARAMETERS

$\mathbf{b}$ correction factor for computing overall mass-transfer, or rate, coefficients
$\mathbf{D}$ distribution parameter, or ratio of concentrations in particle and fluid phases

GREEK LETTERS

α, β valence of ions A, B
$\alpha_1, \alpha_2, \beta_1, \beta_2$ correlation constants for mass-transfer
γ activity coefficient
ϵ ratio of void space outside particles to total volume of contacting zone

ζ	mechanism parameter	app	apparent
Θ	solution capacity parameter, based on mass-transfer calculation	B, C, F, L, S	types of adsorption equilibria
		b	bulk
η	a variable, in Eqs. (110) and (162)	b	boundary value for special case of Eq. (87)
κ_{kin}	general rate coefficient in the reaction-kinetic equation; $k_{kin}Q\rho_b/\epsilon$; hr.$^{-1}$ (or sec.$^{-1}$)	complex	formation of complex ion
		d	distribution
		e	exterior
		f	fluid phase
λ	extent of saturation of fluid phase with reference to initial and final concentrations; $[x_A - (x_A)_o^*]/[(x_A)_o - (x_A)_o^*]$	i	interior
		i	interface
		H	heat transfer
		kin	reaction-kinetic
		m	monomolecular layer
		max	value at the peak of the chromatogram
μ	viscosity of the fluid, lb./hr. ft. (or gm./cm. sec.)	O	over-all
		o	initial or entrance conditions
ξ	a variable, in Eq. (105)	p	particle phase
ρ	density, lb./ft.^{3} (or gm./cm.^{3}), usually the fluid density; ρ_b = bulk-packed density of air-dried particles	pore	fluid phase pore diffusion
		R	reaction unit
		sat	increment during saturating period
		stoic	stoichiometric proportions
Σ	column-capacity parameter for fixed-bed operation, based on mass-transfer calculation	T	total
		trace	trace conditions
τ	time, hr. (or sec.)	∞	value when pores are filled with condensed liquid
ϕ	Thomas's function; see Eq. (106)	∞	temperature of fluid at saturation
χ	internal porosity of solid particles		
ψ	numerical correction factor for solid-phase diffusion, Eq. (86)	SUPERSCRIPTS	
		*	equilibrium
ω	extent of saturation of particle phase with reference to initial and final concentrations; $[y_A - (y_A)_o]/[(y_A)_o^* - (y_A)_o]$	'	second time interval in binary saturation or elution
		'	adsorption equilibrium involving a liquid phase
		†	elution period following complete saturation
SUBSCRIPTS			
1, 2, ...	instantaneous times of measurement	II	equal-valence exchange

REFERENCES

A1. Adamson, A. W., and Grossman, J. J., *J. Chem. Phys.* **17**, 1002 (1949).
A2. Amundson, N. R., *J. Phys. & Colloid Chem.* **54**, 812 (1950).

A3. Anzelius, A., *Z. angew. Math. u. Mech.* **6,** 291 (1926).
A4. Aris, R., and Amundson, N. R., *A. I. Ch. E. Journal* **3,** 280 (1957).
B1. Baddour, B. F., Goldstein, D. J., and Epstein, P., *Ind. Eng. Chem.* **46,** 2192 (1954); Baddour, B. F., and Hawthorn, R. D., *ibid.* **47,** 2517 (1955).
B2. Barrer, R. M., "Diffusion in and through Solids." Cambridge Univ. Press, London and New York, 1941.
B3. Barrow, R. F., Danby, C. J., Davoud, J. G., Hinshelwood, C. N., and Staveley, L. A. K., *J. Chem. Soc.* p. 401 (1947).
B4. Bauman, W. C., and Eichhorn, J., *J. Am. Chem. Soc.* **69,** 2830 (1947).
B5. Beaton, R. H., and Furnas, C. C., *Ind. Eng. Chem.* **33,** 1500 (1941).
B6. Bohart, G. S., and Adams, E. Q., *J. Am. Chem. Soc.* **42,** 523 (1920).
B7. Boyd, G. E., and Soldano, B. A., *J. Am. Chem. Soc.* **75,** 6091, 6107 (1953).
B8. Boyd, G. E., Myers, L. S., Jr., and Adamson, A. W., *J. Am. Chem. Soc.* **69,** 2836, 2849 (1947).
B9. Boyd, G. E., Schubert, J., and Adamson, A. W., *J. Am. Chem. Soc.* **69,** 2818 (1947).
B10. Brinkley, S. R., Jr., Edwards, H. E., and Smith, R. W. Jr., *Math. Tables Aids Computation* **6,** 40 (1952).
B11. Brownell, L. E., and Katz, D. L., *Chem. Eng. Progr.* **43,** 537 (1947); also in Brown, G. G., "Unit Operations," Chapter 16. Wiley, New York, 1950.
B12. Brunauer, S., "Physical Adsorption," Princeton Univ. Press, Princeton, New Jersey, 1943.
B13. Brunauer, S., Emmett, P. H., and Teller, E., *J. Am. Chem. Soc.* **60,** 309 (1938).
B14. Brunauer, S., Deming, L. S., Deming, W. E., and Teller, E., *J. Am. Chem. Soc.* **62,** 1723 (1940).
C1. Caddell, J. R., and Moison, R. L., *Chem. Eng. Progr. Symposium Ser.* **50,** No. 14, 1 (1954).
C2. Cassidy, H. G., "Adsorption and Chromatography," Vol. V of "Technique of Organic Chemistry." Interscience, New York, 1951.
C3. Chilton, T. H., and Colburn, A. P., *Ind. Eng. Chem.* **27,** 255 (1935).
D1. DeVaney, F. D., *in* "Chemical Engineers' Handbook" (J. H. Perry, ed.), pp. 1085–1091. McGraw-Hill, New York, 1950.
D2. DeVault, D., *J. Am. Chem. Soc.* **65,** 532 (1943).
D3. Dodge, F. W., and Hougen, O. A., "Drying of Gases," Edwards, Ann Arbor, Michigan, 1947.
D4. Drew, T. B., and Genereaux, R. P., *in* "Chemical Engineers' Handbook" (J. H. Perry, ed.), pp. 393–394. McGraw-Hill, New York, 1950.
D5. Drew, T. B., Spooner, F. M., and Douglas, J., cited in reference (K4).
D6. Dryden, C. E., Ph.D. thesis in chemical engineering. Ohio State University, Columbus, Ohio, 1951.
E1. Eagleton, L., and Bliss, H., *Chem. Eng. Progr.* **49,** 543 (1953).
E2. Einstein, H. A., D.S.T. dissertation, Eidgenössische Technische Hochschule, Zürich, Switzerland, 1937.
E3. Ergun, S., *Chem. Eng. Progr.* **48,** 227 (1952).
F1. Fredericks, E. M., and Brooks, F. R., *Anal. Chem.* **28,** 297 (1956).
F2. Fujita, H., *J. Phys. Chem.* **56,** 949 (1952).
F3. Furnas, C. C., *Trans. Am. Inst. Chem. Engrs.* **24,** 142 (1930).
G1. Gaffney, B. J., and Drew, T. B., *Ind. Eng. Chem.* **42,** 1120 (1950).
G2. Gilliland, E. R., and Baddour, R. F., *Ind. Eng. Chem.* **45,** 330 (1953).
G3. Glueckauf, E., *Trans. Faraday Soc.* **51,** 1540 (1955).
G4. Glueckauf, E., *J. Chem. Soc.* p. 1302 (1947); *Discussions Faraday Soc.* **7,** 12

(1949); *Trans. Faraday Soc.* **51**, 34 (1955); Glueckauf, E., Barker, K. H., and Kitt, G. P., *ibid.* p. 199 (1949).
G5. Glueckauf, E., and Coates, J. I., *J. Chem. Soc.* p. 1315 (1947).
G6. Goldstein, S., *Proc. Roy. Soc.* **A219**, 151, 171 (1953).
G7. Gregor, H. P., *J. Colloid Sci.* **6**, 20 (1951).
G8. Grossman, J. J., and Adamson, A. W., *J. Phys. Chem.* **56**, 97 (1952).
H1. Hiester, N. K., Ph.D. dissertation in chemical engineering, University of California, Berkeley, Calif., 1949.
H2. Hiester, N. K., Cohen, R. K., and Phillips, R. C., *Chem. Eng. Progr. Symposium Ser.* **50**, No. 14, 23 (1954).
H3. Hiester, N. K., Cohen, R. K., and Phillips, R. C., *Chem. Eng. Progr. Symposium Ser.* **50**, No. 14, 63; also *Chem. Eng. Progr.* **50**, 139 (1954).
H4. Hiester, N. K., Radding, S. B., Nelson, R. L., Jr., and Vermeulen, T., *A. I. Ch. E. Journal* **2**, 404 (1956); *Am. Documentation Inst. Doc.* 4953 (1956).
H5. Hiester, N. K., and Vermeulen, T., *J. Chem. Phys.* **16**, 1087 (1948).
H6. Hiester, N. K., and Vermeulen, T., *Chem. Eng. Progr.* **48**, 505 (1952); *Am. Documentation Inst. Doc.* 3665 (1952).
H7. Hougen, O. A., and Marshall, W. R., *Chem. Eng. Progr.* **43**, 197 (1947).
H8. Hurt, D. M., *Ind. Eng. Chem.* **35**, 522 (1943).
J1. Jacques, G., and Vermeulen, T., *Univ. Calif. Radiation Lab. Rept.* 8029 (1957).
K1. Kasten, P. R., Lapidus, L., and Amundson, N. R., *J. Phys. Chem.* **56**, 683 (1952).
K2. Keulemans, A. I. M., "Gas Chromatography." Reinhold, New York, 1957.
K3. Klinkenberg, A., *Ind. Eng. Chem.* **40**, 1970 (1948); **46**, 2285 (1954).
K4. Klotz, I. M., *Chem. Revs.* **39**, 241 (1946).
K5. Koble, R. A., and Corrigan, T. E., *Ind. Eng. Chem.* **44**, 383 (1952).
K6. Kramers, H., and Alberda, G., *Chem. Eng. Sci.* **2**, 173 (1953).
K7. Kunin, R., and Myers, R. J., "Ion-Exchange Resins." Wiley, New York, 1950.
L1. Langmuir, I., *Phys. Revs.* **2**, 331 (1913); *ibid.* **6**, 79 (1915); *J. Am. Chem. Soc.* **38**, 2221 (1916).
L2. Lapidus, L., and Amundson, N. R., *J. Phys. Chem.* **54**, 821 (1950); **56**, 373 (1952).
L3. Lapidus, L., and Amundson, N. R., *J. Phys. Chem.* **56**, 984 (1952); Amundson, N. R., *Ind. Eng. Chem.* **48**, 26 (1956).
L4. Ledoux, E., *J. Phys. & Colloid Chem.* **53**, 960 (1949).
M1. McCune, L. K., and Wilhelm, R. H., *Ind. Eng. Chem.* **41**, 1124 (1949).
M2. Mair, B. J., Westhaver, J. W., and Rossini, F. D., *Ind. Eng. Chem.* **42**, 1279 (1950).
M3. Mantell, C. L., "Adsorption," 2nd ed. McGraw-Hill, New York, 1951.
M4. Martin, A. J. P., and Synge, R. L. M., *Biochem. J.* **35**, 1385 (1941).
M5. Matheson, L. A., private communication, reported by Tompkins, E. R., *J. Chem. Educ.* **26**, 92 (1949).
M6. Mayer, S. W., and Tompkins, E. R., *J. Am. Chem. Soc.* **69**, 2866 (1947).
M7. Michaels, A. S., *Ind. Eng. Chem.* **44**, 1922 (1952).
M8. Miyauchi, T., Ph.D. dissertation, University of Tokyo, Japan, 1957.
N1. Nachod, F. C., ed., "Ion Exchange: Theory and Application." Academic Press, New York, 1949.
N2. Nelson, R. L., Jr., M.S. thesis in chemical engineering, University of California, Berkeley, Calif., 1951.
O1. Opler, A., and Hiester, N. K., "Tables for Predicting the Performance of Fixed-Bed Ion Exchange and Similar Mass Transfer Processes." Stanford Research Institute, Stanford, Calif., 1954.
R1. Rosen, J. B., *J. Chem. Phys.* **20**, 387 (1952); *Ind. Eng. Chem.* **46**, 1590 (1954).

R2. Rosen, J. B., and Winsche, W. E., *J. Chem. Phys.* **18,** 1587 (1950).
S1. Schumann, T. E. W., *J. Franklin Inst.* **208,** 405 (1929).
S2. Selke, W. A., and Bliss, H., *Chem. Eng. Progr.* **46,** 509 (1950).
S3. Shedlovsky, L., *Ann. N.Y. Acad. Sci.* **49,** 279 (1938).
S4. Sillén, L. G., *Arkiv Kemi Mineral. Geol.* **A22,** No. 15 (1946); Sillén, L. G., and Ekedahl, E., *ibid.* **A22,** No. 16 (1946).
S5. Sips, R., *J. Chem. Phys.* **18,** 1024 (1950).
S6. Spedding, F. H., and Powell, J. E., *J. Am. Chem. Soc.* **76,** 2550 (1954).
S7. Stene, S., *Arkiv Kemi Mineral. Geol.* **18,** No. 18 (1945).
T1. Thomas, H. C., *J. Am. Chem. Soc.* **66,** 1664 (1944).
T2. Thomas, H. C., *Ann. N.Y. Acad. Sci.* **49,** 161 (1948).
T3. Tiselius, A., *Advances in Colloid Sci.* **1,** 81 (1942).
T4. Treybal, R. E., "Mass-Transfer Operations." McGraw-Hill, New York, 1955.
V1. Van Arsdel, W. B., *Chem. Eng. Progr.* **43,** 13 (1947).
V2. Van Deemter, J. J., Zuiderweg, F. J., and Klinkenberg, A., *Chem. Eng. Sci.* **5,** 271 (1956).
V3. Vermeulen, T., *Ind. Eng. Chem.* **45,** 1664 (1953).
V4. Vermeulen, T., and Hiester, N. K., *Ind. Eng. Chem.* **44,** 636 (1952).
V5. Vermeulen, T., and Hiester, N. K., *J. Chem. Phys.* **22,** 96 (1954).
V6. Vermeulen, T., and Huffman, E. H., *Ind. Eng. Chem.* **45,** 1658 (1953).
W1. Walter, J. E., *J. Chem. Phys.* **13,** 229 (1945).
W2. Walton, H. F., in reference (N1).
W3. Weiss, J., *J. Chem. Soc.* p. 297 (1943); Offord, A. C., and Weiss, J., *Discussions Faraday Soc.* **7,** 26 (1949).
W4. Wheeler, A., *Advances in Catalysis* **3,** 250–327 (1951); "Catalysis," Vol. 2 (P. H. Emmett, ed.) pp. 105–165: Reinhold, New York (1955).
W5. Wicke, E., *Kolloid-Z.* **86,** 167, 289 (1939).
W6. Wilke, C. R., and Hougen, O. A., *Trans. Am. Inst. Chem. Engrs.* **41,** 445 (1945).
W7. Wilson, J. N., *J. Am. Chem. Soc.* **62,** 1583 (1940).
W8. Wilson, S., and Lapidus, L., *Ind. Eng. Chem.* **48,** 992 (1956).

MIXING OF SOLIDS

Sherman S. Weidenbaum

Corning Glass Works, Corning, N. Y.

I. Introduction... 211
II. Related Process Steps.. 211
III. State of Mixedness of a Batch of Solids........................ 212
    A. Concept of Degree of Mixing.................................. 212
    B. Sampling Considerations... 213
        1. Method and Location... 213
        2. Spot Sample Size.. 215
        3. Number of Spot Samples...................................... 215
        4. Analysis of Samples... 215
            a. Quantitative... 215
            b. Qualitative... 216
    C. Statistical Techniques... 216
        1. General Comments.. 216
        2. Quantitative Measurements on Each Spot Sample.. 218
            a. The Relative Frequency Distribution............. 218
            b. Definitions... 220
                i. Sample Arithmetic Mean........................... 220
                ii. Sample Variance and Standard Deviation... 220
            c. Statistical Tests of Significance..................... 221
                i. Meaning of a Statistical Test of Significance... 221
                ii. Test of Significance for Means (t test)....... 222
                iii. Test of Significance for Variances (F test)... 222
                iv. Test of Significance for Distributions (**Chi-square** test)........ 224
        3. Qualitative or Semi-quantitative Measurements on Each Spot Sample... 225
            a. Smear Test... 225
            b. Test of Significance for Fraction Satisfactory.. 226
            c. Test of Significance for Streaks per Spot Sample.. 227
        4. Test of Significance for Determining When Equilibrium Is Reached. 228
        5. Confidence Limits... 229
            a. Meaning of Confidence Limits...................... 229
            b. Confidence Limits for the Mean.................... 230
            c. Confidence Limits for the Variance............... 230
            d. Confidence Limits for the Standard Deviation and Their Use in Planning Proper Sample Size for a Prescribed Confidence Range 231
            e. Confidence Limits for Fraction Satisfactory... 235
        6. The Quality Control Chart................................... 235
        7. Remarks on Statistical Techniques....................... 238
        8. Choosing a Degree of Mixing................................ 238

        9. Starting a Quantitative Study.................................. 240
    D. Literature Summary: Degrees of Mixing............................ 240
    E. Summary and Conclusions—State of Mixedness of a Batch of Solids... 258
IV. Theoretical Frequency Distributions..................................... 259
    A. Mixing—Orderly or Random?.. 259
    B. Binomial, Normal, and Poisson Distributions—General Comments..... 261
    C. Binomial and Normal Distributions: Use in Defining and Illustrating
       Mixing; Chi-square; Entropy (W2)............................... 263
    D. Binomial and Normal Distributions: Effect of Sample Size Fluctuation;
       Transformed Variable for Measuring Incomplete Mixing; Chi-square
       (B4).. 264
    E. Theoretical Variance for Random Mixture of Two Materials, Each with
       Its Own Size Distribution (S5)................................. 265
    F. Use of Poisson Distribution in Evaluating Mixtures (A1)............ 268
    G. Continuity of One Phase in a Powder Mixture of Two Phases (F1).... 270
    H. Use of Standard Normal Table with Rate Equations (O9)............. 271
    I. Techniques Necessary for Different Methods of Analysis............ 272
    J. Summary and Conclusions—Theoretical Frequency Distributions...... 273
V. Rate Equations.. 274
    A. The Value of Rate Studies... 274
    B. General Discussion.. 274
    C. Diffusion Analogy... 282
    D. Summary and Conclusions on Rate Equations......................... 285
VI. Equipment.. 287
    A. Description... 287
    B. Performance... 289
      1. Introduction.. 289
      2. Horizontal Rotating Cylinder (O1–O8).......................... 290
      3. Cylinder Rotating at Various Angles with the Horizontal (C3, M2,
         V1)... 298
      4. Horizontal Rotating Cylinder (W2)............................. 304
      5. Cylinder Rotating at Angle with Horizontal (B4)............... 304
      6. Horizontal Rotating Cylinder (O9)............................. 304
      7. Comparison of Several Mixers (A1)............................. 305
      8. Sigma Mixer (S5).. 306
      9. Comparison of Several Mixers with Different Materials (G2).... 307
      10. Finger-Prong Mixer with Materials Having Varying Moisture Content (M4).. 307
      11. Machines to Mix Additives with Soil (S3)..................... 311
      12. Comparison of Muller and Ribbon Mixer (L3)................... 313
      13. Helical Flight Mixer (G3).................................... 315
      14. Twin Shell Blender (Y2)...................................... 315
      15. Comparison of Several Mixers (Y1)............................ 316
    C. Critical Evaluation of Published Performance Data.................. 317
    D. Summary and Conclusions and Related Thoughts—Equipment Performance... 318
VII. Overall Concluding Comments.. 320
    Acknowledgments... 321
    Nomenclature.. 321
    References.. 321

I. Introduction

Solids mixing, although in wide use for a number of years, has received comparatively little rigorous treatment in the literature, compared to other widely used process operations. Quite recently, there has been a spurt of activity to remedy this situation.

Several studies have dealt with ways to evaluate the state of mixedness of a batch of solids. Also, some performance tests of various types of solids mixers have been published. Rate mechanisms have also been dealt with. In addition, there has been some literature concerning the analysis of particle motion during mixing. Each of these areas will be discussed.

The subject matter has been divided into four basic categories which deal with state of mixedness of a batch of solids, theoretical frequency distributions, rate equations, and equipment. Some papers will be discussed in more than one of these categories. Most of the literature deals with batch mixing, although some pertains to continuous mixing, but this method of categorizing has not been followed.

The material contained here, in addition to providing some useful techniques and facts for those who are interested in solids mixing work, can serve as a guide towards research on future problems. Criteria which are considered essential to the best interpretation of a study of mixing performance will be pointed out. Unfortunately, these are not always reported in published studies, and their continued omission can slow up the most fruitful use of such work as building blocks for the future. It is hoped that omissions of data in published literature concerning such things as size and operating variables for commercial equipment, or for that matter, any mixing equipment, will be diminished, and possibly eliminated. Drew (D5) pointed out that diffusivities, which were reported in the literature without frames of reference, imposed serious handicaps on those attempting to use these results for later correlations and to help explain diffusion phenomena.

From time to time questions pertinent to the topic under discussion will be asked, as a technique for stressing important considerations.

II. Related Process Steps

Wherever a solids mixer is necessary, thought must be given to the handling of the batch after mixing. The mixer may turn out an excellent product, but subsequent handling may render it unsatisfactory for the next step in the process. Dumping, shaking during transportation, vibrating, and flow through silos are some of the steps frequently following batch mixing. Each of these offers excellent chances for batch segregation. This phenomenon is probably a lot more widely known and thought about by those using solids mixing in a process, than the comparatively scarce literature on examples of it might indicate. However, batch segregation probably exists and is tolerated in many processes.

There are several reasons for this: (1) It is difficult to obtain an objective picture of the state of mixedness of a solids mixture (this will be discussed in a later section). (2) There are sometimes obstacles to finding out how much of the subsequent process troubles, limitations, and/or defective end products, can be traced to batch segregation in cases where further processing of the mixed batch is necessary. When these factors can be determined, they must be balanced against the cost of remedying the situation, if possible. Sometimes, no obvious solution is available, and it does not appear to be economically feasible to undertake a research and development program to find the solution. With the publication of more information on the mechanisms of batch segregation, this situation will be remedied. (3) It may sometimes be decided that more attention to the proper performance of the existing operations (i.e., better weighing of materials, less loss of dust, less contamination with impurities, etc.) is a more profitable way to reduce process costs and improve the product, with the limited amount of manpower and money for use in this area.

III. State of Mixedness of a Batch of Solids

A. Concept of Degree of Mixing

One of the key questions that comes up whenever a mixture of solids is used, either to make a product directly or to be further processed, is: "Is this batch well enough mixed?" This question has many ramifications because "well enough mixed" may have a different meaning, depending on whether a commercial process or a research project is involved, and also on the particular purpose of the solids mixing operation. Some of the factors that may be considered before answering this question were mentioned in the preceding section. However, it is first necessary to answer the question: "How well mixed is this batch?"

A quantitative expression for the degree of mixing gives an unambiguous picture of the variation in batch composition. To be generally useful, it must be obtained from a study of batch samples, since only in special research projects will analysis of the entire batch be possible. The problem then resolves itself into a matter of how to best get the samples and what to do with them. Each of these subjects will be discussed separately.

B. Sampling Considerations

Several aspects of sampling will be discussed in this section, and related statistical methods will be covered later on. However, before getting into the details, it is well to stand back and first consider some general comments on the whole idea of sampling. Sometimes, obviously poor sampling practices are perpetuated primarily for lack of thoughtful care and attention (e.g., taking one 100-gm. sample from a 140,000-lb. shipment in a railroad car and deciding on this basis that an entire carload is satisfactory). Therefore, it is well as a starting point to ask whether the sampling is arbitrary or truly suitable to the problem to be studied. In reporting results, complete details of the sampling procedure should be given, including the method of taking spot samples; their location, size, and number; the method of analysis; and the fraction of the batch removed for sampling.

1. *Method and Location*

Because of the variety of situations possible, guiding principles rather than detailed rules for removing samples must be given. These are:

a. Removing the sample should cause as little disturbance of the mixture as possible.

b. It is desirable to sample within the mixer as well as after dumping the mixed batch. This will reveal gradients due to segregation and will also prevent confusing of mixer performance with the effects of dumping and "coning." Additional sampling during and after dumping can be carried out as desired to isolate any sources of segregation.

c. Samples should be taken throughout the entire batch volume, rather than from only one part of it.

d. If possible a sampling thief, especially adapted for the particular operation, should be made so that insertion into the batch is possible with a minimum of disturbance. The most desirable procedure is to insert the thief with the holes covered. When it is in position, uncover the sampler holes, trap the sample, cover the holes, and remove the thief.

e. If a sampling thief cannot be used, some special device will have to be made, bearing in mind the above rules. For very tiny spot samples, a special thief which was first used by Maitra and Coulson (M2), is described and shown in detail by Weidenbaum and Bonilla (W2). Larger scale devices, based on this principle, can also be used. For removing spot samples of approximately 0.8 gm. from their sand-salt mixture, a different type of sampling device was used. This sampler (Fig. 2) permitted the removal of a slug of mixture by the application of suction with a precautionary fine screen for preventing particle movement due to the

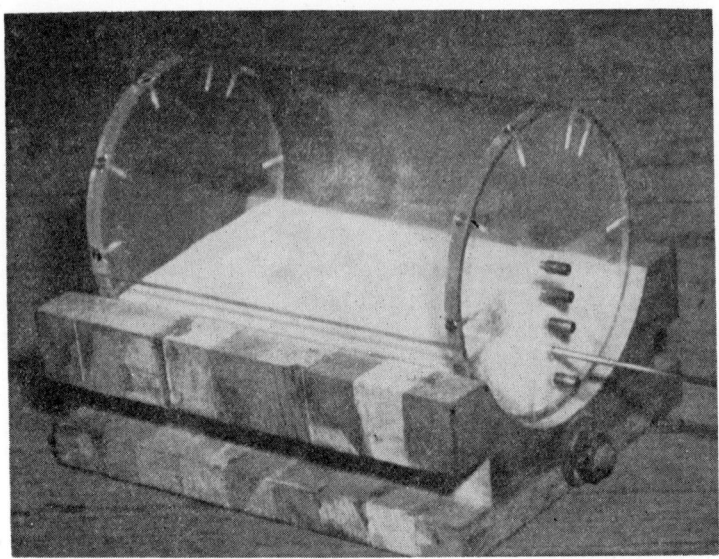

FIG. 1. Sampling thief for very small spot samples (W1).

FIG. 2. Device for taking larger samples (W1).

suction. Pictures illustrating these two methods are shown in Figs. 1 and 2. Sometimes a thin-walled tube can be inserted into the batch, the open end closed off, and the sample thus removed. Upon removal of the tube, the closed-off end is opened and the sample will drop out. These are just a few examples of the improvisation which solids sampling may involve.

2. *Spot Sample Size*

The size of the spot sample will depend upon the use to be made of the mixture, in most cases. A convenient size where packaging or tableting of the mixture takes place is the actual package or tablet size. Where the mixture is to be used in bulk in a process, those familiar with the analyses to be performed on the spot samples and the significance of fluctuations in the mixture composition should be consulted before selecting the sample size. Not too much of the material should be removed during sampling, particularly if more than one set of spot samples is to be taken. A rough rule would be to remove not more than 5% of the mixture, if it is possible to do this while maintaining a meaningful spot sample size.

3. *Number of Spot Samples*

It is not possible to give a single arbitrary number of spot samples that will be optimum for evaluating any solids mixture. The purpose of the mixing operation, as well as the cost of sampling and analysis, must be considered. However, it is rather obvious that the larger the number of properly taken spot samples, the better and more clear-cut the picture of the solids mixture. Statistical methods for the quantitative use of this principle will be given in a later section.

4. *Analysis of Samples*

a. Quantitative. Solids mixing is used for many varied processes. The analytical procedure will depend on the purpose of mixing (which will decide what substances are to be determined, etc.), the facilities available, and economic factors. One of the pieces of information most often desired is the percentage of a certain substance in each spot sample. This determination represents an analytical problem which is outside the scope of this work. However, it is worth mentioning that recent advances in instrumental analysis have made it much easier to give rapid and numerous analyses, which are of great benefit in statistical work. Some of these methods are X-ray fluorescence, flame spectrometry, polarography, and emission spectroscopy. Some references dealing with methods of analysis are included in the bibliography (A2, H3, L5). Whatever method is chosen (gravimetric, volumetric, electrometric, particle counts, optical, etc.), the results of each spot sample analysis will be in numerical form, and can then be treated statistically. It is desirable that the analysis error be very small compared to the variation in the composition (or other property) between spot samples.

b. Qualitative. In some cases it may be possible by means of a nonquantitative test to determine whether any spot sample is satisfactory. For example, smearing a small sample with a spatula to determine if any streaks are visible may be all that is necessary to determine whether the product will be satisfactory. In such cases the desired result of mixing will be a percentage of "satisfactory" spot samples among the total number taken. The statistical treatment of this case differs from that where quantitative analyses are reported for each spot sample.

C. STATISTICAL TECHNIQUES

1. *General Comments*

As a result of the sampling and analyses, a group of numbers will have been obtained which show the compositions at several sampling spots in the mass of mixed batch. This situation is similar to a production line, where a certain amount of product has been made and samples have been taken to determine its quality level. There are well known statistical techniques for working up the data. These techniques have been used for years in fields such as production quality control, tests of significance in experimental work where materials or processes must be compared with one another, card playing and dice throwing, and even in the preparation of information on which insurance companies can base premium rates for different age groups. Among the well-known terms which are involved are the normal (or Gaussian) distribution, the binomial distribution, the Poisson distribution, chi-square, confidence intervals, quality control charts, and tests of significance. These well known and established techniques have a natural application in the quantitative determination of the state of mixedness of a batch of solids. Their use will be discussed here without repetitive derivations which are available elsewhere.

Certain basic statistical nomenclature is essential as a preliminary. The group of measurements is called a "sample" of n observations. (Note: The statistical term "sample" means the whole group of measurements. In order to distinguish between this and any particular measurement at a certain spot, the latter will be called "spot sample.") If a large enough "sample" is properly taken (which includes many "spot samples"), it will give a good picture of the batch (statistical term: population) from which it was drawn. Statistics helps in planning the size and number of spot samples and in working up the data so as to get a maximum of quantitative, objective information. Mathematical statistics and probability have been used to derive relationships for rate equations and other related applications. Familiarity with this aspect of statistics is not necessary for the engineer who merely wants to ascertain facts concerning the extent of mixedness of his batch.

There are many more terms used in statistical work which could be defined and discussed with varying degrees of complexity and mathematical expression. Naturally, it is desirable that further background material concerning these terms be studied wherever possible. It is not absolutely necessary, however, to go into all aspects of the meaning of these terms in order to be able to use them for practical problems in mixing. For example, random sampling is usually desired when taking spot samples.[1] The intuitive feeling, that this means choosing samples so as to avoid a bias in favor of any particular location or material, can serve in obtaining representative samples, even though the mathematical ideas concerning the independence of distribution functions and their relation to the population distribution function, are not used. Again, although discussions of the term "null hypothesis" could figure prominently in the section on statistical tests of significance, this concept has not been mentioned there. Instead, the general idea of a statistical test of significance has been pointed out, particularly as applied to problems of mixing.

This is not meant to imply that supplementary study of the meaning of these points is unimportant. On the contrary, it is definitely recommended where possible. However, use of those techniques that are available need not be delayed because of lack of familiarity with the theoretical background. Where the applicability of a statistical test may be questionable, it can be discussed with someone familiar with its background.

Some further general comments are essential before going into the details of the statistics. The methods that will be discussed should not be thought of as a final enumeration of all statistical methods that can be used to determine the state of mixedness of a batch. They are really just the beginning of an attempt to apply useful techniques in an area that is well suited for them. New statistical tools are constantly being developed. Old established methods are continually being streamlined, and easily usable graphs are continually appearing to replace more tedious analytical methods. The procedures shown below have been

---

[1] In some special cases, however, special location sampling may be preferable when evidence of gradients in a certain direction is being sought.

presented in terms of solids mixing problems, in the hope that this will stimulate their use in that area. They are not meant to replace common sense, nor to substitute some magical mathematical manipulation for a sound intimate knowledge of the particular solids mixing operation and its purpose. They are merely additional tools to aid in utilizing this knowledge. In general, the most productive use of statistical methods in a solids mixing operation is obtained by cooperative work between someone who knows statistics well and someone who is familiar with the solids mixing operation and its purpose. If a single individual is competent in both these areas, the job is that much easier.

So much may be gained in increased efficiency and better product control where solids mixing is concerned, if the proper statistical methods are used, that it is worthwhile to point out some serious obstacles to the employment of statistics in this area. These are:

a. Immediate discouragement on the part of the "practical" man because the mathematical statistician may at first have suggested a procedure that is too impractical. (Such a procedure may involve too many costly analyses, perhaps the mathematical model suggested does not really simulate the actual situation, etc.) These difficulties can almost always be worked out to the mutual advantage of both parties, if the first pitfall due to differences in viewpoint can be overcome.

b. The too frequent tendency of authors, in articles on the use of statistical methods, to become elaborately involved in derivations (some of which have already been derived elsewhere), rather than concentrate on the application of known formulae and methods to mixing problems. To someone unfamiliar with statistics, this may give the impression that some new and untried mathematical ideas are involved, whereas the case may simply illustrate the application of an old and generally accepted statistical method.

c. The relatively few examples in the literature in which statistical methods are specifically applied to solids mixing problems. Although there are numerous cases dealing with statistical methods in mass production manufacturing, this is not as likely to be studied by someone who is groping for a better way to evaluate his solids mixing operation, as a problem dealing with solids mixing. More published examples of the use of statistical methods to evaluate or improve solids mixing operations specifically, would be a great help to their general acceptance and use throughout industry.

2. *Quantitative Measurements on Each Spot Sample*

a. The Relative Frequency Distribution. Returning now to the group of spot samples taken from the mixed batch, if enough spot samples were taken, then a fairly good graph could be drawn, showing "composition" (such as %A) on the abscissa, and "fraction of the spot samples" within a certain composition range on the ordinate, as indicated in Fig. 3. A truly representative graph of this type would give a good picture of the state of mixedness, since it would immediately show how the spot samples varied in composition. However, to get a representative graph, which is called a *relative frequency distribution* in statistical terminology, is a formidable task since a very large number of spot samples would be needed. Except for certain special cases, this is highly impractical. Statistics gives methods whereby objective estimates of the four characteristics, which are needed to completely describe such a distribution,

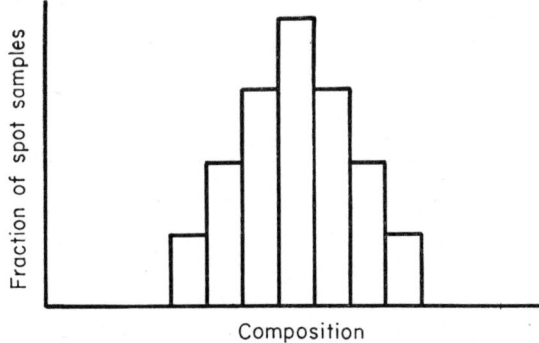

FIG. 3. Relative frequency histogram for spot sample compositions.

can be made with a small number of spot samples. These characteristics, which are called "statistics" are: (1) Some measure of the number about which most of the values tend to cluster. Although many different measures exist (median, mode, geometric mean), the arithmetic mean is the most common. The statistical term for this is a "measure of central tendency." (2) Some measure which indicates the amount of variation among the values. Although the average deviation was formerly very popular, modern statistics has established the superiority of the variance as the most efficient means of extracting information concerning variation within a group of measurements. The standard deviation and range are other terms frequently used. The statistical term for this is "measure of spread" or "measure of variation." (3) Some measure indicating how symmetrical the distribution is with respect to a vertical axis through its mean. The statistical term for this is "measure of skewness." (4) Some measure which indicates how high or peaked the graph is in the neighborhood of the mean. The statistical term for this is "measure of peakedness" or "measure of kurtosis."

For most solids mixing problems, the measure of central tendency and measure of variation will be of greatest use. Therefore these will be defined and tests of significance given for them. Measures of skewness or peakedness can be found elsewhere (H5) if the occasion demands.

Other useful statistical tests, together with their background, will also be given later.

b. Definitions. i. Sample Arithmetic Mean. This well known and widely used term, which will be denoted by $\bar{x}$, is defined as:

$$\bar{x} = \frac{\sum_{i=1}^{n} x_i}{n} \qquad (1)$$

where n = the number of spot samples (statistical term is "sample size")

x_i = the i-th value of x, which is a number representing composition or whatever else is being looked for in each spot sample.

The mean by itself should not be used as a degree of mixing, since, if the batch is properly sampled, the only variation between sample means should be the sampling variation, regardless of how well mixed the batch is. It is often desirable, however, to compute the sample mean in addition to whatever degree of mixing is used. Where the true mean is known (e.g., a known per cent of a certain material has been added in the mixer), this can be used to find out whether the sample mean is significantly different from the true mean. If this significant difference should occur, it would then be necessary to find out whether it were due to improper location and spacing of the spot samples, a bias during sampling (perhaps the sampling method was biased in favor of getting more of a certain material), or some peculiar segregating tendency which favored the concentration of one material in some small area.

ii. Sample Variance and Standard Deviation. The unbiased sample variance, denoted by $\hat{s}^2$, is defined as:

$$\hat{s}^2 = \frac{\sum_{i=1}^{n} (x_i - \bar{x})^2}{n - 1} \qquad (2)$$

[Note: The standard deviation $\hat{s}$ will be obtained by taking the square root of $\hat{s}^2$, although it happens that whereas $\hat{s}^2$ is an unbiased estimate of the population variance, $\hat{s}$ is not an unbiased estimate of the population standard deviation.] (The term unbiased indicates that $n - 1$ is used in the denominator instead of n, in defining variance. In order to avoid repetition from here on, the word unbiased will not be used although all variances and standard deviations will be of this type unless otherwise specified.) Both the standard deviation and variance have their own particular advantages and it is worthwhile to work with both of them. The standard deviation is handier for graphing and discussing mixing performance because it has the same units as the unit of measurement. However, the variance is more desirable for statistical tests of significance comparing variation, since it provides the most efficient means of extracting information concerning variation from a group of measurements, and **F** tables, which are used in statistical tests of significance involving variances, are easily available.

Either the variance or the standard deviation is usually a good first choice for degree of mixing. Sometimes expressions, involving these as well as other terms, are used for special applications, such as determination of the best plotting methods to reveal rate mechanisms.

A computing formula is available for ease in calculating the variance or standard deviation:

$$\hat{s}^2 = \frac{n \sum_{i=1}^{n} x_i^2 - \left(\sum_{i=1}^{n} x_i\right)^2}{n(n-1)} \qquad (2a)$$

c. Statistical Tests of Significance. i. Meaning of a Statistical Test of Significance. After the mean and variance have been computed, it is often desirable to compare them to some "reference" set. Statistical tests of significance give an objective means of doing this. Again, before the details are dealt with, it is essential that certain basic concepts be understood.

A *test of significance* prescribes a method for comparing a characteristic of the test samples with that of the reference samples by means of a term involving both. The probability that as large a numerical value of this term as computed would occur due to chance alone, is then looked up in the proper table. If it is improbable that this large a numerical value would have occurred due to chance alone, then it is concluded that there must be something else besides chance causing the difference between the test value and the reference value, and that therefore the two differ significantly. In the examples that follow, an improbable occurrence is defined as one that would happen by chance less than one time in twenty, in the long run. This means that the tests are run at the 5% level of significance. As is further mentioned under Section III, C, 7, other levels of significance could be chosen.

Choosing a "reference" by which the mean and variance can be compared, is not often an easy job. One possible way of doing this is by determining from a pilot plant run what mixture will be most satisfactory for further processing. A key characteristic of the mix can then be measured in many spot samples. The mean and variance obtained from these spot samples could then serve as "references" for comparison, when it is desired to go to large-scale production in a big mixer. The techniques to be shown can also be used to determine whether significant "unmixing" occurs after mixing, due to handling of the batch, as mentioned in the section on Related Process Steps (Section II).

ii. Test of Significance for Means (**t** *test*). If a known proportion of a certain component has been put into the mixture and the latter has been properly sampled, then the sample mean would not be expected to be significantly different from the known composition, which will be denoted by m. In this test, a method is given for determining objectively whether any observed difference is significant.

$x_1, x_2, x_3 \ldots x_n$ represents a group of measurements on n spot samples.

Step 1. Compute $\bar{x}$, the sample arithmetic mean and $\hat{s}$, the sample standard deviation, as shown previously.

Step 2. Compute $$t = \frac{\bar{x} - m}{\frac{\hat{s}}{\sqrt{n}}} \qquad (3)$$

where n = the number of spot samples.

Step 3. From the **t** tables (H5, pp. 248, 249), determine whether this large a value of **t** would occur more than one time in twenty due to chance. (In statistical terminology: whether the value of **t** is greater than that given in the tables for the 5% level of significance in a two-tail test.) It is necessary to know the degrees of freedom ($= n - 1$) in order to use the tables.

Step 4. If the value of **t** is so large that it would occur by chance less than one time in twenty, conclude that $\bar{x}$ is significantly different from m.

Step 5. This is the most important part of the procedure, namely, the interpretation of the results of the test. If the means under comparison differed significantly, the sampling may have been biased due to location or method. If so, this bias should be corrected before further samples are taken. It should be stressed that statistics alone cannot generally point out exactly what is happening to cause the significant difference. This must come from a close knowledge of the operation.

iii. Test of Significance for Variances (**F** *test*). The need for this might arise as follows: A small-scale pilot plant experiment has shown that a satisfactory mixture can be produced for further processing. All of the work was done by a research group where considerable attention was given to preparing the mixture, since the primary purpose was to ascertain whether the entire process could be used. Time and manpower in preparing the mixture were not limited. Now the process is to be scaled up to production requirements, and it will be necessary to cut the cost per pound of product considerably.

A much larger mixer is needed than that used to mix the say 50-lb. lots for the pilot plant run. Mixing will have to be done in a much shorter time. However, since the mixture in the pilot run has been proven to process successfully, the large-scale, rapid mixing must also yield as good a mixture. The weight fraction of a key constituent in the mixture, which must be well distributed for successful later processing, is determined in a number of spot samples taken from the successful pilot mixture. The variation between spot sample compositions of the large-scale mix should not be significantly greater than the variation between spot sample compositions from the pilot mix. The following procedure indicates how to determine whether the variations are significantly different:

$x_1, x_2, x_3 \ldots x_n$ represent the pilot mixer measurements of which there are n_x.

$y_1, y_2, y_3 \ldots y_n$ represent the large mixer measurements of which there are n_y.

This test is to determine whether $\hat{s}_y^2$ is significantly greater than $\hat{s}_x^2$.

Step 1. Compute $\hat{s}_x^2$ and $\hat{s}_y^2$ as previously shown.

Step 2. If $\hat{s}_x^2$ is larger than $\hat{s}_y^2$, then the larger mixer is satisfactory, and no statistical tests are necessary. If $\hat{s}_y^2$ is larger than $\hat{s}_x^2$, form the ratio

$$F = \frac{\hat{s}_y^2}{\hat{s}_x^2} \qquad (4)$$

Step 3. From the **F** tables (H5, pp. 250–253), determine whether this large a value of **F** would occur due to chance alone more than one time in twenty. (In statistical terminology: whether the value of **F** is greater than that given in the table for the 5% point for the distribution of **F**.) $n_x - 1$ and $n_y - 1$ (denoted as degrees of freedom) are needed to use the tables.

Step 4. If the value of **F** is so large, that it would occur by chance less than one time in twenty, conclude that $\hat{s}_y^2$ is significantly greater than $\hat{s}_x^2$.

Step 5. The reasons for the greater variation must now be sought. Perhaps there is segregation in the mixer, or maybe a longer mixing time is necessary. (Note: It is desirable to make a similar comparison of variation after the mixer is dumped to see whether unmixing occurs.) This type of statistical test can be used to compare the variance among spot samples, taken at any point in the batch handling system, with the reference variance.

iv. Test of Significance for Distributions (**Chi-square** *test*). In Section III, C, 2a, the term *relative frequency distribution* was used to describe a graph showing "composition" on the abscissa and "proportion of spot samples within a certain composition range" on the ordinate. The ordinate values for any composition range are called relative frequencies, because they are expressed as fractions of the total number of spot samples. The mean and variance were mentioned as some of the characteristics needed to describe a frequency distribution, and previous tests dealt with ways to compare them quantitatively with reference values.

In some special cases, it may be desirable to determine whether a set of spot samples comes from a frequency distribution of a certain type, i.e., to compare the experimental frequency distribution with a theoretical one. An example would be the case of a mixture of nuts from which packages of nuts are to be made. If the nuts were mixed in a tumbling mixer then there would be an ideal limiting condition concerning the distribution of different kinds of nuts in the packages, assuming no segregation on packaging. Fifty per cent peanuts might be desired for each package, but the best that could be done by a random tumbling motion would be a certain distribution of "peanut fraction," with 0.50 as the mean. If a number of packages were made up from this mixture, the **chi-square** test would enable us to tell whether the distribution of "peanut fraction" came from a random mixture of peanuts with other nuts. This information would enable an objective decision to be made as to whether the variation in "peanut fraction" from package to package was as low as could be obtained by random tumbling—or whether it was much too large, indicating that some sort of segregation was taking place. The steps for the **chi-square** test are shown below:

Let $o_1, o_2, o_3 \ldots o_i \ldots o_n$ be a set of "observed frequencies" obtained by random sampling from the nut mixture, as previously described. Each of these frequencies is a number whose value shows how often a certain range of "peanut fraction" was obtained.

Let $e_1, e_2, e_3 \ldots e_i \ldots e_n$ be a set of "expected frequencies" obtained by theoretical calculation from the normal distribution. Each of these frequencies is a number showing how often a certain range of "peanut fraction" would be obtained with a certain spot sample size taken from a randomly mixed batch of nuts. The "peanut fraction" range should be chosen so that the expected frequencies are all greater than or equal to 5. The observed frequencies can then be compared with the expected frequencies for various "peanut fraction" ranges. Let $k =$ the number of pairs of observed and expected frequencies to be compared.

Step 1. Compute χ^2 (called **chi-square**):

$$\chi^2 = \sum_{i=1}^{k} \frac{(o_i - e_i)^2}{e_i} \tag{5}$$

Step 2. From **chi-square** (χ^2) tables (H5, p. 246), determine whether this large a value of **chi-square** would occur due to chance more than one time in twenty. The number of degrees of freedom ($= k - 1$) is necessary in order to use the tables. The column heading labeled 0.05 would be used for the "one in twenty" test.

Step 3. If the value of **chi-square** is so large that, it would occur by chance less than one time in twenty, conclude that the observed frequencies represent a distribution which is significantly different from the expected (theoretical) distribution.

Step 4. Now again, the all important question of proper interpretation and use of the results must be considered. If the distributions were significantly different, why did this occur? Perhaps the curved cashew nuts tended to segregate in a certain area. Is there any gradient that can be found by plotting the location and order of the spot sample "peanut fractions"? Statistics does not answer these questions. It simply gives an objective way of pointing out, that there is significantly more variation in "peanut fraction" than there should be for a random mixture. If on the other hand, there is no significant difference, then the nuts have been mixed as thoroughly as possible in a random tumbling operation. There would be no point in looking around for any other type of tumbling mixer in the hope of getting less variation in "peanut fraction" from bag to bag. Instead, if the magnitude of random variation is unacceptable, some other method of nut packaging would have to be considered, such as, perhaps, hand distributing or automatic proportioning of the nuts to give a fixed number of each per package.

3. *Qualitative or Semi-quantitative Measurements on Each Spot Sample*

a. Smear Test. Sometimes, rather than use quantitative analyses for spot samples, a smear test is made. This consists of spreading a spot sample (or a portion from one such sample) in a thin layer on a flat surface by smearing with a spatula, and then visually examining it. Off-color streaks or spots which indicate a lack of complete homogeneity can thus be observed. Samples of a fixed size and a reproducible technique should be used when comparisons are to be made. As will be illustrated in the following two examples, either each spot sample can be judged satisfactory or unsatisfactory on the basis of the smear test (qualitative), or the average number of undesirable streaks per spot sample can be determined by adding up all of the streaks found and dividing by the number of spot samples (semi-quantitative). Where several differently colored streaks are present, the number of streaks of each color per spot sample could be determined, if this is desirable.

b. Test of Significance for Fraction Satisfactory. A case might arise where it is desired to mix differently colored fine materials. Because of the necessity for frequent changes in the materials used in the mixer, it might be desirable not to have any device for breaking up agglomerates in the mixer, even though this means that the final mixture might have a few agglomerates. In other words, in this particular case, the risk of ruining lots due to contamination from previous lots outweighs the advantage of achieving a better mixture by using a device for breaking down agglomerates. However, tests have been made which indicate that there is a limit to the number of agglomerates that can be tolerated, as measured by streaks visible from a smear test. Using a standard size of spot sample, 95% of the smear tests should show a maximum of one streak in order to be considered "satisfactory." The problem now is to determine whether a particular mixer, which is fine from a cleaning point of view, is good enough to meet the above smear test specifications, even though it was not designed for breaking down agglomerates.

In this case the aim of the statistical test is to objectively decide whether a desired "fraction satisfactory" of spot samples has been obtained after mixing.

If the experimentally determined "fraction satisfactory" is greater than or equal to the desired "fraction satisfactory," then there is no need for the statistical test. If it is less than the desired fraction satisfactory, then the following statistical test can be used to make a decision:

Let p = desired "fraction satisfactory" of spot samples, and ϕ = experimentally determined "fraction satisfactory" of spot samples.

Step 1. Compute $$u = \frac{p - \phi}{\sqrt{\frac{p(1 - p)}{n}}} \tag{6}$$

where n = the number of experimental spot samples.

Step 2. Tables of the standard normal curve (H5, pp. 243–245. *Note:* Hoel uses t here for what is called u in Step 1) determine whether this large a value of u would occur due to chance more than one time in twenty. The value of n is not necessary for the use of these tables. The reason for this, very briefly, is that the standard deviation

$$(= \sqrt{p(1 - p)/n}\,)$$

used in computing u, is not an estimate but an exact value, based on a "fraction satisfactory" level equal to p.

Step 3. If the value of u is so large, that it would occur by chance less than one time in twenty, conclude that the desired "fraction satisfactory" is significantly greater than that obtained by sampling the mixer.

Step 4. An unsatisfactory mixture may require longer mixing, or it may be that a more intense type of mixing cannot be avoided if the desired agglomerate breakdown is to be achieved.

  c. *Test of Significance for Streaks per Spot Sample.* In this case two mixers are to be compared to see which will give a product with the fewest streaks per spot sample for a particular mixture. The time necessary to reach the best mixture is not a major factor, so that only the "equilibrium" mixtures for the two mixers are to be compared as to the number of streaks per spot sample. There is no stated lower limit for this number, but these two mixers have been found equally desirable from the cleaning standpoint, and whichever gives the lower number of streaks per spot sample will be chosen.

The significance test for this situation is somewhat similar to the one employed in comparing the sample mean with a known mean, except that in this case two sample means are being compared. First their variances must be compared to see whether they are significantly different. Then, depending upon whether or not the variances are significantly different, one of two alternate tests of significance on the means can be performed. These tests are outlined below:

Let $\bar{\alpha}$ and $\bar{\beta}$ be the "equilibrium" values of streaks per spot sample for mixers A and B respectively. (A statistical test given in Section III, C, 4 shows how the equilibrium values can be objectively obtained.) Let $\hat{s}_\alpha^2$ and $\hat{s}_\beta^2$ be the variances for the spot samples used to compute $\bar{\alpha}$ and $\bar{\beta}$, respectively. First perform an **F** test as previously shown to determine whether $\hat{s}_\alpha^2$ and $\hat{s}_\beta^2$ differ significantly. (Note: This test is not identical with that of the previous example, which was designed to find out whether one variance was significantly *larger* than a given variance; in this case the purpose of the test is to ascertain whether two variances *differ* significantly.)

The remainder of the test will depend on whether or not the variances were found to differ significantly.

CASE I—*Variances do not differ significantly*

Step 1. Compute $$t = \frac{\bar{\alpha} - \bar{\beta}}{\sqrt{\frac{\Sigma(\alpha_i - \bar{\alpha})^2 + \Sigma(\beta_i - \bar{\beta})^2}{n_\alpha + n_\beta - 2}}} \qquad (7)$$

Step 2. From **t** tables determine whether this large a value of **t** would occur by chance more than one time in twenty. The number of degrees of freedom ($= n_\alpha - n_\beta - 2$) is necessary in order to use the tables.

Step 3. If the value of **t** is so large that it would occur by chance less than one time in twenty, conclude that there is a significant difference in smears per spot sample between mixers A and B.

Step 4. Whichever mixer gives the fewer number of smears per spot sample would be better for the job.

CASE II—*Variances differ significantly*

Step 1. Compute $$\nu = \frac{\bar{\alpha} - \bar{\beta}}{\sqrt{\frac{1}{n_\alpha}\hat{s}_\alpha^2 + \frac{1}{n_\beta}\hat{s}_\beta^2}} \tag{8}$$

Step 2. From special tables (A3), determine whether this large a value of ν would occur by chance more than one time in twenty. The ratio

$$\frac{\hat{s}_\alpha^2/n_\alpha}{\hat{s}_\alpha^2/n_\alpha + \hat{s}_\beta^2/n_\beta}$$

must be calculated in order to use the tables. Because the tables are set up for a one-sided test, what is really being tested here is whether $\bar{\alpha}$ is greater than $\bar{\beta}$, rather than whether $\bar{\alpha}$ is different from $\bar{\beta}$.

Step 3. If the value of ν is so large, that it would occur by chance less than one time in twenty, conclude that $\bar{\alpha}$ is significantly larger than $\bar{\beta}$.

Step 4. Assuming the latter conclusion, mixer B should be chosen since it has been able to produce a mixture with significantly fewer streaks per spot sample than the one produced by mixer A.

4. *Test of Significance for Determining When Equilibrium Is Reached*

In the test of significance for comparing streaks per spot sample between two mixers, the "equilibrium" value was referred to. By this is meant the value which does not change significantly with further mixing. The method described below is an objective way of determining whether or not there still is a trend in the value of streaks per spot sample, after a certain period of mixing.

Let $\alpha_1, \alpha_2, \alpha_3 \ldots \alpha_i \ldots \alpha_n$ represent values of streaks per spot sample determined at times $t_1, t_2, t_3 \ldots t_i \ldots t_n$ after it appears that equilibrium has been reached.

Step 1. Compute $$q^2 = \frac{\sum_{i=1}^{n-1}(\alpha_{i+1} - \alpha_i)^2}{2(n-1)} \tag{9}$$

Step 2. Compute $\hat{s}_\alpha^2$

Step 3. Compute $$r = \frac{q^2}{\hat{s}_\alpha^2} \tag{10}$$

Step 4. From r tables (H1, p. 359) determine whether this large a value of r would occur by chance more than one time in twenty.

Step 5. If this large a value of r would occur by chance more than one time in twenty, conclude that there is still a trend, i.e., equilibrium has not been reached. Briefly, the reason for this is that q^2, which is a measure of variation taking into account the order of obtaining α, is significantly different from $\hat{s}_\alpha^2$, which is a measure of variation in which the order of obtaining α is not taken into account.

Step 6. If there is still a trend, get another α for t_{n+1}, and repeat the test with α_2 to α_{n+1}. Continue this procedure until equilibrium is reached. (Note: Calculations can be kept to a minimum by leaving room on the computation sheet for several extra points after equilibrium is suspected.)

Step 7. When equilibrium has been reached, the α's finally used in the test, which indicated that there was no longer a trend, should be averaged to give $\bar{\alpha}_e$, which is an estimate of the equilibrium value of smears per spot sample for the mixer. (Note: The example given above involved computing the equilibrium value of smears per spot sample. The test for determining whether equilibrium has been reached for any other measure of mixing, such as the standard deviation, would be similar.)

5. *Confidence Limits*

   a. *Meaning of Confidence Limits.* In the above example for determining the equilibrium value of α, it was stated that $\bar{\alpha}_e$ was an estimate of this value. The term *estimate* was used because there is variation among the α's used in computing $\bar{\alpha}_e$. It might intuitively be felt that it should be possible to state the limits of a range of values within which one can be confident, that the true but unknown value of streaks per spot sample lies. The term "one can be confident" can be expressed quantitatively by a percentage of the time that the given limits will fulfill their function in the long run. These are known as *confidence limits*. Although the above illustration deals with confidence limits for a mean, they can also be computed for a standard deviation or a variance. The statistical term for the true but unknown value, of which the sample value is an estimate, is *population parameter*. Confidence limits for a sample estimate of a population parameter are limits which will include the true but unknown value of the population parameter a certain specified proportion of the time, in the long run. In the examples that follow 95% confidence limits have been used for illustrative purposes. These will include the true but unknown value of the quantity being estimated nineteen times out of twenty, in the long run. As is further mentioned in Section III, C, 7, other % confidence limits could be chosen.

   It sometimes helps to graph the confidence limits along with the measure of mixing when plotting the latter versus time. This will give a band, rather than a single line for the rate curve, and is a way of showing graphically how sampling variation places limitations on exact statements such as "at 2 minutes the measure of mixing is exactly 1.20." This will be covered in greater detail when Stange's paper (S4) is discussed in paragraph C, 5d of this section. Methods for computing confidence limits are shown below.

b. Confidence Limits for the Mean. The confidence limits for $\bar{\alpha}_e$ are:

$$\bar{\alpha}_e \pm t \frac{\hat{s}_\alpha}{\sqrt{n}} \qquad (11)$$

where the value of **t** is taken from the same **t** table previously mentioned under Tests of Significance for Means (**t** test). As stated there, the number of degrees of freedom ($= n - 1$) is necessary for the use of the tables. For 95% confidence limits, the value of **t** is chosen from the 0.05 probability column when using a two-tail test table and from the 0.025 column when using a one-tail test table. In this case n is equal to the number of α's used to determine $\bar{\alpha}_e$. In the case where confidence limits are to be obtained for a mean $\bar{x}$, n will be the number of values of x taken at one specific time. The confidence limits will be:

$$\bar{x} \pm \frac{t\hat{s}_x}{\sqrt{n}} \qquad (12)$$

c. Confidence Limits for the Variance. Confidence limits for the variance $\hat{s}^2$, require use of *chi-square* tables. These confidence limits are not symmetrical around the variance; instead the 95% confidence limits would be computed as shown below:

Step 1. Using $n - 1$ as the number of degrees of freedom, look up, in the chi-square (χ^2) tables, that chi-square value which is (1) so small that anything less than it would occur by chance less than 2.5% of the time (χ_1^2), or (2) so large that anything greater than it would occur by chance less than 2.5% of the time (χ_2^2). These would be values for $P = 0.025$ and $P = 0.975$, respectively. The values of chi-square between these two limits would occur by chance 95% of the time. (Note: Since in the chi-square tables the nearest columns to the desired $P = 0.975$ and $P = 0.025$ may be $P = 0.98$ and $P = 0.02$, it may be more convenient to use 96% confidence limits to avoid the necessity for extrapolation.)

Step 2. The upper confidence limit of $\hat{s}^2$ is $\dfrac{n\hat{s}^2}{\chi_1^2}$.

Step 3. The lower confidence limit of $\hat{s}^2$ is $\dfrac{n\hat{s}^2}{\chi_2^2}$.

Step 4. If σ^2 is the true but unknown value of the variance which is being estimated by $\hat{s}^2$, then the above discussion concerning its confidence limits can be summarized as:

$$\frac{n\hat{s}^2}{\chi_2^2} < \sigma^2 < \frac{n\hat{s}^2}{\chi_1^2} \qquad (13)$$

d. Confidence Limits for the Standard Deviation and Their Use in Planning Proper Sample Size for a Prescribed Confidence Range. i. Confidence Limits. If σ is the true but unknown value of the standard deviation which is being estimated by $\hat{s}$, then it follows from Eq. (13) that the confidence limits of σ are:

$$\sqrt{\frac{n\hat{s}^2}{\chi_2^2}} < \sigma < \sqrt{\frac{n\hat{s}^2}{\chi_1^2}} \qquad (14)$$

Ordinarily, chi-square tables would be necessary to determine the confidence limits in a manner analogous to that for the variance. However, a recent paper by Stange (S4) has shown a graphical method for computing 90%, 95%, and 99% confidence limits. Stange plots σ/s on the ordinate and the number of degrees of freedom, $\mathcal{K}$, on the abscissa as shown in Fig. 4. (Note: Although Stange refers to $\mathcal{K}$ as number of samples in his first paper (S4), the manner in which he uses the graph in connection with a discussion of degrees of freedom in his other paper (S5) shows $\mathcal{K}$ to actually be degrees of freedom. [When variance is calculated as in equation (2), degrees of freedom $= n - 1$.]) An upper and lower graph is given for each value of confidence limits. What this amounts to, is plotting $\sqrt{n/\chi_2^2}$ vs. $\mathcal{K}$ (lower curve), and $\sqrt{n/\chi_1^2}$ vs. $\mathcal{K}$ (upper curve) to graphically express the inequality:

$$\sqrt{\frac{n}{\chi_2^2}} < \frac{\sigma}{\hat{s}} < \sqrt{\frac{n}{\chi_1^2}} \qquad (15)$$

which is simply Eq. (14) divided through by $\sqrt{\hat{s}^2}$. In Fig. 4, Stange's graph is shown with different nomenclature, which, it is hoped, will make it clearer and immediately usable to those unable to read the original German article. Stange shows how the confidence limits, obtained from a preliminary experiment with an arbitrary number of spot samples, can be used to plan the number of samples for the next experiment so as to give a prescribed confidence range.

These ideas are illustrated with various examples by Stange. In discussing them, standard deviation will be denoted by s, rather than $\hat{s}$, for consistency with Stange's work. Figure 5 simply shows what a rate plot with upper and lower confidence limits looks like. Figure 4 can supply the factors to enable quick computation of confidence limits needed for such a plot.

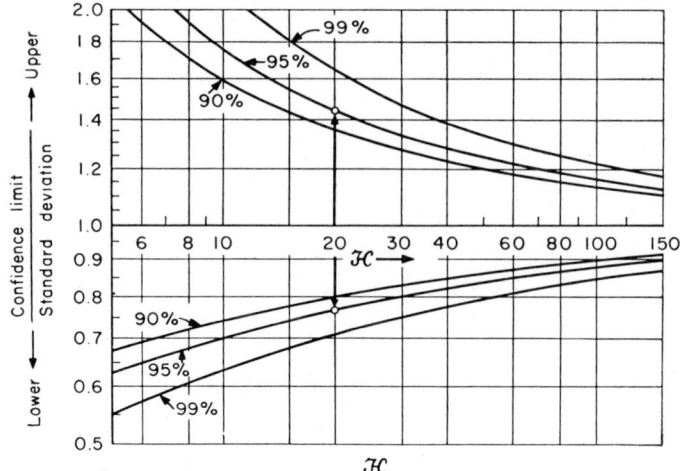

Fig. 4. Confidence limits for the standard deviation as a function of degrees of freedom. This is taken from Stange (S4), with nomenclature changed. The upper graphs are s_u/s vs. $\mathcal{K}$ and the lower graphs s_l/s vs. $\mathcal{K}$, where s_u is the upper confidence limit, s_l is the lower confidence limit, s is the experimental standard deviation, and $\mathcal{K}$ is the number of degrees of freedom. σ, the true but unknown standard deviation, which is being estimated by s, will fall between s_u and s_l nineteen times out of twenty in the long run, for 95% confidence limits.

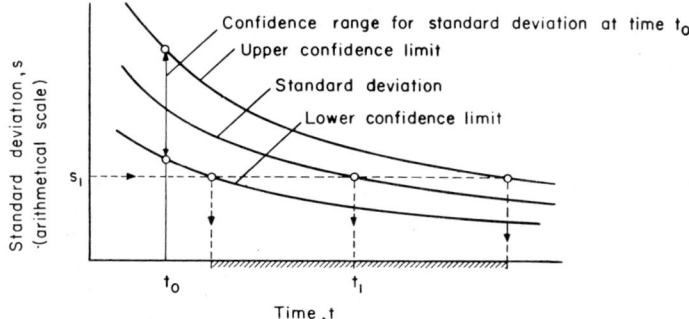

Fig. 5. Illustrative numerical plot of standard deviation vs. time with 95% upper and lower confidence limits. From Stange (S4) with different nomenclature. Cross-hatched area indicates Stange's estimate of confidence range for time, t_1, which corresponds to standard deviation s_1.

Figure 6 illustrates a method given by Stange for estimating confidence limits on the mixing time necessary to achieve a certain specified standard deviation. Lowry's data (L4) for the mixing of 80% ammonium nitrate and 20% TNT by weight to form the explosive *Amatol* is used. (Note: Herdan (H2) contains a discussion of Lowry's report.) Figure 6 indicates how Stange obtains lower (3.8 minutes) and upper (6.0 minutes) confidence limits on the time necessary to achieve the desired value of standard deviation, $s = 1.0\%$. (Note: although the confidence limits for s are based on $\mathcal{K} = 6$, and are, respectively, $2.21s$ (upper—off scale on Fig. 4) and $.645s$ (lower), it would seem that $\mathcal{K} = 5$ should have been used, since six spot samples were taken at each time and $\mathcal{K}$ is the number

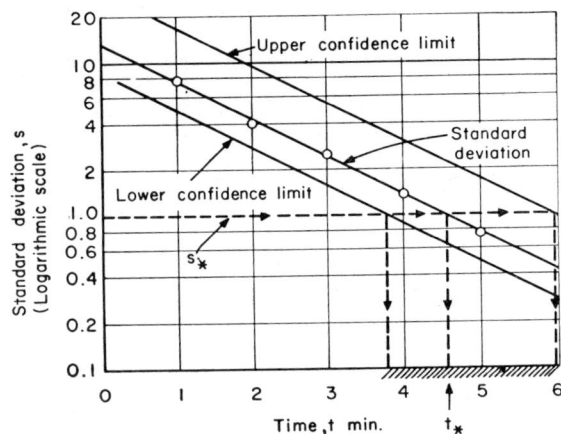

Fig. 6. Logarithmic plot of standard deviation vs. time with 95% upper and lower confidence limits. From (S4) with different nomenclature, but data are originally from (L4). Six spot samples taken at each time, t; t_* = time to achieve desired value of standard deviation, s_*. Crosshatched area indicates estimate of confidence range for t_*.

of degrees of freedom.) Although not stated by Stange, the per cent error involved in this graphical method for estimating confidence limits on the time would be worth knowing.

ii. Sample Size for a Prescribed Confidence Range. If a pilot experiment has shown that a certain sample size ($= n$ spot samples) gives a certain mixing curve of s vs. time ($= t$), then it is possible to use this information in planning the next experiment to give a prescribed confidence range for s, by taking the proper number of spot samples. Stange has shown a graphical method for this operation, which is based on the approximative procedure mentioned previously for getting confidence limits for the mixing time. The limitations mentioned there should be borne in mind when using the following procedure:

Step 1. At the desired time in the pilot curve, compute the slope

$$\frac{A - B}{C + D} = |\dot{s}_*| \tag{16}$$

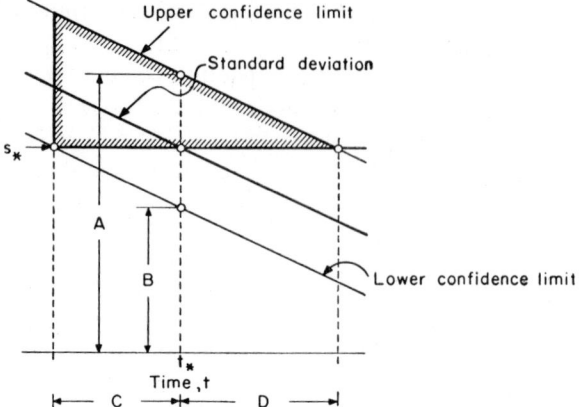

Fig. 7. First part of a graphical method for using a pilot run, in which an arbitrary number of spot samples are taken at each time, to estimate the number of spot samples needed to obtain a confidence interval, the latter being a specified fraction of the desired mixing time, t_*. From Stange (S4) with different nomenclature.

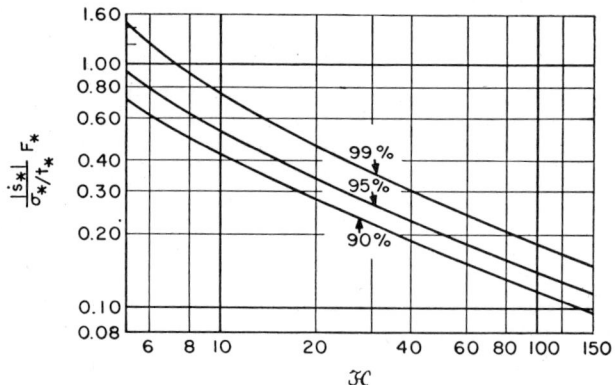

Fig. 8. Second part of graphical method mentioned under Fig. 7 (S4). See Section III, C, 5d.

The nomenclature is shown in Fig. 7. Graphs are assumed to be straight lines in this region.

Step 2. Compute the ratio:

$$\frac{|\dot{s}_*|}{\sigma_*/t_*} F_* \tag{17}$$

where, using Stange's nomenclature (except that time T_*, is changed to t_*),

$F_* = \Delta t_*/t_*$ = desired relative precision of time to achieve desired quality of mixing (95% confidence limits mean $F_* = 0.05$)

σ_* = desired quality of mixing

t_* = mixing time to achieve σ_*

Step 3. From Fig. 8, determine the number of spot samples required. It should be remembered that because of approximations inherent in the above methods this is not a rigorously exact figure.

e. Confidence Limits for Fraction Satisfactory. Confidence limits for "fraction satisfactory" ($= \phi$) may also be computed. These are:

$$\phi \pm u \sqrt{\frac{\phi(1 - \phi)}{n}} \qquad (18)$$

The normal tables mentioned in Section III, C, 3b are also used here to find the value of u for 95% confidence limits. This is equal to 1.96.

6. *The Quality Control Chart*

In addition to the analytical techniques previously described, which permit an objective decision as to whether a solids mixture is satisfactorily mixed, there are other techniques which have long been used in manufacturing processes and are applicable to solids mixing studies. The quality control chart is a well known graphical method for this purpose. Some recent papers on solids mixing which make use of quality control methods are Herdan's discussion of industrial mixing (H2), and the paper of Adams and Baker (A1). Before going into these, some general comments on quality control may serve as an introduction.

The quality control chart for the sample mean is a graph showing a center line which is a grand average mean, and upper and lower lines which are control limits. These control limits can be calculated in a manner similar to that employed to determine confidence limits on the mean. There are tabulated shortcuts available for their computation. These shortcuts involve the use of the range, rather than the standard deviation, and make use of the fact that the standard deviation is a known function of the range under certain conditions. The range is denoted by R—the difference between the highest and lowest values of a spot sample property in a group of spot samples.

Control charts can also be made for the range itself. Since the range is an easily computed measure of variation among a group of spot samples, this kind of control chart would be applicable to mixing work, where it is desired to find out how the variation among spot samples decreases with mixing. A book relating to statistical quality control can be read for further background information (G1). However, for the purposes of illustrating how the techniques can be applied to a solids mixing problem, examples which have appeared in solids mixing literature will be used here without delving into the theoretical background. The American Society for Testing Materials, *Manual on Quality Control of Materials* (A4), goes into considerable detail on the control chart method of analysis and presentation of data.

Herdan (H2), using data from Lowry's report (L4) on the incorporation of *Amatol*, illustrates a method for following mixing via a control

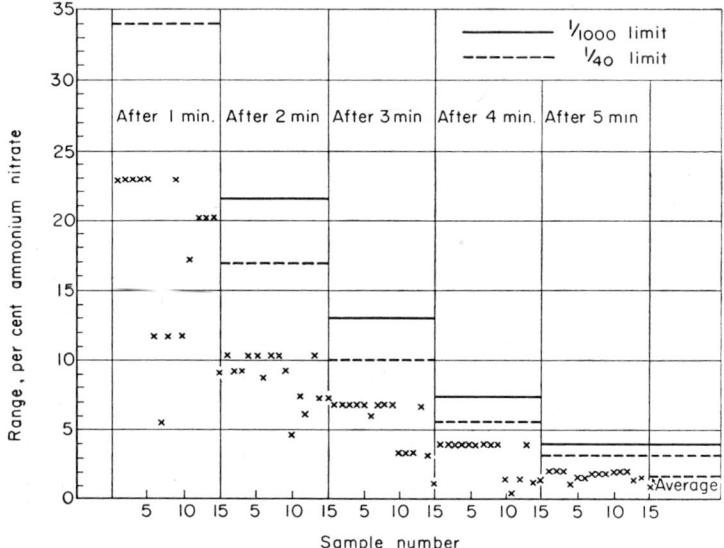

FIG. 9. Quality control chart for range (H2). (Courtesy Elsevier Publishing Co., Amsterdam.)

chart. He has rearranged the original data so as to get 15 groups of four spot samples each, at each time of mixing. (Note: The statistical term for each group of four spot samples is a "rational subgroup.") For each spot sample, the per cent ammonium nitrate has been determined. Actually, Lowry took six samples at each time of mixing, and these were combined in 15 different ways by Herdan to give 15 groups of four, but this rearrangement is not involved in the idea of a quality control chart; it was made only to obtain numbers with which to illustrate the method.

Mixing is considered complete when the standard deviation among rational subgroups, of size $n = 4$, reaches a certain specified value. Because of the functional relationship between standard deviation and range, an equivalent criterion would require that the range reach a certain value. Since sampling variation is present, instead of a single value of the range, control limits, based on an average of range values for the 15 groups of four, are given. If all the values of range based on groups of four fall within these control limits, then mixing has been completed.

Figure 9 illustrates the use of the quality control chart for the range in connection with the previously mentioned problem on the incorporation of Amatol. The $1/1000$ and $1/40$ control limits are shown, and ranges outside of these limits would occur by chance one time in a thousand and one time in 40, respectively.

As an example of how such control limits may be calculated consider the 15 groups of four spot samples, each taken after 5 minutes of mixing. The average range is denoted by $\bar{w} = 1.64$. Table I, from the Herdan

TABLE I
FACTORS USED IN THE QUALITY CONTROL CHART TECHNIQUE[a]

Numbers in sample (n)	d_n	$D_{0.025}$	$D_{0.001}$	$D'_{0.025}$	$D'_{0.001}$
2	1.128	5.62	8.24	2.81	4.12
3	1.693	5.09	6.99	2.17	2.99
4	2.059	4.99	6.65	1.93	2.57
5	2.326	5.00	6.52	1.81	2.36
6	2.534	5.02	6.47	1.72	2.22
7	2.704	5.06	6.45	1.66	2.12
8	2.847	5.11	6.45	1.62	2.04
9	2.970	5.14	6.46	1.58	1.99
10	3.078	5.19	6.47	1.56	1.94

[a] From Herdan (H2). (Courtesy Elsevier Publishing Co., Amsterdam.)

book, gives factors used in the quality control chart technique. Of these d_n, $D_{0.025}$, and $D_{0.001}$ are not directly involved in this example, since they are concerned with cases where control limits for the standard deviation are to be computed. By multiplying the average range ($\bar{w} = 1.64$) by the factors $D'_{0.025}$ and $D'_{0.001}$ the respective upper $1/40$ and $1/1000$ control limits can be obtained. These are 3.17 and 4.22. The lower control limits will be zero in both cases. As can be seen in the column headed "After 5 min.," all of the ranges fall within these limits. Thus, the quality control chart has shown that the present mixture is "in control" within the prescribed limits.

Adams and Baker (A1) have illustrated how other kinds of quality control charts were used to evaluate dry blending equipment. Since these charts were based on the Poisson distribution, a discussion of them will be postponed until theoretical distributions have been covered.

7. Remarks on Statistical Techniques

It should be stressed that the above discussion is merely an introduction to some of the many available statistical techniques. To cover them thoroughly would require considerable additional background material and many more examples. It is hoped, however, that the simple examples given will stimulate both investigation of the possibilities of this field by those interested in solids mixing, and particularly, discussions of its applications between the mixing engineers and statisticians. Some comments concerning cooperative work between the engineer and the statistician appear in a recent article (M3).

The choice of "5%" level of significance or of "95%" confidence limits is an arbitrary one used for illustrative purposes. Although these criteria are frequently employed, other values can also be used and may be more suitable for specific applications. The Discussion of the paper by Adams and Baker (A1) includes some material on this topic.

8. Choosing a Degree of Mixing

The many varied approaches in different kinds of mixing studies preclude giving one single measure of mixing which will be ideal for all cases. Weidenbaum and Bonilla (W2) have mentioned several considerations which will affect the degree of mixing.

These include (1) unit size of the end product; (2) whether a little too much of one component in the end product would be undesirable, even if over-all variation between samples were low; (3) whether a random mixture of several solids or a coating of one solid with another is to be achieved; and (4) whether composition gradients within the mixer due to segregation are to be determined. As these authors have indicated, each investigator must ask whether the degree of mixing for a particular case is arbitrary, whether it is truly suitable for the problem at hand, and also whether the most efficient use is being made of the information contained in the spot samples. They further state (p. 31J of reference W2): "When any degree of mixing is used, complete details of sampling procedure should be reported, including size, number, and location of samples, method of their removal, fraction of mixer contents removed, and method of expressing compositions."

Although the above comments may seem obvious, some examples from the literature will indicate the value of emphasizing them. Weidenbaum (W1) has discussed a statement made by Coulson and Maitra (C3)—comparing their method of computing degree of mixing with that used by Hixson and Tenney (H4)—and has pointed out the dangers inherent in comparing degrees of mixing which were meant for entirely different purposes. Coulson and Maitra estimated the per cent mixed by withdrawing approximately thirty samples of about 150 particles each from different positions in the mixing drum, examining the samples with an eye lens, and so determining how many of them had "approximately the same composition as in the whole system." Call this number n. The mixture was then defined as $n/30 \times 100$ per cent mixed. The authors gave no further details on what was meant by "approximately the same composition as in the whole system." Concerning their method of sampling, they stated (M2): "Since the theory relies on a statistical examination of the mixture, 30 samples are insufficient to give adequate accuracy. In general, three tests were carried out under as far as possible identical conditions and in this way the mean of 90 readings could be obtained for the extent of mixing at any time." Coulson and Maitra then stated that their method of expressing degree of mixing is to be preferred "to the mixing index of Hixson and Tenney (H4) where only a small number of comparatively large samples are taken."

Coulson and Maitra's degree of mixing was intended to measure extent of mixing in a batch of solids which were sampled at intervals, starting with the partially mixed loaded state. The Hixson and Tenney mixing index was used to give a measure of the degree of uniformity of a mixture of solids and liquids which were in a state of dynamic equilibrium due to agitation. For the case where liquid is present in excess, their degree of mixing is defined as per cent mixed $= S/S_0 \times 100$, where S and S_0 represent the per cent by weight of a solid in a sample and in the total mixture, respectively. The average of the percentages mixed, of a number of samples taken from the vessel, was considered the measure of the degree of uniformity of mixture.

Further discussion of these two cases indicates the danger of overall comparison of degrees of mixing when such indices are used for different purposes. In Hixson and Tenney's studies, assumptions made as to system symmetry, due to use of a centered, top-entering agitator in a flat-bottomed cylinder, led to taking all of the samples on one side of the agitator shaft. Solids might be expected to concentrate near the bottom when the agitator did not keep them uniformly dispersed. Segregation would occur if the agitator were stopped. The mechanics of taking a sample without too much disturbance of the "suspension regime," and the method of analysis, presented problems peculiar to this type of system. In Coulson and Maitra's work, a sloping, particulate solids mixture in a drum was being studied. System symmetry with respect to mixing patterns of the different components was not obvious nor were the concentration gradients known, should segregation occur. The materials being mixed would not tend to rapidly segregate when the mixer was stopped.

The removal of small samples without too much disturbance of the mixture was possible, via the very useful sampling thief which Coulson and Maitra devised.

The ambiguity of the term "approximately" in Coulson and Maitra's definition was pointed out above. The Hixson and Tenney mixing index uses an average of the per cent mixed, based on samples at definite locations. Thus the value of this measure of mixing will be heavily dependent upon the location of the samples. A very pertinent comment made by Chilton (C1) in the discussion of the Hixson and Tenney paper (H4) states: "In regard to the 'mixing index' used by the author, would not this figure come to 100% if enough samples were taken? Would not a factor which would represent the deviation of the samples from the average be a better measure of the uniformity of the suspension produced by the agitator?"

Another example illustrating the need for clarity is a discrepancy in an illustrative problem given by Brothman *et al.* (B5). After deriving a rate equation theoretically, they illustrated its use by means of a blending problem, which was stated as follows: "Consider a conical tumbler blender in which it is proposed to blend 5 cu. ft. of carbon black with 3 cu. ft. of calcium carbonate. A fair sampling of the contents at the end of ten spins of the mixer device indicates that where one-cubic-inch samples are withdrawn, 0.2 of the samples contain a minimum of 1 mg. of calcium carbonate." Without considering the rest of the problem, reflection on this statement will show, that if originally three-eighths of the mixture (by volume) were calcium carbonate, a fair sampling after ten spins would be expected to reveal that more than 0.2 of the one-cubic-inch samples withdrawn would contain a minimum of 1 mg. of calcium carbonate.

9. *Starting a Quantitative Study*

It is extremely difficult, as has been pointed out previously, to suggest a single ideal criterion for a degree of mixing. However, should someone be groping for a place to start a study of solids mixing involving quantitative measurements on spot samples, either the standard deviation or the variance is suggested for simplicity, efficiency in utilizing data, unambiguousness, and adaptability, if desired, to later mathematical manipulations. The sample mean should also be computed, and information on sampling and analysis should be clearly stated.

D. Literature Summary: Degrees of Mixing

A variety of methods have been used to determine and express the degree of mixing of a solids mixture. These are summarized below together with short resumés of the papers from which they were taken. In order to facilitate reference to the original papers, each method for expressing degree of mixing is given in the original author's own terminology, which is defined when it is introduced. The general format to be followed is: author, description of contents of paper, and method used to determine and express degree of mixing or term for evaluating uniformity of mixture. In cases where the definition of degree of mixing has not been separately published by the original author but is mentioned in someone else's published paper, it will be listed after the latter although the original author will be given.

1. *Oyama*

(Note: Because much of this work is in Japanese, it will be given a more detailed discussion than some of the other papers, which are more easily accessible to those who only read English. Some of the material is taken from the thesis of Weidenbaum (W1) which includes a translation by Chai Sung Lee of Oyama's photometric method for measuring degree of mixing.)

a. In a series of papers dealing with "the motion of granular or pulverous materials in a horizontal rotating cylinder," Oyama covered the following subjects:

1. Relations between rpm of the cylinder and the state of motion of the materials (O1)
2. The types of mixing and the determination of mixing velocity (O2)
3. The effect of the cylinder diameter on the state of the motion of materials (O3)
4. The effect of a flight [baffle] on the state of the motion of materials (O4)
5. Packing and mixing of binary system (O5)—(Note: previous work had been done with particles of uniform size and shape.)
6. Theoretical study of motion of granular materials (O6)
7. Study of power consumption (O7)

A paper entitled "Studies on Mixing of Solids: Mixing [in a] . . . Binary System of Two [Particle] Sizes by Ball Mill Motion" (O8) is available in English and covers some material from (5) above as well as fragments from some of the other works.

b. For Mixing Particles of Uniform Size and Shape but Differently Colored. In work with particles of uniform size and shape, Oyama used photographic methods to determine what he called the mixing grade or "mixing effect." In his procedure, light was passed through a film obtained by photographing the end of the cylinder as mixing proceeded (O2). The light went through a properly sized slit to a photoelectric bulb which generated a current. This current actuated the fiber of a fiber-potentiometer and the motion of the fiber was recorded on a moving-picture camera. Thus it was possible to record a curve which was related, as explained above, to the changes in what Oyama called the "photographic density," as mixing proceeded. A formula is given for the intensity of light (i) received by the film from a group of particles:

$$i = \frac{1}{t}\left(\frac{o}{o'}\right)^{\frac{1}{\gamma}} \tag{19}$$

where t = time of exposure
    o = degree of obscureness (or absorbity)
    o' = degree of obscureness (or absorbity) of the original fog on the negative
    γ = a constant depending on the film and developing conditions (given as 1.53 by Oyama).

Thus, i can be measured and, as will be shown, is used to obtain a number which, in turn, can be utilized in an expression for the degree of mixing. The graph of i vs. different portions of the negative (gotten by photographing the cylinder end) has two main parts before mixing: one due to the black particles which have been loaded on the bottom, and one due to the white layer on top; a sharp jump occurs in the graph as the light passes through the borderline between the two layers. This is

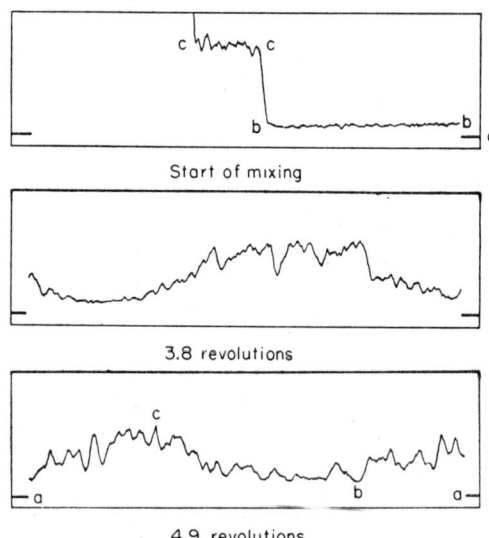

FIG. 10. Tracings of photographs from Oyamo (O2), showing change in pattern of

shown in Fig. 10. As mixing proceeds, the difference between the two extremes will become smaller while the frequency of small peaks and depressions will increase (Fig. 10). Note that in the picture corresponding to 4.9 revolutions, the difference between b and c is less than that for the initial state. Also, the large number of small peaks and valleys as compared to the initial state indicates that alternate layers of the differently colored particles are beginning to form.

Oyama defined his degree of mixing as follows:

$$\text{Degree of mixing} = 1 - \frac{i_{\max.,c}}{i_{\max.,c_0}} \qquad (20)$$

where $i_{\max.,c}$ = light intensity, i, corresponding to maximum concentration difference, and $i_{\max.,c_0}$ = light intensity, i, corresponding to maximum concentration difference at the start.

Figures 11 and 12 show graphs of i vs. number of revolutions for two different speeds (16-rpm and 80-rpm). In each case, the upper graph is for c, the lower one for b, and the one in the middle represents $c - b$. This difference, divided by the initial difference (at 0 revolutions) and subtracted from 1, was Oyama's degree of mixing. Realizing that the frequency of peaks and valleys was also of importance in indicating the extent of mixing, Oyama stated that if the maximum concentration difference were the same, a greater frequency would indicate better mixing. He interpreted the change in frequency as indicative of positional

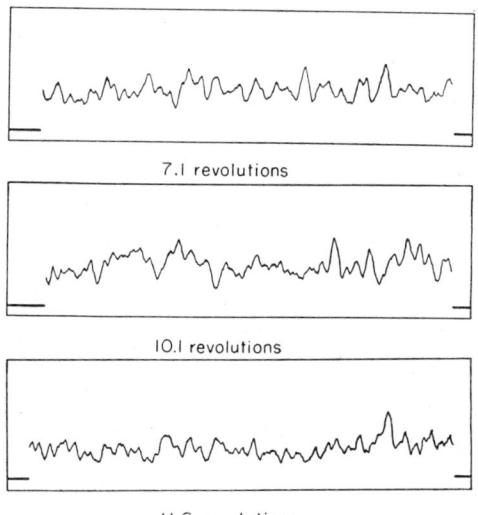

variation of light intensity i, across the cylinder end-face, as mixing progresses.

changes of particles throughout the mix, and called this the "uniform" or "total" mixing effect. The changes in maximum concentration difference can be attributed mainly to the slip, or friction among particles at localized positions, and this effect, referred to as "local" mixing, could

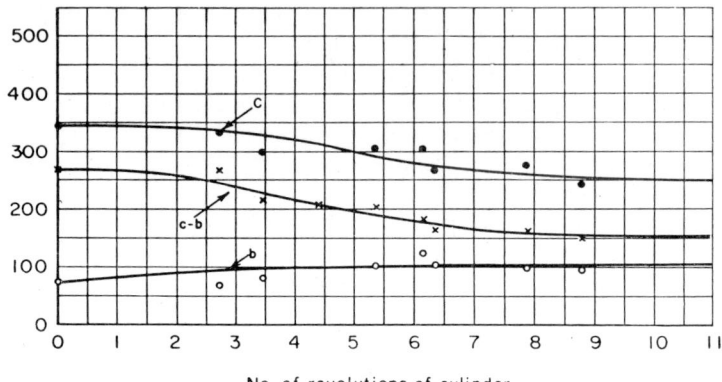

Fig. 11. Light intensity i, vs. number of revolutions. In Section III, D, 1b the procedure is described for using this graph to compute degree of mixing (O2). Cylinder speed = 16 rpm.

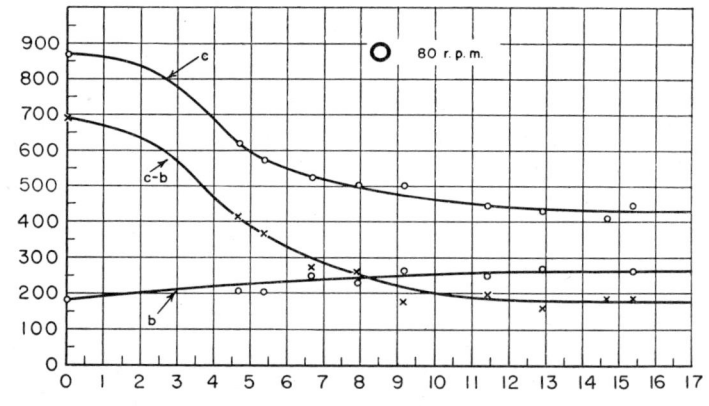

Fig. 12. Light intensity i, vs. number of revolutions. Cylinder speed = 80 rpm (O2).

be related to intensity. His degree of mixing is a measure of the latter type of mixing.

    c. For Mixing Particles of Two Different Sizes. Drawing on the background of packing studies provided by Furnas (F2, F3), Westman and

Hugill (W3, W4), and Kasai (K1), Oyama (O8) used the specific volume of the mixture in a term for degree of mixing for mixtures of two different particle sizes. His method involved a comparison of the specific volume of the mixed batch, with that obtained when the two materials were packed as closely as "practically attainable." The graph of specific volume *vs.* per cent composition by weight was computed for the latter case from experimentally determined data of Westman and Hugill, who

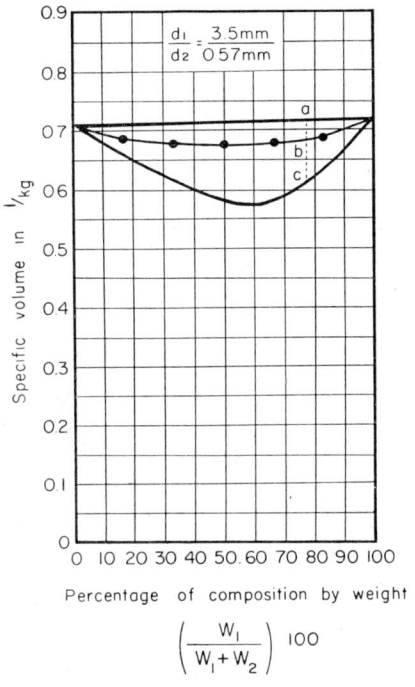

Fig. 13. Specific volume *vs.* per cent composition by weight. Used by Oyama (O5, O8) to compute per cent of mixing for a mixture of two different size particles.

used a specially designed machine to determine the minimum volume attainable in a mixture of particles of different sizes. Details of the calculation methods are given in Oyama's paper (O8) (in English). The graph of specific volume *vs.* per cent of composition by weight was also determined after mixing various weight ratios of the two materials in the cylinder. (Note: Although this is not stated in the paper, it is presumed that the figures are given for the batch after a long period of mixing, since only one figure is given for each weight per cent of W_1.) A third graph of specific volume *vs.* weight per cent, W_1, was drawn to correspond with the case where the two components were introduced separately into the cylinder without mixing; the total specific volume here would be the sum of the specific volumes of the two granular materials (i.e., apparent volumes).

These graphs are shown in Fig. 13. The per cent mixing for any weight ratio $W_1/(W_1 + W_2)$ was found as follows: (1) Determine the difference ($= ac$) between "non-mixed" specific volume and "closest packed" specific volume; (2) Determine the difference ($= ab$) between "non-mixed" specific volume and "mixed" specific volume; (3) Per cent mixed or "grade of mixing" $= ab/ac \times 100$.

2. *Beaudry* (B3)

The author presents a means for determining how well a continuous blender is operating, as compared to how well it is capable of operating. This takes into account the type of blending cycle used, the ratio of blender volume to batch volume, and for certain cycles, the ratio of inflow rate to outflow rate.

$$\textit{Blending Efficiency (B.E.)} = \frac{[(V_b/V_p)_{\text{actual}} - 1]}{\gamma - 1} \times 100 \qquad (21)$$

where for $(V_b/V_p)_{\text{actual}}$:

V_b = variance among batches before blending

V_p = variance among batches actually obtained after blending.

Variance is defined as $\dfrac{\Sigma(c_i - c_{\text{av.}})^2}{N}$

where

$c_{\text{av.}}$ = the average value of property A
c_i = the value of property A for the i-th batch
N = the number of batches sampled
γ = limiting blending ratio = V_b/V_p, where V_p in this case is the variance with perfect blending, which is computed theoretically taking into account the factors mentioned above.

3. *Lacey* (L1)

Early fundamental work on the degree of mixing was performed by this author, and involved very laborious counting of all particles (0.2 in. diam.) in the mixture. Experimental work was carried out.

$$M = \sqrt{\frac{\bar{\alpha}\bar{\beta}}{n}} \qquad (22)$$

where $\bar{\alpha}$ = fraction of A particles in a mixture of A and B
$\bar{\beta}$ = fraction of B particles in a mixture of A and B
n = number of particles in a cell (mixture completely divided into cells).

Plot M vs. $1/\sqrt{n}$: "Very probably a family of curves of characteristic shape would be obtained for each combination of mixing machine and materials" (L1).

4. *Buslik* (B10)

This paper deals with the derivation and application of a formula for computing standard deviation (σ_g) of the per cent of a given size in random samples from a granular material with a known size distribution:

$$\sigma_g{}^2 = \frac{G(100 - G)\bar{w}_g + G^2(\bar{w} - \bar{w}_g)}{W} \tag{23}$$

where $\bar{w}_g$ = average particle weight in the size fraction being considered. A range of sizes is actually included, but this size fraction is called the w_g size. (Constant density is assumed.) Methods for computing $\bar{w}_g$ to various degrees of precision are given by Buslik.

G = per cent by weight of the mixture of particles of size w_g
W = total sample weight
$\bar{W}$ = average weight of all the particles of all sizes in the entire mixture. Calculated from

$$\bar{W} = \tfrac{1}{100}(AW_a + BW_b + \cdots + GW_g + \cdots)$$

where $A, B, C, \ldots$ are the percentages by weight of the sizes $w_a, w_b, w_c, \ldots$ in the mixture.

5. *Bonilla and Crownover, and Bonilla and Goldsmith*

Results from these unpublished theses were mentioned by Weidenbaum and Bonilla (W2).

Reports of tests on a batch mixer with horizontal screw-shaped blades of opposite pitch, moving material in a trough with an approximately semi-cylindrical bottom. Graphs of $-\log M$ vs. $\log r$ made under different operating and loading conditions (r = number of revolutions of the mixer):

$$M = -\log \frac{a.d.}{2c(1-c)} \tag{24}$$

where $a.d. = \dfrac{\sum\limits_{i=1}^{N} |c_i - \bar{c}|}{N}$, which equals the mean deviation of the weight fractions of the "key constituent." (About five 5-gm. samples were taken.)

c_i = weight fraction of key constituent in the i-th spot sample
$\bar{c}$ = arithmetical mean value of c for the N spot samples
N = number of spot samples
c = weight fraction of the key constituent in the entire batch.

6. *Brothman, Wollan, and Feldman* (B5)

These workers are mainly concerned with developing a theoretical rate equation for mixing, and interpreting this equation. No experimental data are given.

P_t = proportion of v units of volume which contain some specified minimum amount of component A, where the desired goal is to effect a permeation of x units of A in a system of volume V. (Thus $v = V/x$.) The authors' rate equation is adjusted to take into account the effect of taking samples of a different size than v (say size V_0). The proportion of the total number of cubes into which the mixture has been divided, containing at least one element of the surface of separation, is used in developing the rate equation; but the above definition, in terms of the specified minimum amount of one component, is necessary for practical use of the equation experimentally.

7. *Coulson and Maitra* (C3, M2)

Experiments were performed with an inclined rotating drum to determine the effect of several equipment and material variables on rates of mixing and segregation tendencies. Also a rate equation was postulated and tested with the experimental data.

X, the percentage unmixed, was used in the rate plots. It was experimentally determined as follows: 30 samples were withdrawn from different positions in the drum and examined with an eye lens. If n samples were found to have "approximately the same composition" as the overall mixture, then the per cent mixed was defined as $n/30 \times 100$, and $X = (100 -$ per cent mixed$)$.

8. *Visman and Van Krevelen* (V1)

The authors replotted Coulson and Maitra's data on probability paper. They preferred this method of plotting to that given by Coulson and Maitra.

$100 - Z$ = percentage unmixed; it is experimentally determined as follows: A number of samples are removed. If a sample does not show any visually observable mixing when examined with a magnifying glass, it is considered unmixed. The number of these unmixed samples divided by the total number of samples and multiplied by 100, is $100 - Z$.

" . . . samples which show visible mixing are defined as samples with more than 10% and less than 90% coal or salt. The samples which do not show visually observable mixing are thought to contain less than 10% coal or salt."

9. *Stange* (S5)

A formula is presented for calculating the theoretical variance of samples taken from a randomly mixed batch of two materials (P and Q), each of which has its own size distribution:

$$\sigma_z{}^2 = \frac{PQ}{g}[P\bar{\nu}_Q(1 + C_Q{}^2) + Q\bar{\nu}_P(1 + C_P{}^2)] \qquad (25)$$

where P and Q are fractions by weight of the mixture components. $\bar{\nu}_P$ and $\bar{\nu}_Q$ are average particle weights for P and Q. These are measures of the *degree* of fineness of the particles. C_P and C_Q are variation coefficients for P and Q, referring to the frequency distribution of particle weights for these substances. (Note: variation coefficient equals standard deviation divided by the mean.) These are measures of the *uniformity* of fineness of the particles. g is the sample weight.

10. *Stange* (S4)

Graphical methods are presented for computing confidence limits for standard deviations that are used in plotting rate-of-mixing curves. The author gives an example of the method's use in comparing graphs of throughput *vs.* power consumption for different mixers. He also shows how to graphically compute sample size for a prescribed confidence range. (See Section III, C, 5d.)

11. *Danckwerts* (D2, D3)

In one paper (D2), Danckwerts discusses methods of measuring "goodness of mixing" which will be described. In another (D3), he gives a more general discussion on the scale of scrutiny of mixtures, and also discusses other mixing topics, which include minimum work required for mixing, scaling up from models, mechanisms of mixing, and back-mixing in continuous flow systems. His papers cover other systems besides solid-solid mixtures.

Danckwerts expresses the "goodness of mixing" by two statistically defined quantities, the *scale* and the *intensity of segregation*. He states that his treatment is suitable chiefly for mixtures where the smallest particles capable of independent movement are very small compared to the size of the portions which will normally be taken for use or analysis. For his analysis, he assumes that the mixture is uniform in texture; that is, it cannot be divided into two parts of equal size in which the mean concentration or the scale or intensity of segregation differ significantly. He further states: "This is the most important limitation on the practical value of the definitions and tests which will be proposed." He emphasizes the fact that "large scale segregation, caused for example by sedimentation, or dead space in a mixer, is of great practical importance but its study cannot conveniently be combined with that of the small-scale characteristics, or texture, . . . "

He considers that mixing occurs by two independent processes which produce distinguishable results: (1) the breaking-up process which reduces sizes of clumps and (2) the "interdiffusion" process which obliterates differences of concentration between neighboring regions of

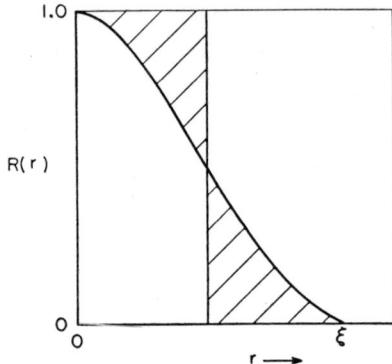

FIG. 14. Illustrative graph of correlogram (D2). Shaded areas are equal.

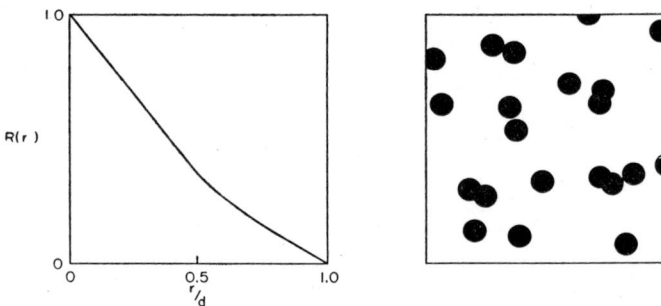

FIG. 15. Correlogram. Diameter of circles is d; $S/d = 0.42$. An analogous random collection of spheres of diameter, d, will give $S/d = 0.38$ (D2).

the mixture. He therefore postulates two quantities for describing the degree of mixing—namely the "scale of segregation" and the "intensity of segregation," both of which he defines by statistical methods. (Note: He generalizes these methods for mutually soluble liquids, *fine powders*, or gases.)

a. Scale of Segregation. Danckwerts considers the scale of segregation as analogous to the "scale of turbulence" used in the statistical theory of turbulence. He defines it as:

$$R(r) = \frac{\overline{(a_1 - \bar{a})(a_2 - \bar{a})}}{\overline{(a - \bar{a})^2}} = \frac{\overline{(b_1 - \bar{b})(b_2 - \bar{b})}}{\overline{(b - \bar{b})^2}} \quad (26)$$

where a_1 and a_2 (and b_1 and b_2) are concentrations measured at two points in the mixture a distance r apart, and $\bar{a}$ and $\bar{b}$ are the mean concentrations of a and b in the mixture as a whole. $R(r)$ is called the coefficient of correlation between the values of a (or b) at points separated by a distance r.

He then interprets the graph of $R(r)$ vs. r, known as a *correlogram*, pointing out how, as r increases, the correlation coefficient $R(r)$ will drop towards zero if there is no large scale segregation or regular periodicity in the mixture. Figure 14 illustrates a "correlogram," and correlograms for certain cases are shown in Figs. 15 and 16. Note that in the case of

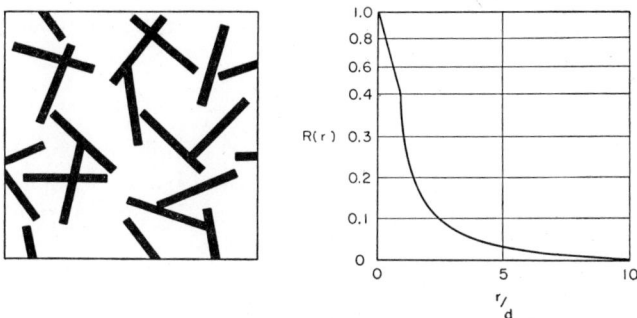

FIG. 16. Correlogram. Width of strips is d, length of strips is $10d$; $S/d = 1.1$ (D2).

the circles (Fig. 15), when the distance between points r, is equal to the diameter of the circle d, i.e., $r/d = 1$, the correlation coefficient has dropped to zero. In the case of the elongated strips (Fig. 16), when the distance between points, r, is about two to three times the width of the strip, the correlation coefficient drops sharply to less than 0.1; when r is equal to the length of the strip, it drops to zero.

Danckwerts mentions two measures of scale of segregation, (1) the *linear scale*, S, defined as the area under the correlogram:

$$S = \int_0^\infty R(r)dr = \int_0^\xi R(r)dr \quad (27)$$

where ξ = value of r for which $R(r)$ falls to zero; and (2) the volume scale, V, defined as 2π times the area under the curve of $r^2 R(r)$ vs. r:

$$V = 2\pi \int_0^\infty r^2 R(r)dr = 2\pi \int_0^\xi r^2 R(r)dr \quad (28)$$

The relationship between S and V depends on the shape of the clumps and hence of the correlogram. For a random collection of spheres of diameter, d, analogous to Fig. 15, the V/S ratio is 4.7. In order to use the above theoretical relationships it is necessary to measure the scale of segregation. Danckwerts gives several ways of doing this and the computational methods for each. These involve (1) measuring concentration at a large number of points, (2) in batch-mixing, measuring the concentrations at two fixed points continuously while the mixture moves past them, (3) measuring the total content of one of the components along a straight line joining two points, (4) determining V by analyzing a number of samples of volume v taken from the mixture, and (5) other methods adaptable to continuously measuring S, as a fluid mixture flows through a pipe.

A familiar experimental technique used in solids mixing is method 4, which involves analysis of a number of samples of size v. Danckwerts computes the volume scale of segregation V, from this analysis as follows:

$$V = \frac{\sigma_K^2}{2v\sigma_a^2} \qquad (29)$$

where σ_K^2 is the variance among the spot sample measurements (the content of A in each sample of size v is denoted by K). σ_a^2 is the variance of point concentrations of component A from the over-all mean concentration $\bar{a}$. This requires determination of point concentrations by special techniques, of which Danckwerts discusses a few (electrical, optical, following the progress of a chemical reaction). Although the ideas are very interesting, their adaptation to a dry solids mixing process would require considerable investigation of practical means of measuring point concentrations in a solids mixture. Of particular importance in solids mixing, the measuring techniques must not so disturb the mixing patterns, that a picture is obtained which is not truly representative of the mixer in normal operation without the interference of measuring instruments.

b. *Intensity of Segregation.* This is defined as:

$$I = \frac{\sigma_a^2}{\bar{a} \cdot \bar{b}} \equiv \frac{\sigma_b^2}{\bar{a} \cdot \bar{b}} \equiv \frac{\sigma_a^2}{\bar{a}(1-\bar{a})} \equiv \frac{\sigma_b^2}{\bar{b}(1-\bar{b})} \qquad (30)$$

where σ_a^2 is, as previously mentioned, the point to point variance in concentration of a.

Although Danckwerts presents I "*in toto,*" it is actually a ratio of variances.

σ_a^2 is the variance mentioned above and $\bar{a} \cdot \bar{b}$ is the variance of the unmixed materials. ($\bar{a}$ and $\bar{b}$ are respectively the mean concentrations of a and b in the mixture as a whole.) Danckwerts states: "In general, I reflects not the relative amounts of A and B, nor the size of the clumps, but the extent to which the concentration in the clumps departs from the mean. If B is present in large excess, the value of I will depend primarily on the extent to which the clumps of A in the mixture have become diluted by B." The methods for measuring $\sigma_a{}^2$ have been discussed previously. In addition, Danckwerts gives an alternate method involving measurement of the chemical reaction rate which might be useful under certain conditions of mixing liquids A and B.

12. *Herdan* (H2)

In his chapter on "Industrial Mixing" in *Small Particle Statistics* (H2), Herdan covers ways of measuring degree of mixing, including the quality control chart. He also gives a discussion and explanation of the paper by Brothman, Wollan, and Feldman (B5).

Methods for using quality control charts, for both standard deviation and range, are discussed and illustrated with reworked data from Lowry's report (L4). This is discussed in Section III, C, 6. Numbers 13 and 14 are also mentioned in Herdan (H2) but no reference is given.

13. *Lexis*

Herdan states that the ratio given "was first introduced by Lexis for the purpose of ascertaining whether a universe from which samples had been taken was homogeneous, or whether the probability of the event in question differed significantly from place to place, or from time to time." (H2)

$Q = R/r$, where R = observed standard deviation, and r = theoretical standard deviation, calculated under the assumption of complete randomness.

14. *Charlier*

Herdan states that ρ^2 "was designed for the same purpose as Lexis' Q, viz., as a measure of heterogeneity in a denumerable population."

$$\rho^2 = \frac{S_t{}^2 - \sigma^2}{M^2} \qquad (31)$$

where $S_t{}^2$ = observed variance at time t
σ^2 = theoretical variance for a random mix
M^2 = mean number or mean proportion of the component in question
ρ^2 = coefficient of perturbation or disturbance.

15. *Kramers*

The work of this author was not published separately but mentioned by Lacey (L2). (See 16 below.)

Here
$$M = \frac{s_0 - s}{s_0 - s_r} \tag{32}$$

where s_0 = theoretical value of s for unmixed material = $\sqrt{P(1-P)}$.
[Note: Lacey (L2), p. 259, gives $P(1-P)$, but there should be a square root sign, as shown.]

s = standard deviation among samples of size n

s_r = theoretical standard deviation for the completely random mixture ≡ $\sqrt{P(1-P)/n}$, where n = number of particles per spot sample.

16. Lacey (L2)

This author discusses several papers on various aspects of solids mixing, aiming to coordinate existing ideas, develop the theoretical aspects, and show their relationship to the few systematic practical investigations that have been considered.

He uses variance instead of standard deviation in Kramers' type of degree of mixing (see 15). This gives:

$$M = \frac{s_0^2 - s^2}{s_0^2 - s_r^2} \tag{33}$$

which Lacey states " . . . is more satisfactory statistically, since s^2, unlike s, has additive properties." He has further developed the expression to:

$$1 - M = \frac{s^2 - s_r^2}{s_0^2 - s_r^2} \tag{34}$$

which he called " . . . a fundamental equation for expressing the state of a mixture."

17. Oyama and Ayaki (O9)

Using a horizontal rotating cylinder and several different sand-sand systems, Oyama and Ayaki studied rate equations for mixing and the effect of rotational speed and the volume ratio of sand to mixer. Different methods of loading were also tried.

Variance of the spot sample compositions, denoted by σ^2, was called the degree of mixing of the whole mixture. Like Lacey (L2), Oyama and Ayaki considered this to be composed of two parts, σ_m^2 and σ_r^2. σ_m^2 is the variance due to local composition variations which would be independent of sample size, and σ_r^2 is the variance due to sample size.

When σ^2 is used in a rate equation and integrated to give

$$\ln(1-M) = -\phi t$$

M is the same mixing index defined by Lacey. ϕ is the coefficient of mixing velocity.

18. *Weidenbaum and Bonilla* (W2)

A fundamental study of solids mixing is given, with experimental data for a rotating horizontal cylinder containing salt and sand. Theoretical random equilibrium mixture, chi-square test, and sampling considerations are discussed. Also rate equations and segregating effects are covered.

$$\text{The degree of mixing} = \frac{\sigma}{s} \tag{35}$$

where $\sigma = \sqrt{p(1-p)/n}$, and p is the true particle fraction of sand in the mixture, or the mean of the binomial distribution

$$s = \sqrt{\frac{\sum_{i=1}^{N}\left[\left(\frac{x}{n}\right)_i - \left(\frac{\bar{x}}{n}\right)\right]^2}{N}}$$

$(x/n)_i$ = particle fraction of sand in the i-th spot sample
$(\bar{x}/n)$ = arithmetical mean particle fraction of sand for the N spot samples
n = total number of particles in a spot sample
N = number of spot samples taken from the mixture at any spot sampling.

Graphs of s vs. number of revolutions are plotted. Also, the use of the usual *chi-square* test is illustrated for comparing distributions to determine whether a batch is randomly mixed. In addition, a table is given for classifying mixtures according to the relative frequency of occurrence of chi-square values.

19. *Blumberg and Maritz* (B4)

Theoretical concepts of complete and incomplete mixing are discussed from a statistical point of view. Experimental data are given for mixing two differently colored, but otherwise identical batches of sand. Topics that are covered include the *chi-square* test and the number of samples needed to indicate completion of mixing.

For testing whether a batch is *completely mixed*, the usual *chi-square* test is used. However, for measuring the degree of mixing of an *incompletely mixed* batch, the following term was used:

$$\phi = \nu \sum_{i=1}^{k} (z_i - \zeta_0)^2 \tag{36}$$

where ν = the average number of particles per spot sample
$z_i = 2 \arc \sin \sqrt{x_i}$
$x_i = r/n$ which equals the fraction of red particles in a spot sample of n particles
$\zeta_0 = 2 \arc \sin \sqrt{0.5}$, because the fraction of red particles in the entire mixture is 0.5.

Since ϕ is also a chi-square variable, the probability that a value as large as the computed value of ϕ would occur due to chance can be looked up in chi-square tables. If this is less than the level that has been set (say one time in twenty), then mixing is considered to be incomplete. If it turns out, however, that this large a value of ϕ would occur by chance more than one time in twenty, conclude that mixing is complete, i.e., a randomly mixed batch has been obtained.

20. *Michaels and Puzinauskas* (M4)

This paper describes a study of the mixing of powdered dextrose with water-wet kaolinite in a small finger-prong mixer. The rate of homogenization, and the energy consumption during mixing are determined for clay-water systems of varying water content and total volume, other conditions being held constant. Optimum mixture volumes are given for rapidity of mixing and energy utilization.

$$I_v = \frac{D_v}{D_{v_0}} \sqrt{\frac{\sum_0^n (C_A - C_A^m)^2}{n(1 - C_A^m)(C_A^m)}} \tag{37}$$

where I_v = Uniformity Index

$$D_v = \sqrt{\frac{\sum_0^n (C_A - C_A^m)^2}{n(C_A^m)^2}}$$

D_{v_0} = value of D_v at no mixing, which can be shown to equal $\sqrt{1 - C_A^m/C_A^m}$

C_A = concentration of A in a random sample of volume v

n = the number of samples

C_A^m = the mean concentration of A in the mixture.

(Note: In their work, Michaels and Puzinauskas use $n = 6$ and $v = 1$ cc.)

21. *Adams and Baker* (A1)

Graphical and other methods are given for evaluating dry-blending equipment, by making use of accepted statistical laws. The authors discuss the effect of changes in the relative proportions of the two ingredients on variations in composition of the final mixture. Also presented, are experimental data comparing four types of blenders.

Three types of tests are discussed. The first two tests require graphs of the number of blacks vs. the sequence number of samples (the mixture consists of black and natural polythene colored particles). One test looks for the number of samples outside of certain control limits or confidence lines; and the other looks for the number of consecutive samples on one side of the mean. The third test requires computing $b = s^2/\bar{x}$, where $s^2 = \dfrac{\Sigma(x - \bar{x})^2}{N - 1}$; x is the number of black granules in a sample, and N is the number of samples. From a graph, the probability of obtaining this large a value of b from a Poisson distribution is obtained. This probability is used as a measure of blending in a manner analogous to that of a statistical test of significance.

22. *Smith* (S3)

Quantitative data are shown in this paper to emphasize the necessity for good mixing in treatment of soils with additives. The mixing machines, necessary to achieve this aim are also discussed.

$$R = \frac{\sigma_0}{\sigma} = \left[\frac{n\mu_a(1 - \mu_a)}{\sum_{i=1}^{n}(x_i - \mu_a)^2} \right]^{0.5} \quad (38)$$

where n = number of spot samples taken from a mix containing a mean fraction of additive μ_a
x_i = indicated content of any sample
σ_0 = theoretical standard deviation at "zero" mixing = $\sqrt{\mu_a(1 - \mu_a)}$

σ = actual standard deviation at any time = $\sqrt{\dfrac{\sum_{i=1}^{n}(x_i - \mu_a)^2}{n}}$.

23. *Gray* (G2)

The results of tests on the mixing performance of some of the common types of dry-solids mixing equipment are described by this author. A specially constructed reflectivity probe, with a light and photocell behind a glass window, is used. The intensity of light reflected from a layer of particles outside the glass window of the probe gives, via the photocell meter, an indication of the composition at the point in front of the glass window, since light and dark particles are mixed. The standard deviation s, of probe meter readings taken at several spots in the mixture, is used as a measure of the uniformity of composition.

Bullock (B8) later pointed out that the reflectivity probe is limited in its usefulness because it can not sufficiently well differentiate between mixes containing uniformly dispersed small agglomerates, and mixes in which these agglomerates have been broken down into ultimate particles.

24. *Yano, Kanise and Tanaka* (Y2)

These investigators studied the mixing of anhydrous sodium carbonate and polyvinyl chloride powders, of $-100/+200$ Tyler mesh, in two V-type mixers of ¼-liter and 2-liter working capacity, respectively. The variables include rpm of the mixer, volume per cent loaded, volume ratios of the feed, and method of loading.

The degree of mixing used by Yano et al. is the standard deviation, defined as follows:

$$\sigma = \sqrt{\frac{1}{n}\sum_{i=1}^{n}(C_i - C_0)^2} \qquad (39)$$

where C_i is the volume fraction of Na_2CO_3 in the i-th spot sample, and C_0 is the known volume fraction of Na_2CO_3 in the over-all mixture.

25. *Sakaino* (S1)[2]

An expression is developed for computing the mean size z, of randomly dispersed agglomerates, from the variance S^2, obtained with a certain sample size, R. The author postulates a rate equation for solids mixing, which he feels may also be applicable to the homogenizing process in glass melting.

The degree of mixing is defined as μ,

where
$$\mu = \frac{1}{z} = \frac{pq}{S^2 R} \qquad (39a)$$

p is the particle fraction of one type of particle in the mixture, and q is $1 - p$. S^2 and R are defined above.

E. SUMMARY AND CONCLUSIONS—STATE OF MIXEDNESS OF A BATCH OF SOLIDS

1. The day to day control of industrial mixing operations to obtain a desired product, with properties or composition within specified limits, is a frequently occurring problem. A variety of statistical techniques are

---

[2] Prof. Sakaino kindly furnished a more detailed summary than was available in the English Abstract of (S1).

available for quantitatively determining the adequacy of the mixing operation. Included among these are: statistical tests of significance, confidence limits, and quality control charts. These can be adapted to both quantitative and qualitative measurements on spot samples.

2. There can be considerable variations from one process to the next in the required scale of uniformity. Choice of a suitable standard for a degree of mixing should take this into account.

3. To avoid ambiguity, complete details of sampling procedure should be available when spot samples are removed to measure a degree of mixing.

4. The most intelligent use of statistical techniques is made by combining them with a sound knowledge of the mixing operation and the purpose of its end use.

5. A variety of methods from the literature, for determining and expressing the degree of mixedness of a batch of solids, have been summarized. These vary from photometric methods, involving the use of a motion picture camera, to methods involving comparisons of sampling data with reference theoretical distributions.

6. A good general measure of the extent of mixedness is the standard deviation among spot samples removed from the batch. The sample mean should also be reported.

IV. Theoretical Frequency Distributions

A. Mixing—Orderly or Random?

For a long time, the picture of mixing shown in Fig. 17, had been widely accepted as a guide to ideas on mixing. Such a picture would indicate that all spot samples taken from the mixed batch should give the same exact composition (or other measurement). If this were not the case, the variation between spot samples might be attributed to errors due to sampling or analysis; or perhaps to a feeling that the mixer was just not doing a perfect job, the latter statement implying that things were not quite as good as they could be. It has now been recognized and pointed out several times, that a mixture such as shown in Fig. 17(b) is not the equilibrium state of a mixing operation where tumbling motion causes the mixing. Instead, since random tumbling is involved, the mixing process is recognized as producing a random arrangement of particles, if no segregating effects are present.

The fact that the end result of a tumbling operation is not an orderly system was stated by Lacey in 1943 (L1). He considered a solids mixture from the point of view of individual particles, calling an "ideal" mixture of A and B one in which the degree of dispersal of the components is the one most likely to exist in a mixing machine after equilibrium has been established. By means of Figs. 18(a) and 18(b), he showed the difference between ordered and random arrangements of particles.

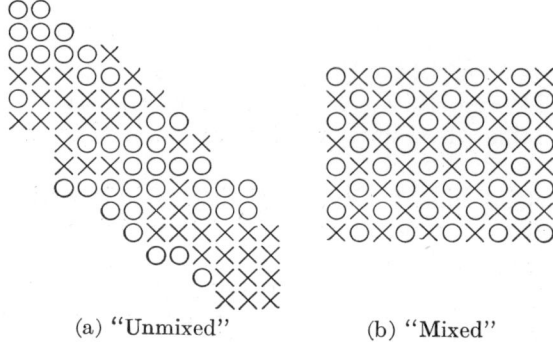

(a) "Unmixed"   (b) "Mixed"

FIG. 17. Illustration entitled "The Effect of Mixing" from (P1). (b) is not the end result of random mixing, since the probability of getting state (b) by a random tumbling operation is zero for all practical purposes.

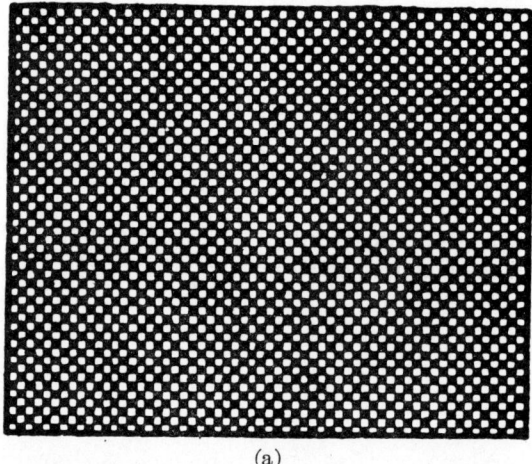

(a)

FIG. 18(a). Ordered arrangement of particles (L1).

Buslik (B10) stated that the term *random* does not describe a single specific arrangement but rather the way in which the arrangement was achieved. He showed how these ideas could be quantitatively developed to enable computation of the weight per particle necessary to achieve a certain specified uniformity of mixture. Also, he derived and illustrated the use of the formula shown in Section III, D, 4 for computing the standard deviation of the per cent of a given size in random samples from a granular material with a known size distribution.

Several others have dealt with various aspects of the development of this idea of a theoretical distribution of spot sample compositions from a randomly mixed batch. Before further discussion of the outstanding

Fig. 18(b). Random arrangement of particles (L1).

contributions of these investigators, the basic statistical concepts which underlie their work will be examined.

B. Binomial, Normal, and Poisson Distributions—General Comments

The distribution of spot sample compositions of a certain size, taken from a randomly mixed batch of A and B, can be calculated theoretically. The methods of calculation are standard statistical techniques, and several papers have shown how various aspects of these basic ideas can be applied to solids mixing. Most of the calculations and discussion center around three distributions: binomial, normal, and Poisson.

The physical picture is as follows: a batch of A and B particles have been mixed until equilibrium has been reached. There are no segregating or clustering tendencies; therefore, A is considered to be randomly distributed throughout the mixture. If this is true, then *the probability of finding a particle of type A at any point in the mixture is a constant equal to the proportion of that kind of particle in the whole mixture.* This key statement immediately identifies the distribution of spot sample compositions (expressed in terms of number of A particles or "particle fraction A'') as a binomial distribution, provided only a relatively small amount of material is removed for sampling. (Note: Rigorously, the multinomial hypergeometric distribution applies (W1), but if only a small amount of material is removed for sampling the assumption of constant probability is valid and for practical purposes the binomial distribution applies.) The general equation for the binomial distribution is:

$$P(x) = \frac{n!}{x!(n-x)!} p^x (1-p)^{n-x} \tag{40}$$

where $P(x)$ is the probability of obtaining x successes in n independent trials of an event, for which p is the probability of finding a success in a single trial. In the case of a random mixture of particles of A and B, n would be the total number of particles in each spot sample, p would be the fraction of A particles in the entire mixture, and $P(x)$ would be the probability of finding x particles of type A in a spot sample of n particles. $P(x)$ can also be thought of as the proportion of spot samples which will contain x particles of type A. The spot sample composition may also be expressed as "particle fraction A," rather than the number of A particles. In this case x/n is used instead of x, but the relation is not changed, i.e.,

$$P\left(\frac{x}{n}\right) = \frac{n!}{x!(n-x)!} p^x (1-p)^{n-x} \tag{41}$$

The mean of the distribution is p and the variance, σ^2, is $p(1-p)/n$. To compute values of $P(x/n)$, when n is fairly large, say 100, is a time consuming task; yet spot samples with an n of this order of magnitude are usual, even with a very small spot sample, and techniques for computing $P(x/n)$ are necessary for these cases. Therefore, approximations are used as follows:

When p is ≤ 0.5 and $np > 5$ at least, experience indicates (H5) that the normal distribution, which is a continuous function, is a good approximation to the binomial distribution, which is a discrete function. The normal distribution is:

$$f\left(\frac{x}{n}\right) = \frac{1}{\sigma \sqrt{2\pi}} e^{-\frac{1}{2}\left(\frac{\frac{x}{n}-m}{\sigma}\right)^2} \tag{42}$$

Since this is an approximation of the binomial distribution, the mean m, and standard deviation σ, are p and $\sqrt{p(1-p)/n}$, respectively.

Hoel (H5) states that "when p is very small, even though n is large, the normal approximation to the binomial distribution may be poor; consequently some other form of approximation is needed." This would cover the case when $np < 5$. The Poisson distribution function is such an approximation and is defined as:

$$P\left(\frac{x}{n}\right) = \frac{e^{-m}m^x}{x!} \quad (43)$$

where $m = np$.

With the above three distributions, it is possible to compute theoretical frequency distributions for the composition of samples of size n, for the case where A is randomly mixed with B. As will be shown in the following discussion, this basic theme has been developed in several different directions.

C. Binomial and Normal Distributions: Use in Defining and Illustrating Mixing; Chi-square; Entropy (W2)

Weidenbaum and Bonilla made use of the normal approximation to the binomial distribution in defining mixing. They illustrated its course,

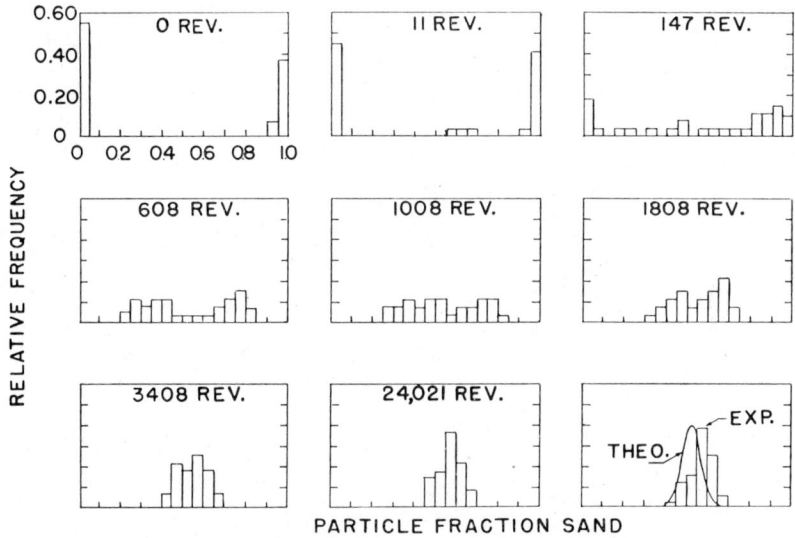

Fig. 19. Change in relative frequency histogram as mixing proceeds (W2).

as shown in Fig. 19. [Several of these items are discussed in greater detail in Weidenbaum's thesis (W1).] The limiting theoretical distribution was computed for $p = 0.53$ and $n = 133$, which were the particle fraction of sand, and the average number of particles per spot sample, respectively. By computing the theoretical standard deviation for this case, they showed that their mixing system—consisting of a small cylinder with salt and sand particles—did not give a random mixture, but instead had an equilibrium state of segregation, as will be further described in the Section on "Equipment Performance." They also illustrated the use of the chi-square distribution to (1) compare the experimentally determined frequency distribution of spot sample compositions with that for a randomly mixed batch, and (2) as a method of rating the extent of mixedness. Also, they have introduced the concept of the entropy of the system increasing as it changes from an orderly loaded state to a random mixture. They define the change in entropy ΔS, as:

$$\Delta S_{1-2} = k'' \ln \frac{P_2}{P_1} \tag{44}$$

where P_1, P_2 are the probabilities (or relative frequencies of occurrence) of chi-square values for states 1 and 2, respectively, and k'' is a constant.

D. BINOMIAL AND NORMAL DISTRIBUTIONS: EFFECT OF SAMPLE SIZE FLUCTUATION; TRANSFORMED VARIABLE FOR MEASURING INCOMPLETE MIXING; CHI-SQUARE (B4)

Blumberg and Maritz mixed a closely sized batch of sand, half dyed red and half dyed blue. With this highly idealized case, they showed that a randomly mixed batch could be obtained in a drum rotated at 55 rpm at a 30° angle with the horizontal. By this experimental work, they confirmed the fact that where segregating forces are not present, the limiting or equilibrium mixture is a random arrangement of particles, from which the number r of particles of one kind (red in this case), in spot samples of size n, are normally distributed. They also showed, by a statistical proof, that although n might vary from one spot sample to the next, if the variance of N ($= S_n^2$) is small compared to the square of the mean ($= \bar{n}^2$), the proportion of red particles in the sample ($x = r/n$) will still be approximately normally distributed. In statistical terminology, they showed that if, instead of the equation

$$P(x) = \frac{n!}{x!(n-x)!} p^x (1-p)^{n-x}$$

the function $P(x,n) = P(x)P(n)$ were used, then assuming that the distributions of both r and n were approximately continuous, the distribution of x would be approximately normal ($x = r/n$). They later show this to be true experimentally for a ratio of $S_n^2/\bar{n}^2 = 0.005$. After their section on the completely mixed state (which concerned the random mixture), Blumberg and Maritz consider incomplete mixing. They state, "When mixing is incomplete, what appears to be important fundamentally is not the degree of mixing but rather the approach of the mixture towards the completely mixed state." Previously, for the completely mixed state, they had used a simple chi-square test to determine whether a random mixture had been achieved. This chi-square test involved a comparison, of the observed frequency distribution of x_i with the theoretical calculated frequency distribution for a random mixture. For the unmixed state, however, although a chi-square test is used as a measure of mixing, it is not based on x_i, but instead on $z_i = 2 \arcsin \sqrt{x_i}$. By introducing this transformation, they sought to obtain a variable z_i, which would be approximately normally distributed with means ζ_i and variances $1/\nu$. They also did some experimental work from which they concluded that 10 spot samples at each state of mixedness were sufficient to plot a graph of degree of mixing vs. time, which would accurately indicate when mixing was completed.

E. Theoretical Variance for Random Mixture of Two Materials, Each with Its Own Size Distribution (S5)

In a paper entitled, "The Degree of Mixing of a Random Mixture as a Basis for the Evaluation of Mixing Experiments," Stange gives a formula for calculating the theoretical variance of samples taken from a randomly mixed batch of two materials (P and Q), each of which has its own size distribution (Eq. 25).

In order to determine $\bar{\nu}_P$ and $\bar{\nu}_Q$, the relative frequency distribution for particle weights must be known for each component. The variation coefficients are also determined from these distributions. Examples of the use of this formula are shown in the paper.

Special cases which enable simplification of the equation are given, e.g., the coefficients of variation are not needed in the simplified expression for the case where the initial distributions consist of almost like particles, since their squares are "small" compared with 1. Stange also shows how the confidence limit graph given in his other paper (S4) can be used to give a confidence band around the theoretically computed standard deviation for a randomly mixed batch. If the sample variation is within this confidence band, then there is considered to be no further significant change in the extent of mixedness. Variation among sample means would be accounted for by the sample size and prescribed confidence limits. Stange also gives experimental evidence to show that two substances which do not have any tendency to segregate during mixing will form a random mixture for which one can compute the standard deviation theoretically. In this case, the formula used to compute the theoretical standard deviation was a simplification of Eq. (25), because the range of particle sizes was such that the ratio of the largest particle weight to the smallest was 3.4. In Stange's paper, it was stated that if this ratio is ≤ 3.5, the following simplification of Eq. (25) can be used:

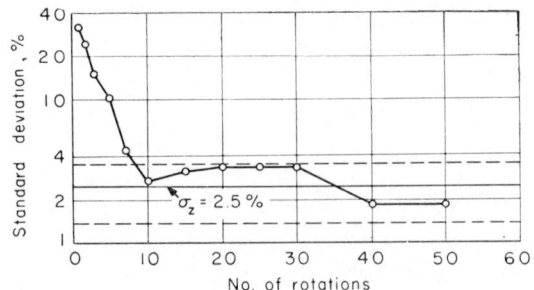

Fig. 20. Standard deviation s, *vs.* number of rotations of the mixer for experiment 1 of Table II. σ_z is the theoretical value of standard deviation computed from Eq. (45). Eight samples taken at each number of rotations. Broken lines are 90% confidence limits on σ_z for seven degrees of freedom (S5).

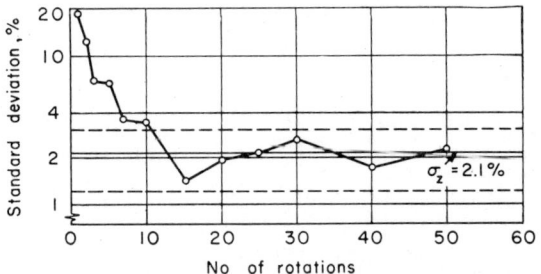

Fig. 21. Same as Fig. 20 for Experiment 2 F of Table II (S5).

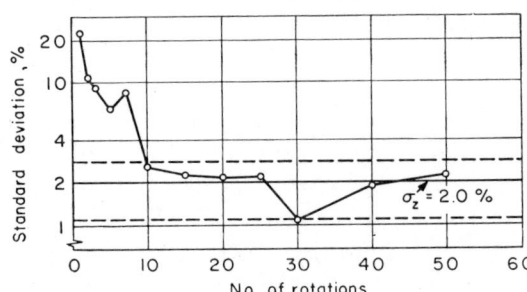

Fig. 22. Same as Fig. 20 for Experiment 3 F of Table II (S5).

$$\sigma_z^2 = \frac{PQ}{g}(P\bar{\nu}_Q + Q\bar{\nu}_P) \tag{45}$$

(See Section III, D, 9 for nomenclature.)

By applying this formula and performing experiments with the systems shown in Table II, Stange obtained the three graphs shown in Figs. 20 through 22. Further details are given in the section on Equipment Performance.

TABLE II
Data Concerning Mixtures Graphically Tested for Randomness by Stange[a] (S5)

Experiment No.	Total wt. G[gm.]	Component P					Component Q					Weight of sample, g [10^{-3} gm.]	Quality coefficient of the random mixture $\sigma_z^2 \times 10^4$	Standard deviation from theoretical values P, Q σ_z, %
		Particle size [mm.]	Absolute wt. G_P[gm.]	Relative wt. P	Average particle wt. $\bar{\nu}_P$[10^{-6} gm.]		Particle size [mm.]	Absolute wt. G_Q[gm.]	Relative wt. Q	Average particle wt. $\bar{\nu}_Q$[10^{-6} gm.]				
1	250	0.4 0.6	150	0.60	177		0.4 0.6	100	0.40	190		71	6.25	2.50
2 F^b	250	0.2 0.3	175	0.70	24.2		0.4 0.6	75	0.30	190		65	4.53	2.13
3 F^b	250	0.1 0.15	175	0.70	2.8		0.4 0.6	75	0.30	190		66	4.26	2.06

[a] Stange states: "The experimental results are from the thesis of R. Schaeffer, *Mixtures of Granular Substances: Comparisons between the Variations of Samples by Calculation and Experimentation with Molding Sand Mixtures* (Technical University at Karlsruhe, 1953)."
[b] The addition of F in experiments 2 and 3 signifies an addition of 2 wt. % distilled water, in order to prevent separation of the mixture.

F. Use of Poisson Distribution in Evaluating Mixtures (A1)

Adams and Baker have pointed out some properties of the Poisson distribution that were particularly useful in enabling them to test and compare blenders for a polythene blending problem. This involved blending a small quantity of master batch granules with natural granules, the master batch containing additives such as pigments and anti-oxidents necessary for certain uses. These properties are: (1) the Poisson distribution mean and variance are the same ($= m$), and (2) when $m \geq 20$, the normal distribution serves as a good approximation to the Poisson distribution.

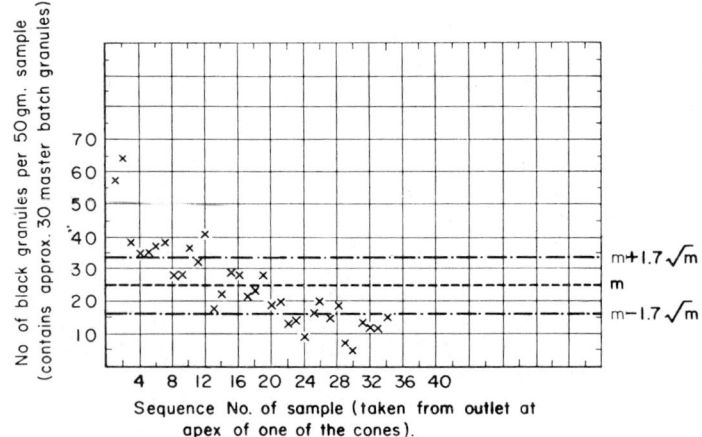

Fig. 23. Illustration of how confidence limits computed from a Poisson distribution can be used to evaluate blender performance. Double cone blender, 500 turns. Natural polythene (rough 4 mm. cubes, sp. gr. 0.92) mixed with master batch containing carbon black (3 mm. cubes, sp. gr. 1.2). Conclusion: batch not randomly mixed since probability of this graph occurring with a randomly mixed batch is less than 0.01 (A1).

Three tests were used for assessing the efficiency of blending. The first was a variation of the quality control chart or confidence band technique, which was used on individual samples rather than on means of rational sub-groups. Ninety, ninety-five, and ninety-nine per cent control lines were drawn for a mean m, which was calculated from the weight of the master-batch added, the weight of natural granules used, and the volume and density of the master batch granules. The use of the average number of master-batch granules per sample (denoted by $\bar{x}$) as an estimate of m was suggested as an alternate method which would not significantly affect the results. This test required $m \geq 20$ and $N \geq 10$, where N is the number of samples taken from a batch. Examples of such graphs are shown in Figs. 23 and 24; 90% control lines are shown.

The second test, which required $N \geqq 20$, is a graphical method for detecting trends in sample compositions when the latter are plotted versus sample sequence. Seven successive results on one side of the mean m, indicate that there is a trend in the sample compositions rather than a purely random arrangement. The authors advise a repetition of this test to confirm whether this trend exists. If the trend is still indicated the batch is not well blended.

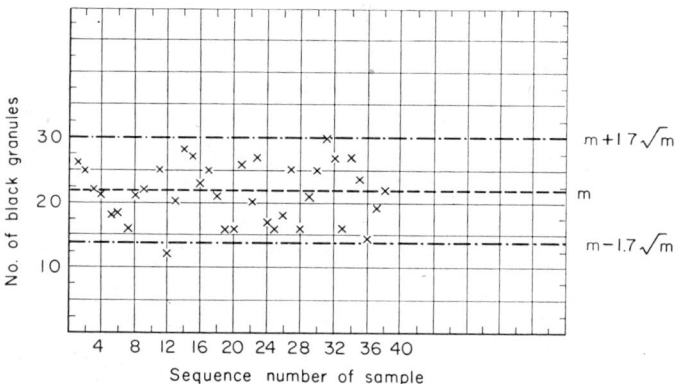

Fig. 24. Same type of graph and equipment as in Fig. 23, but with different materials. Natural polythene (sp.g. 0.92) mixed with natural polythene with black compound (sp.g. 0.922). Probability = 0.7. Conclusion: batch is randomly mixed (A1).

The third test involves use of the ratio $b = S^2/m$, where S^2 is the variance and equals $\dfrac{\Sigma(x - \bar{x})^2}{N - 1}$. In practice, m is replaced by $\bar{x}$ (x is the number of black granules in a sample from a mixture of natural and black granules). The probability of obtaining as large a value of b as computed, from a Poisson distribution, is plotted versus values of b. This graph is shown in Fig. 25 for sample sizes of 20 and 30, and can be used as follows:

Step 1. Calculate b from the sample compositions.

Step 2. From Fig. 25, determine the probability of obtaining this large a value of b from a Poisson distribution.

Step 3. If this probability is less than 0.05 (one chance in twenty), conclude that the samples did not come from a Poisson distribution. Thus the theoretical limiting mixture has not been reached.

This test is similar to the statistical tests of significance described earlier. In this case, the test of significance is being performed on $S^2/\bar{x}$. Instead of 0.05, some other level of significance could be used if desired.

The authors mention the fact that probability can be taken as a measure of the quality of blending. This is somewhat analogous to the Weidenbaum and Bonilla method for classifying the quality of a mixture in accordance with the probability of occurrence by chance of the chi-square value that is obtained when the mixture spot sample distribution is compared with the theoretical normal distribution (W2).

Adams and Baker illustrate a method for determining the carbon black concentration of the master batch, and also the per cent of the master batch that must be added to the natural polythene in the pre-

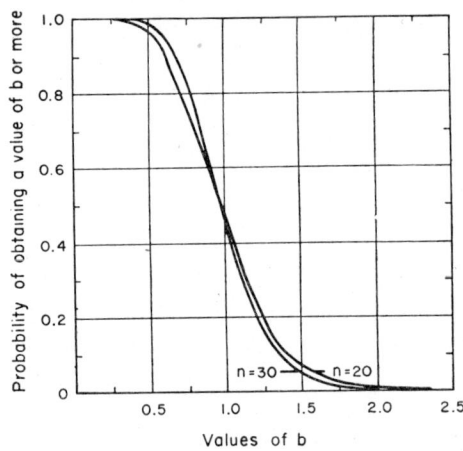

Fig. 25. Graph of probability, of obtaining a value of b or greater from a Poisson distribution, vs. values of b. $b = S^2/\bar{x}$, where S^2 = sample variance and $\bar{x}$ = sample mean (A1).

blend, in order that certain specifications for variation of additive in the final compound will be met. They do this by developing and illustrating the use of an equation involving density of polythene, density of carbon black, volume per granule, average concentration of carbon black, and permitted tolerance. In deriving the equation, they use the principle that the number of granules of master batch in a weight, w, of blended granules, will be distributed according to the Poisson distribution; thus the variance of this quantity can be calculated theoretically.

G. Continuity of One Phase in a Powder Mixture of Two Phases (F1)

In a paper dealing with the analysis of the continuity of one phase in a powder mixture of two phases, Forscher presents another approach to the use of the concept of a random mixture as a criterion of mixing.

His mathematical analysis deals with the problem of determining how much of one powder A, must be added to another powder B, to obtain continuity of powder A in a random mixture of the two. When more than the minimum is added, more continuity will result. Continuity is expressed as the probability that pairs of particles of powder A have radii, the sum of which just equals the distance between their centers. Forscher's mathematical analysis leads him to conclude that for an aggregate of equal-size spheres A is probably discontinuous if the ratio of A to B is below ⅕. Both A and B are continuous when A/B exceeds this ratio and eventually B becomes discontinuous if B/A falls below ⅕. In the discussion given, powder A is a good conductor of electricity, and powder B, a poor conductor. The degree of continuity of powder A was assumed proportional to the over-all conductivity of the compound powder mixture.

This approach has interesting potentialities as a research tool for studying mixing. Although the assumptions inherent in Forscher's article require a rather idealized system, it may be possible to determine from conductivity measurements whether very fine powders are randomly mixed with one another. This could be done, by calculating at approximately what concentration continuity could be achieved, and then determining whether this were obtained, or whether the concentration needed for continuity was far in excess of that theoretically required. Particularly where the aim of mixing is to achieve some sort of special electrical property, this approach would appear to have potentialities.

H. Use of Standard Normal Table with Rate Equations (O9)

Oyama and Ayaki's development of their rate equation yields an expression:

$$\epsilon = \int_{-z_1}^{z_1} f(x)dx \tag{46}$$

where ϵ = probability of a spot sample composition x, lying within the range between $a - \delta$ and $a + \delta$,

and
$$f(x) = \frac{1}{\sqrt{2\pi}\,\sigma} e^{-\frac{1}{2}\left(\frac{x-a}{\sigma}\right)^2} \tag{47}$$

The authors describe how this relation can be used to compute mixing time for the case where the equation

$$\ln(1 - M) = -\phi t \tag{48}$$

gives a straight line.

The essentials of this method are:
1. Set the specifications to be achieved by mixing. Thus, if a is the

average composition of the whole mixture, the specifications would state that a certain fraction of spot samples taken from the mixed batch must be between $a - \delta$ and $a + \delta$.

2. Determine z_1, which is the specification expressed in multiples of the standard deviation σ, by using standard tables of the normal distribution. The area under the normal curve between $a - \delta$ and $a + \delta$ must equal the fraction of spot samples specified in step 1. This value, which equals ϵ, is just another way of looking at "the probability of a spot sample composition x lying within the range between $a - \delta$ and $a + \delta$." Thus, z_1 is the "normalized" standard deviation corresponding to this area.

3. Compute $\sigma_s = \dfrac{\delta}{z_1}.$ (49)

4. Use this to compute $1 - M$ in:

$$1 - M = \frac{\sigma^2 - \sigma_r^2}{\sigma_0^2 - \sigma_r^2} \tag{50}$$

(See Section III, D, 17 for further nomenclature.)

5. Using Eq. (48), compute t, the time necessary to achieve the desired degree of mixing.

Primarily, this is still only an interesting theoretical idea since ϕ, which is needed for this calculation, was obtained only for a few systems of differently colored but otherwise identical sands, loaded in a certain manner and mixed in a horizontal rotating cylinder. When two different sands were mixed, the straight line relationship did not hold, and thus no over-all ϕ could be obtained. Also, it should be borne in mind that this method assumes spot sample compositions will be normally distributed at whatever specification is set for the mixture.

I. Techniques Necessary for Different Methods of Analysis

The experimental work of Weidenbaum and Bonilla (W2), Blumberg and Maritz (B4), Adams and Baker (A1), and Oyama and Ayaki (O9) involved particle counting in order to test whether or not a batch was randomly mixed. If weight fraction rather than particle fraction were used as the method for expressing spot sample composition, the standard deviations based on these two different methods of analysis, might be expected to differ. Weidenbaum (W1, pp. 131–133) has shown quantitatively that the magnitude of this standard deviation difference would depend upon the magnitudes and constancies of the weight fractions, and also upon the magnitudes and constancies of the differences between the weight fractions and particle fractions. These latter differences, in turn, are contingent upon the number of particles per gram of each of the two substances being mixed, and the relative proportions of the two materials in the spot sample. Hatch's equations relating count and weight distributions, which are summarized by Dallavalle (D1), may also be of use in this area.

Adams and Baker mention some variations of the particle count technique for cases where counting is tedious. One entails cutting down the number of counts, which involves using the square root of the count as the variable rather than the count per sample. The other involves spreading the sample over a flat surface and placing a transparent covering with grid markings above it, and then counting the number of squares which contained none of the additive. They then used the transformation of Blumberg and Maritz to obtain a new variable which is approximately normally distributed. It is stated that in both cases, tests can be developed with these new variables, similar to those which are illustrated in their paper (A1) using the number of master batch granules per sample as the variable.

J. Summary and Conclusions—Theoretical Frequency Distributions

1. It has been shown, both theoretically and experimentally, that with certain systems of particulate solids in which precautions were taken to prevent segregation, a randomly mixed batch was achieved from a mixing process.

2. The distribution of spot sample compositions from such a mixture can be computed from the binomial distribution, and closely approximated by either the normal or Poisson distributions, depending on the relative amounts of the two materials in the mixture and the size of the spot samples taken.

3. Several methods for using the above principles to evaluate mixing machines have been developed and illustrated. To a certain extent, the above ideas have been applied to systems where the two types of particles differ in size distributions.

4. Examples have been given illustrating criteria for deciding whether a specific mixing operation turns out a random mixture. Also the principles have been applied to planning how best to prepare quantities of coarse particulate ingredients for blending in a mixture to meet specifications as to variability.

5. In order to increase the usefulness of this type of approach, techniques should be developed to extend the application, of both the experimental and calculation methods, to much finer mesh materials. Some

equations for doing this are available, but these are still in early exploratory stages.

6. A recent paper mentions an approach—involving continuity of one ingredient throughout a mixture—which although purely theoretical, offers a potential method for ascertaining whether random mixtures of fine particles have been achieved. The conductivity of the mixture could be studied in such work.

V. Rate Equations

A. The Value of Rate Studies

The attempt to find rate mechanisms for solids mixing is important for several reasons. First, and most obvious, is the fact that if a method were available for calculating how long it would take to mix materials in various mixers, the amount of time necessary for testing could be reduced considerably. At present, there is no general way of calculating the optimum mixing time accurately from such data as have been reported. Another value of rate studies is that they throw light on the basic mechanisms of the mixing operation, thereby pointing to ways in which mixing machines and techniques can be improved.

In the summaries of rate studies which follow, both these considerations should be borne in mind.

B. General Discussion

Solids mixing rate theory has been discussed in papers by Brothman et al. (B5), Coulson and Maitra (C3, M2), Visman and Van Krevelen (V1), Weidenbaum and Bonilla (W2), Weidenbaum (W1), Lacey (L2), and Oyama and Ayaki (O9). In all of these papers, some sort of a rate equation has been postulated. These are tabulated in Table III, which is an expanded version of a similar one from Weidenbaum and Bonilla (W2).

Lacey (L2) points out that simply finding an exponential law of mixing rate in practice, must not be taken by itself as substantiation of a theory without considering its range of agreement. He indicates how theories for the rate of mixing can be developed by assuming one or more of the following three mechanisms: (1) convective mixing—transfer of groups of adjacent particles from one location in the mass to another, (2) diffusive mixing—distribution of particles over a freshly developed surface, and (3) shear mixing—the setting up of slipping planes within the mass. Later, in discussing the analysis of incomplete mixtures, he states that $1 - M$ (see Section III, D, 16) should be independent of sample size for diffusive mixing, but greatly dependent on sample size for convective mixing. The importance of using the most meaningful term in a

TABLE III
Rate Equations and Plotting Methods Proposed by Various Investigators[a]

Investigators	Original rate equation proposed in derivation of final equation[b]	Final equation given[b]
Brothman et al. (B5)	$y_{t+1} = y_t + \phi(1 - y_t)$	$P_t = 1 - e^{-kS_p(1-e^{-tc})}$
Coulson and Maitra (C3)	$\dfrac{dS}{dt} = k(S_0 - S)$	$\ln \dfrac{100}{X} = kt$
Bonilla and Crownover, and Bonilla and Goldsmith [mentioned in (W2)]	Not given	Data plotted as $-\log \dfrac{a.d.}{2c(1-c)}$ vs. $\log r$
Visman Van Krevelen (V1)	Not given	Data plotted as $(100 - Z)$ vs. t on linear probability paper using the probability scale for $(100 - Z)$
Weidenbaum and Bonilla (W2)	$\dfrac{d\left(\dfrac{\sigma}{\bar{s}}\right)}{dt} = k'\left[\left(\dfrac{\sigma}{\bar{s}}\right)_{eq.} - \left(\dfrac{\sigma}{\bar{s}}\right)\right]$	$\ln\left[\dfrac{1}{\left(\dfrac{\sigma}{\bar{s}}\right)_{eq.} - \left(\dfrac{\sigma}{\bar{s}}\right)}\right] = \ln\left[\dfrac{1}{\left(\dfrac{\sigma}{\bar{s}}\right)_{eq.} - \left(\dfrac{\sigma}{\bar{s}}\right)}\right]_0 + k't$
Weidenbaum (W1) and Lacey (L2)	Diffusion equation—see Table IV	
Oyama and Ayaki (O9)	$\dfrac{\partial \sigma^2}{\partial t} = -\phi(\sigma^2 - \sigma_r^2)$	$\ln \dfrac{\sigma^2 - \sigma_r^2}{\sigma_0^2 - \sigma_r^2} = -\phi t + c'$
Sakaino (S1)	$-\dfrac{dz}{dt} = c(z - z_f)$	$\ln\left(\dfrac{z - z_f}{z_0 - z_f}\right) = -ct$

[a] This is an extension of a similar table given by Weidenbaum and Bonilla (W2).

[b] Nomenclature: t = time; ϕ, k, k', c' = constants; y_t = proportion of the maximum theoretically possible surface of separation, S_p, that has been developed in t units of time; $c = \log 1/1 - \phi$; S = interfacial area of surface per unit volume of the mix; S_0 = maximum surface per unit volume that can be achieved with the given system; r = number of revolutions of the mixer; $\left(\dfrac{\sigma}{\bar{s}}\right)_{eq.}$ = value of $\dfrac{\sigma}{\bar{s}}$ for theoretical random equilibrium mixture = 1; z = mean size of "agglomerate" which is being distributed throughout the batch $(-dz/dt$ = rate of decrease of size of unit); z_0 = initial value of z; z_f = equilibrium value of z. As previously given, the definition of c applies to Brothman et al. However, c is also a constant in Sakaino's equation. Also for Bonilla and Crownover and Bonilla and Goldsmith, c refers to weight fraction of the key constituent in the entire batch. $a.d.$ is mean deviation and σ and s are standard deviations defined in Section III, D under the various authors who use them. P_t, X, and Z are also explained in Section III, D.

rate plot is thus stressed. Weidenbaum and Bonilla (W2) also emphasized this point by showing how the integrated form of the simple first-order rate equation $dA/dr = K(A_p - A)$, can give different plots simply by changing the definition of A, which is a term expressing degree of mixedness. A_p is the value of this term for perfect mixing (definitions of the latter also vary), r is the number of revolutions of the mixer and K is a constant.

The Brothman et al. (B5), and Coulson and Maitra (C3, M2) rate theories have been analyzed in several other articles (W2, L2, H2). As Table III indicates, these papers start with essentially the same original rate equation concerning the rate of change of the surface of separation,

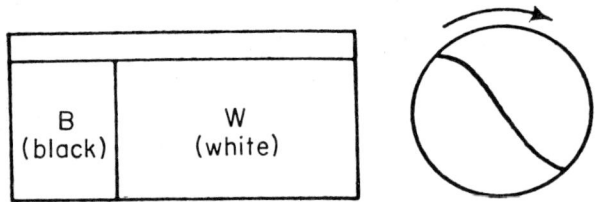

Fig. 26. Starting point for pure diffusive mixing (L2).

but they end up with different forms of the final equation. The lack of data for testing the theory by Brothman et al., plus the obviously curved initial portions in the Coulson and Maitra plots of log $100/X$ vs. t, still leave much to be desired before either of these theories can be accepted. The work of Weidenbaum and Bonilla (W2), although indicating a straight line initially in a plot of $\ln(1/1 - \sigma/s)$ vs. t, gives an irregularly shaped portion thereafter due to segregating effects, thus limiting the area of usefulness of this rate equation. Their method of loading, which was similar to that shown in Fig. 26, probably affected the shape of their rate plot, since the original rate equation would more likely be followed in such a system, than in an irregularly loaded one. However, this is an idealized situation which probably does not arise in most commercial mixing applications. It should be noted that Weidenbaum and Bonilla's original equation, although similar in form to those of Brothman, et al., and of Coulson and Maitra, differs from them in that it does not involve surface of separation, but instead uses a ratio involving the standard deviation among spot samples from a randomly mixed batch, divided by the experimentally determined standard deviation.

Visman and Van Krevelen (V1) replotted the data of Coulson and Maitra on linear probability paper, using $100 - Z$ (which they called the per cent unmixed) on the probability scale, and time t, on the arithmetic scale. Probability paper changes one of the scales, so that the cumulative normal distribution curve appears as a straight line. Thus, if there were a straight line from such a plot, its equation would be:

$$(100 - Z) = \frac{1}{\sqrt{2\pi}} \int_{-\infty}^{t} e^{-\frac{y^2}{2}} dy \tag{51}$$

where y is a dummy variable. Differentiating this with respect to t gives

$$\frac{d(100 - Z)}{dt} = \frac{1}{\sqrt{2\pi}} e^{-t^2/2} \tag{52}$$

which is the rate equation giving a straight line on such a graph.

Blumberg and Maritz (B4), while not covering rate theory, provided

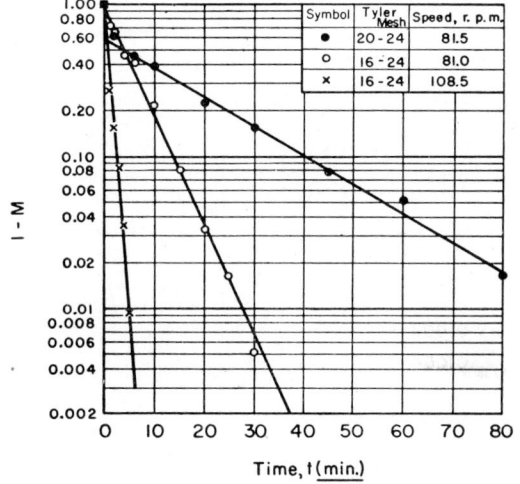

FIG. 27. Log $(1 - M)$ vs. time. Horizontal cylindrical mixer rotated at indicated speeds. Sōma standard sand of indicated mesh sizes was mixed. Half was dyed red, the other half blue, and initially the colored sands were loaded on opposite sides of the mixer. Analysis performed by particle counts. Mixer 30% full (by volume); total charge = 6 kg. (O9). Spot samples taken of approximately 108–118 particles each.

data on a system of two otherwise identical, but differently colored sands, which were mixed. Lacey has shown that if their degree of mixing, ϕ is replaced by $(\phi - s_r^2)$, then a plot of $- \log (\phi - s_r^2)$ vs. time will give a comparatively straight line over the range of 100 to 1. (All terms are defined in Section III, D, 15, 16, 19.)

Oyama and Ayaki (O9) performed experiments in a small cylinder, with otherwise identical but differently colored sands, in which they tested the rate equation

$$\frac{\partial \sigma^2}{\partial t} = -\phi(\sigma^2 - \sigma_r^2) \tag{53}$$

which was integrated to give

$$\ln \frac{\sigma^2 - \sigma_r^2}{\sigma_0^2 - \sigma_r^2} = \ln(1 - M) = -\phi t + c' \tag{54}$$

(Note: The background for M is given in Section III, D, 16 under Lacey.)

They found that graphs of $\log(1 - M)$ vs. t were straight lines for the cases where identical (except for color) sands were mixed, indicating

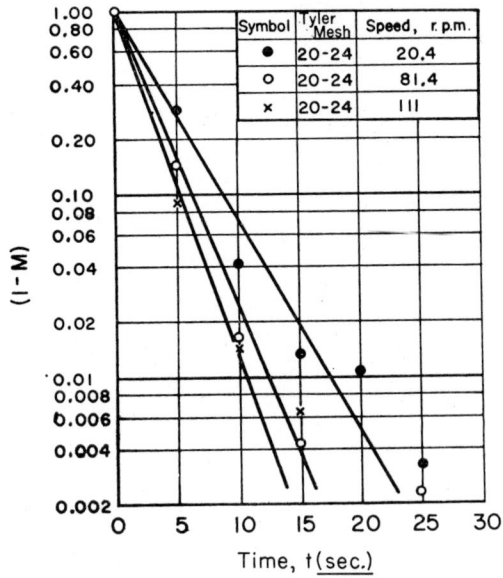

Fig. 28. Log $(1 - M)$ vs. time. Horizontal cylindrical mixer rotated at indicated speeds. Sōma standard sand of 20–24 Tyler mesh was mixed, 20% differently colored from the other 80%; initially loaded one on top of the other. Analysis by particle counts; spot samples taken of approximately 118 particles each. Total charge = 6 kg. (O9).

that their proposed rate equation applied here. Figures 27 and 28 show such graphs. The data for Fig. 28 are rather sketchy.

However, where two different sands were mixed, this rate equation did not hold, as shown in Fig. 29.

Michaels and Puzinauskas (M4) studied the mixing of powdered dextrose with water-wet kaolinite in a small finger-prong mixer. Using the Uniformity Index, I_v, shown in Section III, D, 20 as the degree of mixing, the mixing rates and energy requirements were determined for systems of different water contents and total volumes.

The authors stated that graphs of I_v vs. θ (time) on logarithmic coordinates approximate straight lines in the interval $10 > I_v > 1$, which can be expressed by $I_v = k\theta^n$, the exponent n varying between -0.7 and -1.2. Although they realize that this is insufficient to prove any of the rate theories which have been proposed, they feel that the trends of the Uniformity Index with time are qualitatively in agreement with the equation of Brothman, *et al.*, predicting that the logarithm of mixture uniformity will be an exponential function of time.

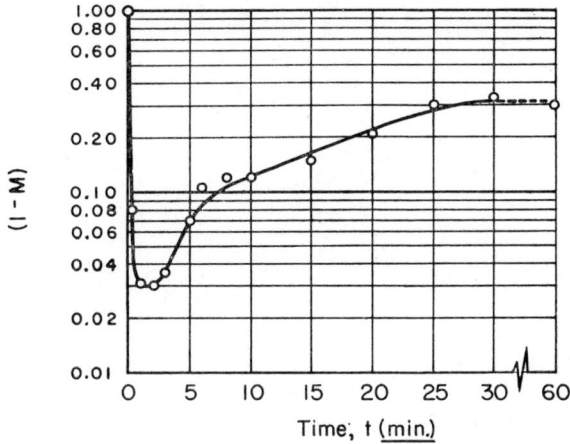

Fig. 29. Log $(1 - M)$ vs. time. Horizontal cylindrical mixer rotated at 54.5 rpm. Equal weights of 80–100 Tyler mesh Toyouro standard sand and 35–42 Tyler mesh Chigasaki sand were mixed, initially loaded on top of one another. Mixer 25% full (by volume). Total charge is 5 kg. Samples of about 0.45 gm. each were taken (O9).

Sakaino (S1) has attempted to utilize the same type of rate equation for explaining both solids mixing rate theory and the homogenizing process during glass melting. Drawing on data from Tooley and Tiede (T1), he uses their plot of density spread *vs.* melting time to illustrate his theory. His equation, as shown in Table III, deals with the rate of decrease in size of agglomerates, of size z, which are distributed throughout the mixture. "Apparent sample size" is denoted by R/z, where R is the actual sample size. It is used in computing the limiting variance for a pseudo-perfect mixture, which is considered a mixture in which agglomerates have reached their limiting size and are randomly dispersed. In the case where agglomerates reduce to individual particles and the sample size is so large that the variance due to theoretical random variation may

TABLE IV
ANALOGY FOR HEAT CONDUCTION, MOLECULAR DIFFUSION, AND DIFFUSIVE MIXING

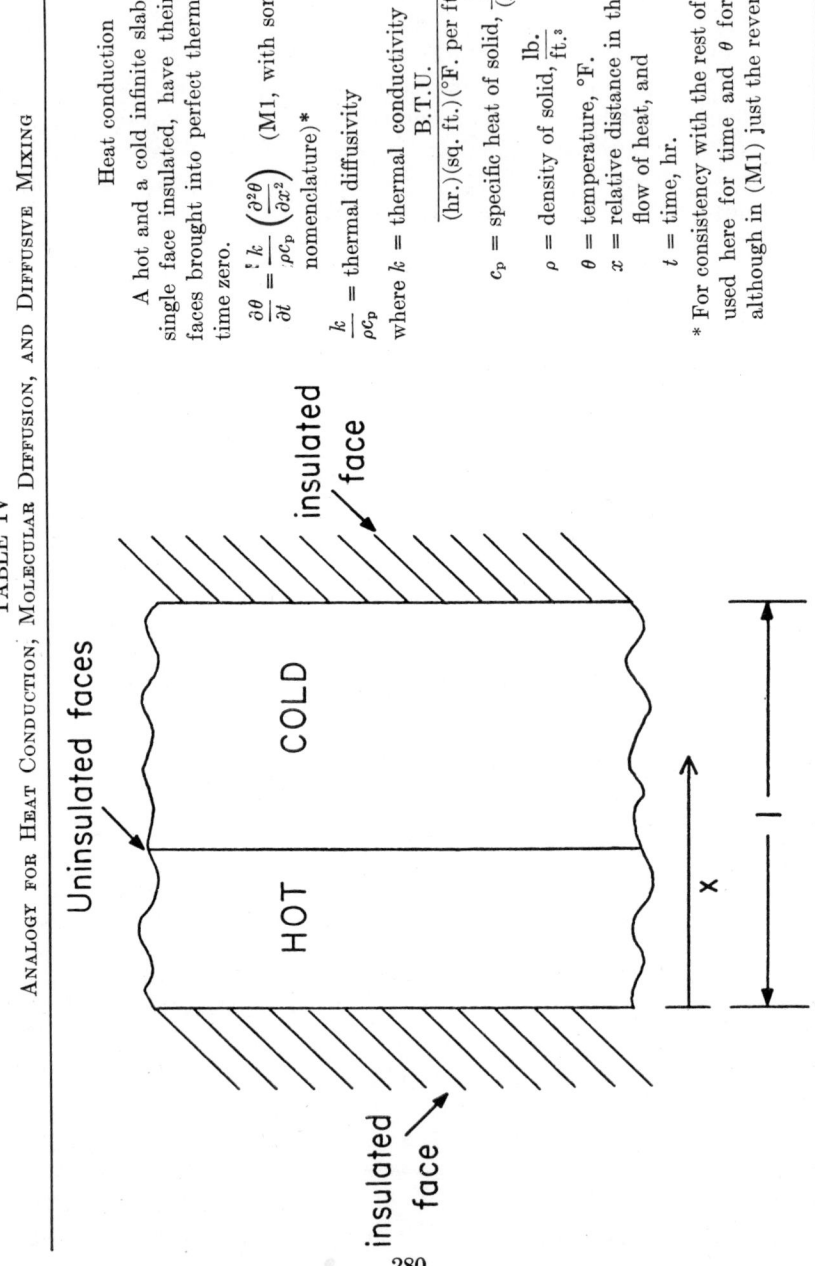

Heat conduction

A hot and a cold infinite slab, each with a single face insulated, have their *uninsulated* faces brought into perfect thermal contact at time zero.

$$\frac{\partial \theta}{\partial t} = \frac{k}{\rho c_p}\left(\frac{\partial^2 \theta}{\partial x^2}\right) \quad \text{(M1, with some changes in nomenclature)}*$$

$\dfrac{k}{\rho c_p}$ = thermal diffusivity

where k = thermal conductivity of the solid, $\dfrac{\text{B.T.U.}}{(\text{hr.})(\text{sq. ft.})(°\text{F. per ft.})}$

c_p = specific heat of solid, $\dfrac{\text{B.T.U.}}{(\text{lb.})(°\text{F.})}$

ρ = density of solid, $\dfrac{\text{lb.}}{\text{ft.}^3}$

θ = temperature, °F.

x = relative distance in the direction of flow of heat, and

t = time, hr.

* For consistency with the rest of the table, t is used here for time and θ for temperature although in (M1) just the reverse is done.

TABLE IV (*Continued*)

Molecular diffusion (W1)*

A thin diaphragm separating gases A and B is suddenly removed, and the gases are allowed to diffuse.

$$\frac{\partial y_A}{\partial t} = D \frac{\partial^2 y_A}{\partial b^2}$$

y_A = mole fraction A
b = relative distance in the direction of diffusion of A molecules, as shown
D = molal diffusivity
t = time

* (D4) gives basic reference material on diffusion.

Diffusive mixing (W1, L2)

A mixing cylinder is loaded with particulate solids and separated by a partition as shown.* The partition is removed and the cylinder is rotated around its horizontal axis.

$$\frac{\partial p}{\partial t} = D \frac{\partial^2 p}{\partial x^2}$$

where p = per cent black,
t = time,
x = relative distance in the direction of movement of black particles, as shown
D = mixing diffusivity.

* In (W1) the partition is in the center of the cylinder.

Lacey's nomenclature is used here for ease in following the discussion of Eqs. (55–58) which he gave (L2).

In Weidenbaum's terminology (W1):
y_A (= particle fraction of component A) replaces p and b replaces x.
In Eqs. (56–58), P is the value of x at the initial boundary, or the final per cent black in the overall mixture.

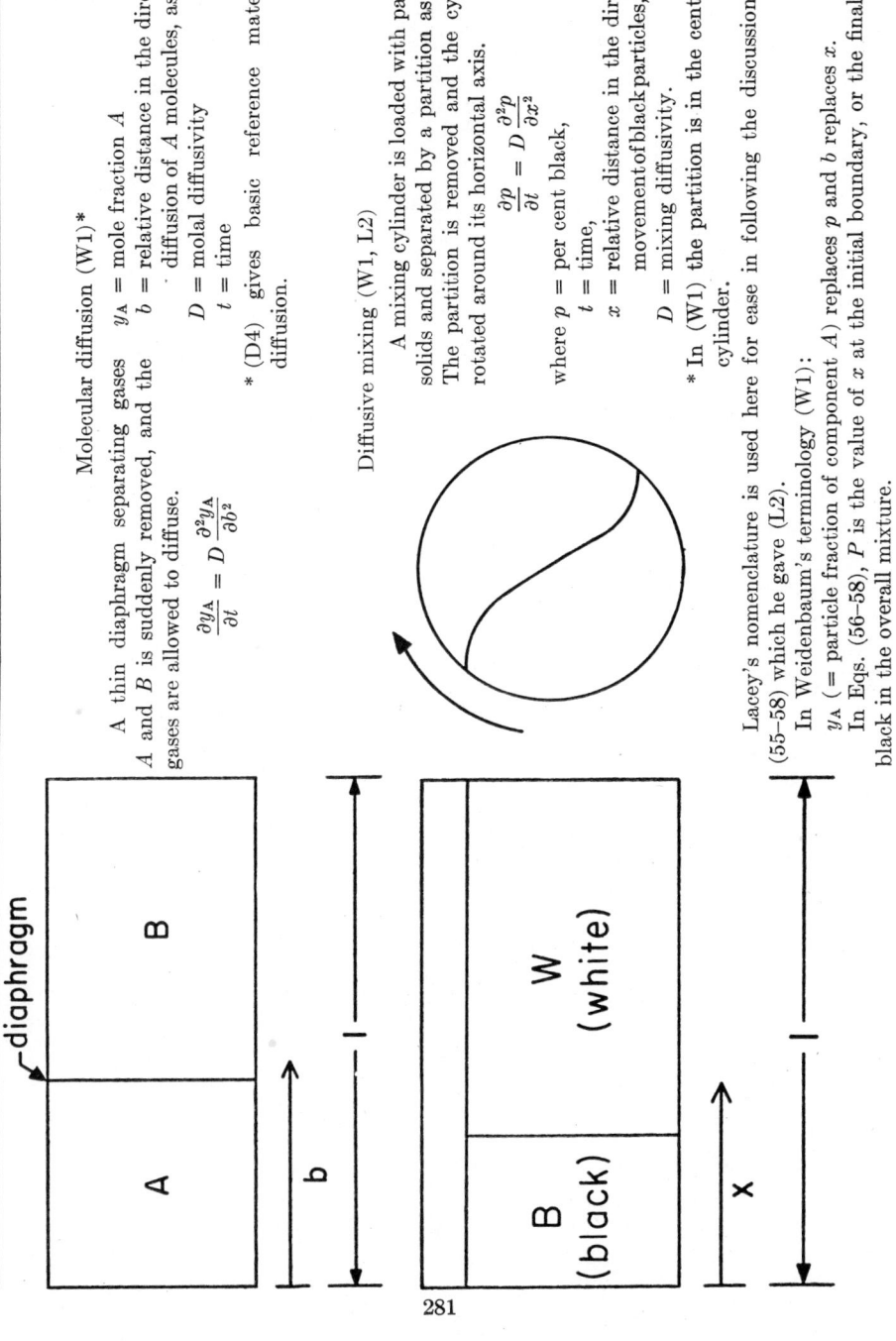

be neglected compared to the experimentally determined variance, Sakaino's final rate equation becomes log $S^2 = kt$ (k = a constant). He shows a straight line plot of log S vs. time for Lowry's data (L4), the latter being taken from Herdan (H2). This is the same as the standard deviation line in Stange's plot (S4) which is shown in Fig. 6.

C. Diffusion Analogy

Considering mixing in an analogous manner to molecular diffusion has led Weidenbaum (W1) and Lacey (L2) to investigate the use of Fick's second law of diffusion:

$$\frac{dy_A}{dt} = D \frac{d^2 y_A}{db^2} \tag{55}$$

as a rate equation for the mixing of solid particles.[3] The ideal case, which has been separately considered by each of them, is that where a mixing cylinder is loaded, as shown in Fig. 26. If the mixer is revolved as shown, then the motion of the particles of the two substances being mixed can be considered as analogous to the motion of molecules of two different gases, which are allowed to diffuse after a partition separating them has been removed. The above equation is for the case where diffusion is considered in the plane perpendicular to the axis of rotation of the cylinder. In order to more clearly illustrate the analogy with other unit operations, Table IV illustrates how similar equations can be applied to heat conduction, molecular diffusion, and diffusive mixing.

Lacey has shown how the solution of the differential equation for the stated boundary conditions gives:

$$(p - P) = \frac{2}{\pi} \sum_{q=1}^{q=\infty} e^{-q^2 \pi^2 T} \frac{\sin (qP\pi)}{p} \cdot \cos (qx\pi) \tag{56}$$

where $T = Dt/L^2$, L is the actual length of the mixer, and q is an integer. (Table IV gives further nomenclature.)

From Eq. (56), the variance can be theoretically computed as:

$$s^2 = \frac{2}{\pi^2} \sum_{q=1}^{q=\infty} \left(\frac{\sin qP\pi}{q} \right)^2 e^{-2q^2 \pi^2 T} \tag{57}$$

Therefore a plot of $-\log s^2$ vs. T should be a straight line if the original equation was valid. Lacey has plotted this graph, as shown in Fig. 30. The initial portion in all cases is curved, which Lacey attributes to initial

---

[3] See Table IV for nomenclature.

transients that are indicative of the distance that the black particles have to travel to the middle of the mixer. The transients assume more importance at low values of P, but the graphs all have the same slope for the

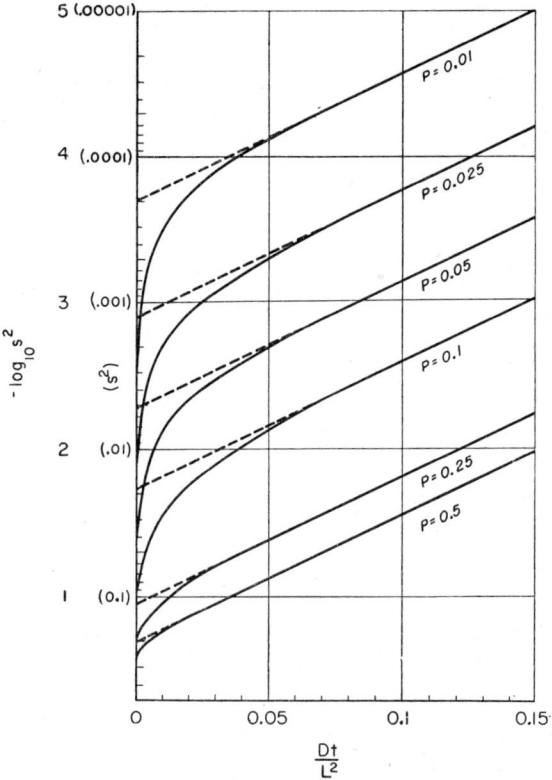

FIG. 30. Logarithmic plot of sample variance s^2, computed according to Eq. (57), vs. Dt/L^2 (L2). P = proportion of black particles in the overall mixture, D = diffusion coefficient, t = time, L = length of mixer.

linear portion, showing that the final rate of mixing does not depend upon composition. Using a simplified equation for the post-transient period:

$$s^2 = \frac{2}{\pi^2} e^{-2\pi^2 T} \sin^2 P\pi \qquad (58)$$

and dividing s^2 by $\sin^2 P\pi$ gives a corrected s^2, denoted by s_c^2. When this is plotted against T, the straight line portions all eventually reduce to one line, although there is still divergence in the transient region (Fig. 31).

It should be noted that Oyama and Ayaki (O9) arrive at an expression similar to that of Eq. (58), by developing the idea of mathematically formulating the probability, that any particle would transfer from one location to another in the mixer at any particular rotation. They mathematically summed up the product of probability × concentration for the N cells into which they had divided the mixture volume, in order to get concentration at any time. By then assuming that the particles being mixed have the same physical properties, and that the mechanism of mixing is independent of the locations of cells in the mixer, the authors

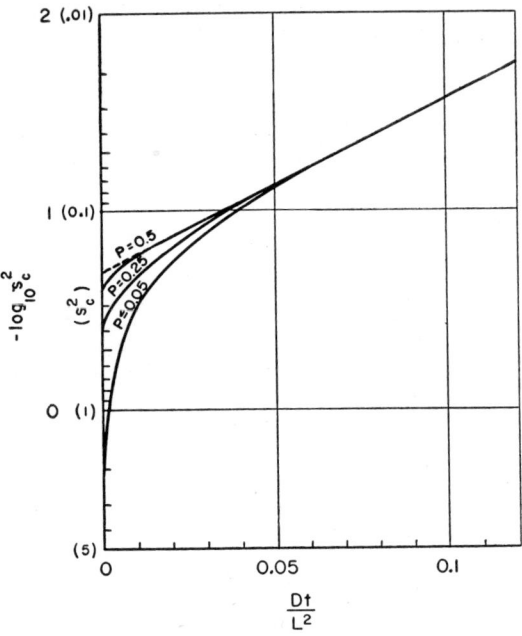

FIG. 31. Same as for Fig. 30 except that "corrected" variance, s_c^2, is used in place of s^2. (See Section V, C.) This shows the elimination of the effect of P, except in the transients (L2).

simplified the mathematical expression and obtained the equation for a horizontal rotating cylinder initially loaded as shown in Fig. 26. They assumed the motion of the particles in the mixer to be similar to what they referred to as "rodinal one-dimensional Brownian motion in a finite domain."

Lacey (L2) concludes from this that the straightness of a plot of this type for any machine is an indication of how uniformly the machine in question acts upon its contents. He states that straightness of such a plot implies "that either D is independent of position, or else there is good circulation which evens out variations in D; and thus that the design is an optimum for the type of mechanism in use." He further concludes, "If there is curvature, this indicates that there are regions of poor mixing (or even 'dead' spots), which control the mixing time, owing to the absence of good circulation. Such regions can then be looked for and the defect remedied."

Lacey gives some sketchy data from Blumberg and Maritz (B4), and a private communication from Kramer which, although somewhat indicative a compatibility with diffusion theory, still leave several phenomena that require more investigation and experimentation before this theory can be confirmed. In applying diffusion theory it is well to bear in mind Weidenbaum's comments (W1) concerning the solution of Eq. (55) for any boundary conditions. He states that it " . . . has in its derivation two assumptions which are probably not true for the physical situation in most cases. They are: (a) 'equimolal' counterdiffusion of A and B. (b) D is a constant independent of concentration." Mole fraction is replaced by particle fraction when using the diffusivity concept for solids mixing. He (W1) further states that the diffusion analogy " . . . might eventually lead to a way of summing up the effects of the particle properties which affect the rate of mixing, i.e., the D might be used as an ultimate variable in a correlation of the variables involved in mixing particulate solids."

Extension of these ideas on diffusion, to cases where there are differences in diffusivity between the materials, would be of interest. Barrer's chapter (B2) on the solution of the diffusion equation may be of use here.

D. Summary and Conclusions on Rate Equations

1. Coulson and Maitra's rate equation (C3), although resulting in straight lines in the latter parts of the rate plots for certain systems, was not followed during their initial portions.

2. The rate equation of Visman and Van Krevelen (V1) gave two straight line portions in their rate plots for certain systems, the change in rate being attributed to a de-mixing effect setting in after a certain time of mixing.

3. Weidenbaum and Bonilla's rate equation (W2) was followed at the beginning of mixing, but not all the way to the end of mixing. The authors explain this as due to random motion being the initial driving force with the later appearance of an additional driving force, namely, a segregating tendency in the particular system they used.

4. Oyama and Ayaki's first-order type of rate equation (O9) was followed when two substances, identical except for color, were mixed. The major mechanisms of mixing were diffusion and shear, in this case. However, when two different substances were mixed, the equation was

only followed at the beginning, later deviation from it probably being due to segregating tendencies.

5. The equation of Brothman, Wollan, and Feldman (B5) was not supported with any experimental evidence when first postulated. Lacey's further analysis of this equation (L2), which he illustrated with a hypothetical graph, led him to conclude that the apparent rate constant should increase with increasing proportions of the constituent being analyzed for, and with decreasing particle size. He considers the fact that Coulson and Maitra's rate constants follow this pattern as some confirmation of the Brothman theory. Michaels and Puzinauskas' rate plots (M4) lead them to feel that the trends of changes in the Uniformity Index with time are qualitatively in agreement with the Brothman, *et al.* equation, which predicts that the logarithm of mixture uniformity will be an exponential function of time. (Note: Shearing forces would probably be very important in the mixer used by Michaels and Puzinauskas.)

6. Weidenbaum and Bonilla's data indicate that a first-order rate equation can give either a straight line or a curve depending on the term chosen for the degree of mixing. Lacey's logarithmic rate plot of Blumberg and Maritz' data indicates that a straight line over a wide range is obtained if the variance of a random mixture is subtracted from their expression for the degree of mixing, whereas a curved line is obtained if that expression alone is used.

7. Lacey's plot of a theoretically calculated variance based on a diffusion equation, *vs.* time, has an initial curved portion after which a straight line is obtained. Lacey concludes that the straightness of this latter part is an indication of how uniformly the machine in question acts upon its contents; curvature indicates that there are regions of poor mixing which control the mixing time, owing to the absence of good circulation.

8. The rate studies thus far described have dealt with mixers in which the predominant mechanisms of mixing were diffusion and shear, rather than large scale turbulence. Rate studies in a machine where the latter was predominant would be of interest.

9. Bearing in mind the three types of mixing previously mentioned, namely, convective, diffusive, and shear mixing, and the wide variety of results obtained in the various rate studies, the following picture emerges: In any but an extremely idealized system, numerous driving forces are at work to cause mixing. When the materials are loaded, groups of similar particles will be thrown around. At the same time, they will be broken up into smaller groups. The means of distributing the larger groups of particles to the different parts of the mixing vessel may involve shearing or convective mechanisms, whereas the smaller scale mixing

which causes the clusters to decrease in size, involves a diffusion-type mechanism. When the larger groups of particles have been distributed throughout the vessel, the most rapid mixing will then be obtained, if the scale of turbulent movement of particles is the same as the scale of segregation.

With large-scale and small-scale random movement going on, and considerable differences in the initial method of loading the material, any single rate mechanism will not be able to adequately cover all cases. In addition, the forces acting upon the particles will affect the rate of mixing. Some of these are gravitational and centrifugal forces, frictional forces between particles and between particles and equipment, and surface forces on particles such as electrostatic forces. Differences in properties between particles being mixed may cause segregating tendencies which affect the rate of mixing. Rate data thus far mentioned apply only to the specific systems of machine and materials used. While some of the above conclusions indicate their possible usefulness as a guide to improving the performance of mixing machines, they are not enough to enable any general theoretical calculations of mixing times. However, as more data on different systems and machines are acquired so that the above equations may be further tested, more light will be thrown on the controlling mechanisms for different machines mixing different materials. This information will aid in analyzing mixing motion and improving mixer design by enabling the determination of the major mixing mechanisms in different types of mixing systems. It would be interesting to see what type of rate equations best describe the mixing action of various types of mixers with different materials, and how closely this classification compares with Scott's (S2) grouping of mixers according to the category of predominating action. By determining which rate mechanisms best describe the action of various mixers, it would be possible to group the mixers according to predominating mechanisms. The rate constants for any type of mixing action could then be compared in order to determine which of the mixers in that category gave most rapid mixing.

VI. Equipment

A. DESCRIPTION

The one phase of solids mixing which has been treated at length in the literature is the description of equipment. In textbooks, reference books, and review articles, pictures of equipment are shown and tables or charts given, which list suitable equipment for various applications. Also, details of machines, diagrams of their shapes, and practical construction data are available, including approximate power requirements.

Should one want any further information on a specific type of mixer shown, he need only write to the manufacturer for a detailed bulletin concerning construction, operation, and applications of that mixer. In many cases a description of the motion during mixing is also given. Some

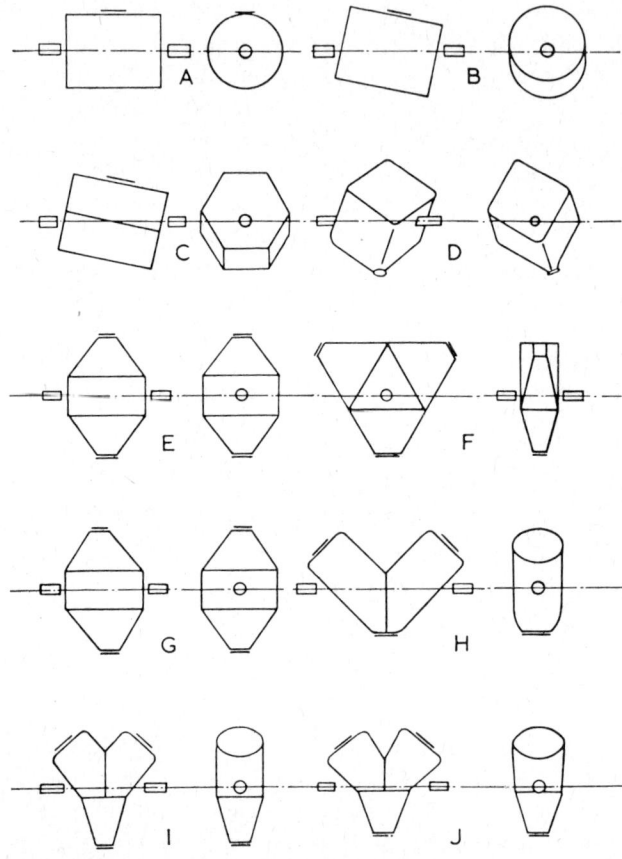

Fig. 32. Diagrammatic outlines of powder mixers (A1). A, horizontal rotating cylinder; B, cylinder—off center axis of rotation; C, similar to B but hexagonal shape; D, rotating cube; E,G, double cone; F, pyramid; H, twin shell; I,J, "Y" cone.

of the standard references describing solids mixing equipment are the Chemical Engineers' Handbook (P1) (very brief) and the Encyclopedia of Chemical Technology (K2). The latter lists several types of dry solids mixers and has a table listing types of machines that can be used for various specific applications. Brown (B6) has discussed five types of tumbling mixers and tabulated various construction details. For a descriptive summary of a large number of pieces of dry mixing equipment, Quillen's article (Q1) is available. Work (W5) has discussed some of the considerations involved in solids mixing and prevention of subsequent segregation. In particular, he has discussed the concept of systems mixing whereby, instead of a specific mixer, various process steps, such as dumping in rotation, and grinding, are used to further the mixing of ingredients. Scott (S2) has described several kinds of mixers, indicating which of the several mechanisms mentioned by Lacey (L2) are predominant in them. Wornick (W6) has summarized characteristics of several dry solids mixers in connection with their use in premixing micro-ingredients for feed-stuffs. Young and Snaddon (Y3) have described an adjustable angle mixer which they found of use in mixing laboratory samples preparatory to analysis. Bullock (B7) has discussed the types of material for which muller type mixers are applicable and described the action of mullers. He cautions against trying to use the muller for materials which are too fluid or too sticky. However, he states that if used in the proper ranges, mullers are well suited to breaking down aggregates. Pierce (P2) has mentioned a vertical helical screw mixer with planetary movement in a conically shaped vessel. In the discussion of the paper by Adams and Baker (A1), several new shapes are illustrated (Fig. 32), and the mixing action of these plus other shapes is discussed. It is unnecessary to duplicate here descriptions of equipment which are readily available in the above sources.

B. Performance

1. *Introduction*

In contrast to equipment description, the subject of equipment performance has few available published references. This has probably been due, at least in part, to the lack of quantitative methods for testing, which requires certain theoretical and experimental tools. Also, because a machine can be used to turn out a product which can either be further processed or sold at a profit, it might be thought of as a "satisfactory" machine. However, it may still not be giving the best mixture attainable; the rest of the process may be making up for its deficiency, possibly at a decreased profit. Higher off-specification losses than necessary may be occurring in the finished product, without realization that the solids mixer is contributing to them. As has previously been mentioned, there are often many steps following the mixing operation, so that irregularities in the final product may not be traced to improper mixing unless they become so serious that the plant may have to shut down if their cause is not found.

Burton (B9) points out that unmixing is nothing new, giving as an example what he calls " . . . the first recorded instance of its importance in industry," which was in 1910. It relates to a problem that the Rumford Chemical Works had concerning the control of baking powder uniformity. After considerable fruitless checking of weighing accuracy, research discovered the cause of variation to be improper mixing in the 5000-lb. capacity mixers. This example deals with the mixer itself. Burton gives other examples concerned with segregation after mixing due to free fall; one of these, he reports, caused the loss of a chief engineer's job in a munitions plant producing electronic cores for military equipment. "Absolutely uniform and perfect magnetic requirements" were demanded in the cores, which were made of thermosetting resin and powdered iron of high purity. Evidently the trouble-shooting did not go back far enough: a free fall from the mixer had been causing segregation which resulted in non-uniformity. Still another example of unmixing was given by Burton, this case leading to non-uniformity in the height of material in packages. These phenomena emphasize the need for starting with a good mixture and making sure that it stays that way. Certainly the first step would appear to be the choosing of mixing equipment which will not have inherent tendencies to cause segregation of the components and which will give the most uniform product most rapidly, making most efficient use of power. The following paragraphs will aim to present a coordinated picture of such work as has been done to further our knowledge along these lines.

2. *Horizontal Rotating Cylinder* (O1–O8)

A comprehensive study of solids mixing in rotating cylinders was made by Oyama between 1933–1940. Later, in 1956, Oyama and Ayaki (O9) did some further work concerning mixing in a rotating cylinder. Some of the highlights of the above papers, most of which are in Japanese, will be discussed below.

a. State of Motion of Materials. Oyama observed the state of motion of materials in a rotating horizontal cylinder under various conditions of speed and volume per cent loaded, using black and white sand of particle diameter equal to about 1.3 mm. (between 12 and 16, U.S. mesh size). He described the various states of motion by the terms "cascade," "cataract," and "equilibrium." The *cascade* state, which he reported for low speeds, consisted of particles rolling down the inclined surface after leaving their circular paths. At a certain speed, the motion changed to the *cataract* state, which consisted of particles close to the cylinder walls adhering to it until they reached a certain height, whereupon they were thrown onto the inclined surface and rolled down the remainder of the way. The state at which this changeover occurred was called the *critical* state. When the speed was increased sufficiently, the particles finally reached a state such that they remained on the cylinder wall until they fell upon the bottom edge of the inclined surface, thus bypassing the process of rolling down the inclined surface. In this state, called the *equilibrium* state, the particles in motion were considered to always keep their own fixed paths in flight and circular motion. The best operating conditions existed between the *equilibrium* and *critical* states, the particular optimum speed for Oyama's system being 80 rpm. Oyama gave an empirical relationship, by which the rotational speed of the mixer which would give a certain state of motion could be computed as shown below:

$$N = \frac{C}{D^{0.47} X^{0.14}} \tag{59}$$

where N = speed in rpm
D = inner diameter of cylinder in meters
X = per cent of mixer volume occupied by batch;
and C = a constant as follows: 54 for the *critical* state, 72 for the *equilibrium* state, and 86 for the state in which the whole mass of particles was almost in contact with the inner wall of the cylinder.

It is interesting to see how Oyama's empirical relationship compares with the type of relationship derived from model theory. This can be done by assuming that the ratio of centrifugal to gravitational forces is the only thing that would be changed by varying only the cylinder size (frictional forces, electrostatic forces, etc., would not change). The Froude number = $N^2 L/g$ expresses the ratio of centrifugal to gravitational forces. Therefore, to produce the same state of motion in two different cylinders (say 1 and 2), their Froude numbers, V^2/Dg or $N^2 D/g$, should be the same.

Therefore
$$\frac{N_1^2 D_1}{g} = \frac{N_2^2 D_2}{g} \tag{60}$$

or
$$N_1/N_2 = \sqrt{D_2/D_1} \tag{61}$$

where V = linear velocity, ft./sec., and g = acceleration of gravity, 32.17 ft./sec.^{2}

Thus the speed required to give a certain state of motion should be inversely proportional to the 0.50 power of the cylinder diameter. Oyama's relationship shows that speed is inversely proportional to the 0.47 power of the cylinder diameter, and in addition, is inversely proportional to the 0.14 power of the volume per cent loaded.

The above relationship might serve as a "guesstimate" for scaling up any type of tumbling mixer, where centrifugal and gravitational forces are of primary importance. It would be necessary to use the proper characteristic dimension for the various sizes of geometrically similar equipment.

b. Mechanisms of Mixing. Oyama listed two mechanisms of mixing for the cylinder, based on analytical analysis and visual observations. These he called "local" mixing and "whole" or "total" mixing. The former he attributed to the scattering of particles as they roll down the surface in a "cascade" motion, whereas the latter was ascribed to positional changes of particles by circulation throughout the mixture. In terminology later used by Lacey (L2), these would correspond to the "diffusion" and "shear" mechanisms, respectively, for Oyama's system. By observing that the state of motion with the lowest dynamic angle of repose also gave the slowest mixing, Oyama pointed out the importance of the diffusion mechanism in promoting mixing.

His photographic method of measuring degree of mixing gives a visual means of illustrating the "diffusion" and "shear" mechanisms. Figure 10 mentioned in Section III, D, 1b showed a maximum intensity difference and a frequency difference on the various photographs. The drop in maximum intensity difference with mixing means a breakup of large groups of particles of one type, and the increase in frequency means a greater distribution of these groups throughout the mixture. These illustrate, respectively, the effects of the diffusion and shear mechanisms previously described.

In a paper by Oyama and Ayaki (O9), the authors point out the perfectly reasonable fact that radial mixing in such a cylinder is extremely rapid in comparison with longitudinal mixing, so that the controlling factor in determining mixing time is the coefficient of longitudinal mixing.

c. Performance Tests. Graphs from some of Oyama's tests with a cylindrical mixer of 20 cm. diameter and 40 cm. length, are shown in Figs. 33 and 34. This was for a system of black and white sand of approximately 1.3 mm. diameter (12–16 U.S. mesh size). The total weight of the mixture in the cylinder was 8 kg. The degree of mixing was measured photometrically, as described in Section III, D, 1b. For this system, the optimum speed is seen to be approximately 80 rpm. Equation (59) would permit calculation of the range of speeds between the "critical" and "equilibrium" states, which would include the optimum operating speed.

In a more recent paper, Oyama and Ayaki (O9) reported the results of some experimental work in connection with testing a rate equation (see Section V, B). Using a cylinder, 20.4 cm. in diameter and 39.6 cm. in height, they mixed several combinations of sand—sand systems. Employing 20–24 mesh, differently colored Sōma sand, they measured the coefficient of longitudinal mixing, ϕ_L, vs. rotational speed, with 6 kg. of

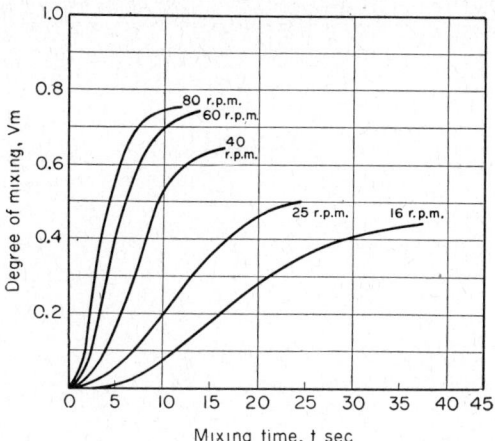

Fig. 33. Relation between the degree of mixing and mixing time. From (O2, p. 581). Horizontal rotating cylinder. Degree of mixing determined photometrically as described in Section III, D, 1b; 12–16 U.S. mesh, black and white sand. Total weight of mixture is 8 kg.

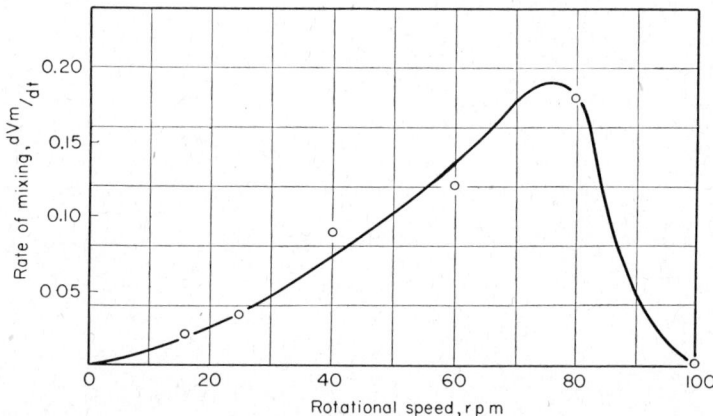

Fig. 34. Relation between the rate of mixing and rotational speed of mixer. Same materials and equipment as for Fig. 33 (O2). Values of rate of mixing are obtained from the slope of the linear portions of Fig. 33.

material in the mixer, corresponding to 30% loaded by volume. The degree of mixing was based on quantitative particle counts rather than photometric methods. The results, shown in Fig. 35, again indicate an optimum speed, although it differs from that shown in Fig. 34 due to differences in experimental conditions and the methods of measuring degree of mixing. Also, using a speed of 81.4 rpm, the curve in Fig. 36 was

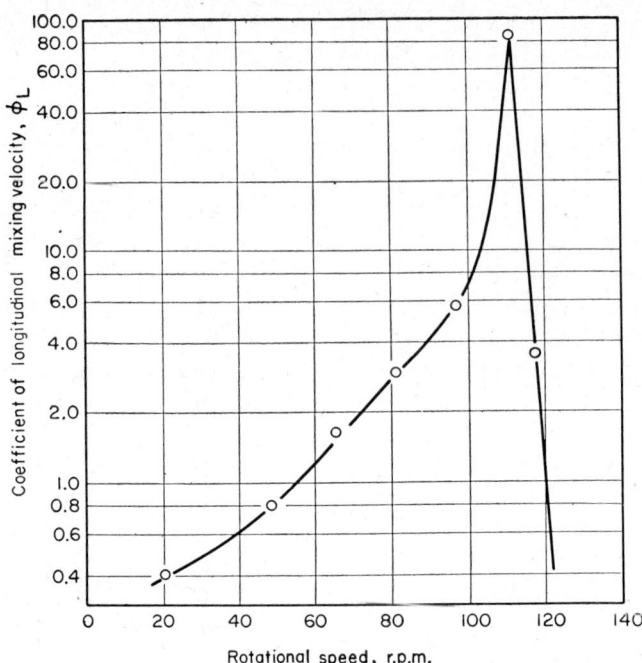

Fig. 35. Effect of rotational speed on values of ϕ_L. Horizontal rotating cylinder. 20–24 mesh sand. Degree of mixing from particle counts on spot samples (O9).

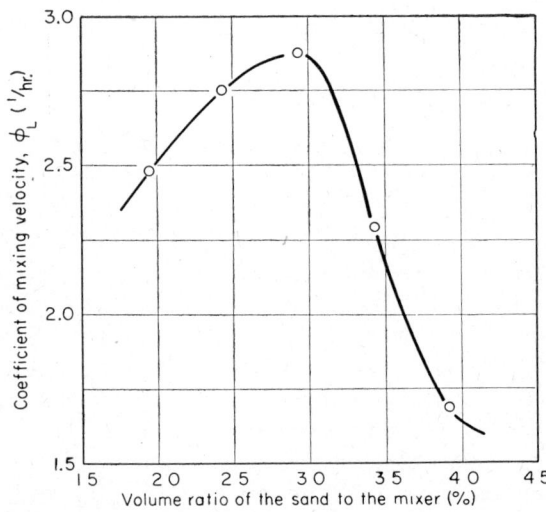

Fig. 36. Relation between volume ratios of the sand to the mixer and values of ϕ_L. Same equipment and method of measuring degree of mixing as used for Fig. 35. Speed = 81.4 rpm (O9).

determined. This indicates that the optimum volume ratio of sand to mixer was about 30% for the conditions shown.

d. Mixing of Solids of Two Different Sizes. A very interesting series of experiments by Oyama (O5) showed that the equilibrium state, when solids of different sizes are mixed in a horizontal rotating cylinder under certain conditions, is associated with a definite type of segregation. In fact, even if a well-mixed batch is put into the cylinder, it will end up in the segregated state after mixing. When Oyama mixed black and white limestone, he found that there were alternately arranged strips of black and white particles perpendicular to the horizontal axis of the cylinder, when the mixer had been stopped. These strips, which were visible on the inclined surface of the bed in the cylinder, were reproducible for the same conditions and appeared to get sharper with longer mixing.

The strips became clearer with decreasing weight ratio of the two components (W_1/W_2), where $d_1 = 3.5$ mm. (diameter of white limestone particles), and $d_2 = 0.57$ mm. (diameter of black particles). He reported that they were easily produced at low speeds of the cylinder, even with a larger weight ratio W_1/W_2, and that the frequency of their appearance decreased after this, but again increased at the speed corresponding to the equilibrium state. In Fig. 37, are shown pictures of these strips for certain conditions of rpm, d_1/d_2, and W_1/W_2. The total charge was $W_1 + W_2 = 6$ kg.

These results certainly would indicate, that when two differently sized solids are mixed in a cylinder, some sort of segregation is likely to occur. Testing for the above type of effect where such a system is in use, would be strongly advisable. As has been pointed out in the Introduction to Section VI, B the results of segregation may result in trouble later on in the process.

e. Effect of Baffle (or Flight) in a Cylindrical Mixer. In another part of his work, Oyama (O4) studied the effect of a flight (or baffle) in a cylindrical mixer, using the same photometric method previously mentioned to measure degree of mixing. His graphs indicate that when an unbaffled cylinder was run at other than the optimum speed (80 rpm in this case), much was to be gained in improved mixing by the use of a baffle plate, provided it was not too high in relation to bed depth. At the optimum speed, the unbaffled cylinder gave more rapid and uniform mixing than for any baffled condition. In cases where the baffle can improve mixing, the h/R_f' ratio is important; here h is the height of the baffle, and R_f', the maximum bed depth. Oyama reported that the best results were obtained when this ratio was less than 0.9. An example given for a specific case where the baffle improved mixing is shown in Fig. 38. Note that h/R, rather than h/R_f', is used on the graph. The bed depth was constant for all runs.

296   SHERMAN S. WEIDENBAUM

These results indicate that a baffle would be most advantageous when its use causes the motion in the cylinder to be as close as possible to that existing at the optimum speed in the unbaffled cylinder.

(Note: In the discussion of the paper by Adams and Baker, Johnson (J1) mentions a horizontal rotating drum, approximately 21 in. long and about 1 ft. in diameter, which was used for fundamental studies. He notes that without baffles, when the drum was one-third full, mixing was far from perfect, even after 81 hours. A specific kind of baffling,

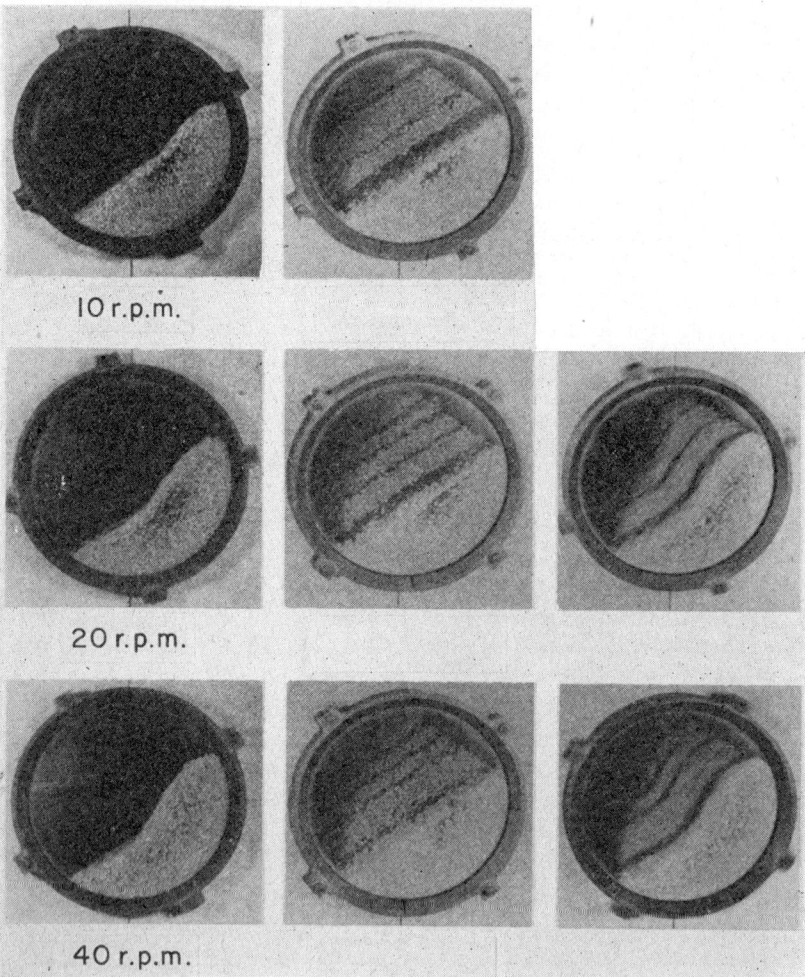

FIG. 37. Photographs illustrating segregation of a mixture of two different sizes limestone the coarser one. Cylinder rotated at indicated speeds. $d_1/d_2 = 3.5$ mm.

which is not elaborated on, managed to improve mixing so that a fairly wide specification could be met.)

f. On Power Consumption of Horizontal Rotating Cylinder (O7). Oyama measured the power consumption under various mixing conditions for certain systems. Some of his results are illustrated in Fig. 39, which gives plots of power in kg.-m./sec. *vs.* rpm. Each graph covers three different diameter cylinders at a certain volume per cent loaded. A major point is that the maximum rate of mixing occurred at a different speed

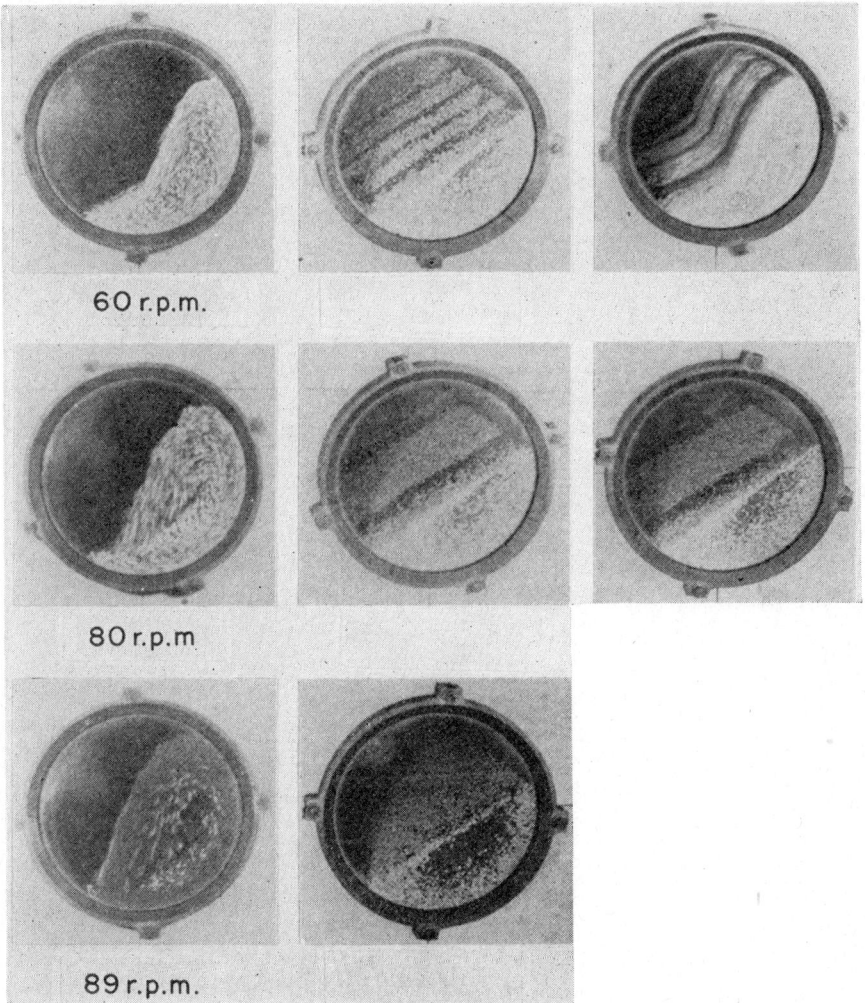

of particles into strips upon mixing. Black limestone is the finer component, and white 0.57 mm., $W_1/W_2 = 3/3$, $W_1 + W_2 = 6$ kg. (O5).

from that which required the most power. Thus, highest power consumption does not necessarily mean highest rate of mixing. Specifically, the "equilibrium" state of mixing required the highest power consumption although the maximum rate of mixing occurred *between* the "critical" and "equilibrium" states.

3. *Cylinder Rotating at Various Angles with the Horizontal* (C3, M2, V1)

Using a small drum mixer, 6 in. diameter and 9 in. long, Coulson and Maitra investigated the effect of several equipment and material varia-

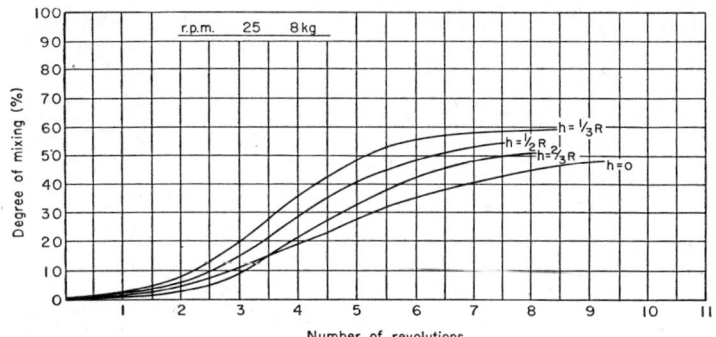

FIG. 38. Relation between degree of mixing and number of revolutions for various conditions of baffling. 20 cm. × 40 cm. horizontal cylinder rotating at 25 rpm. Black and white particles of about 1.3 mm. diameter mixed. Degree of mixing measured photometrically. Total weight of solid is 8 kg. h = height of baffle plate, R = radius of cylinder (O4).

bles on rates of mixing and segregation tendencies. As mentioned under Rate Equations, they used graphs of $\ln 100/X$ vs. kt to plot the data, ignoring the initial curved portions of these graphs. The k's, obtained from the latter straight-line portion of these graphs, were determined for various systems. From preliminary experiments with black and white glass spheres in a glass drum they concluded that the maximum quantity of material that can be effectively mixed is that which forms an elliptical surface with one apex at point A of the drum shown in Fig. 40. They also stated that if a free surface such as A_1A_1 is formed, then virtually no mixing will take place. The effect of several other variables will be separately discussed. Before this, however, some qualifying remarks are necessary in order to point out the limitations of these results.

First of all, the variables were changed one at a time using an arbitrary condition as standard, and then a number of rate constants were

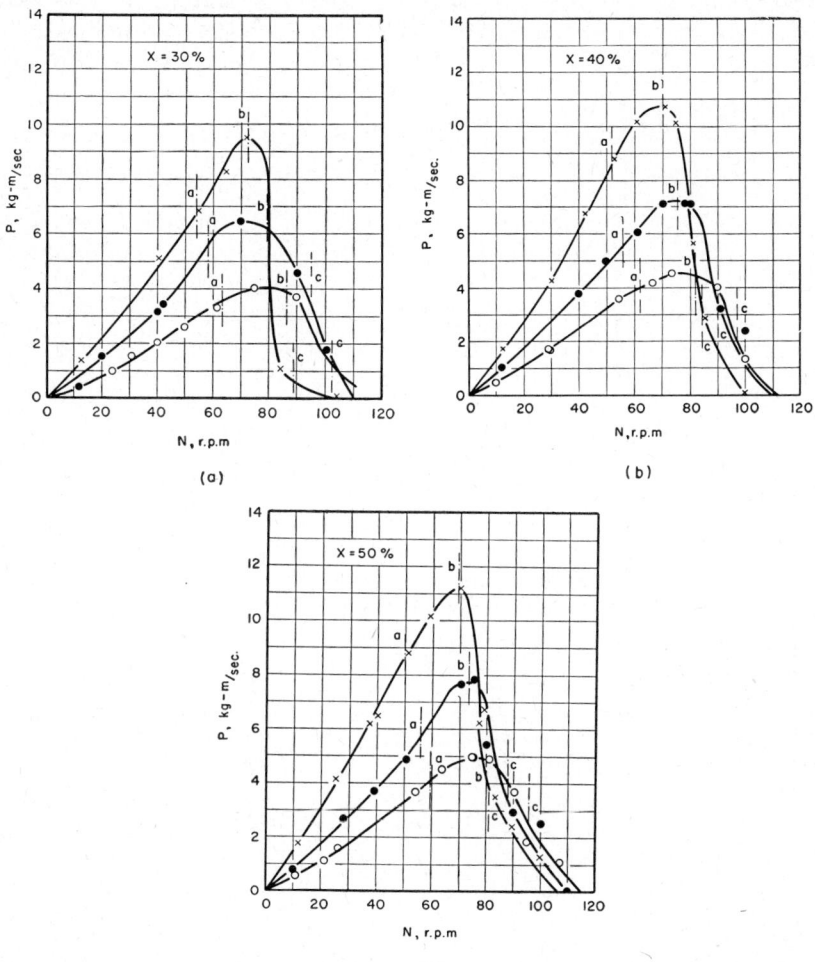

Fig. 39. Power consumption *vs.* rotational speed of a rotating horizontal cylindrical mixer. Limestone was mixed, particle size about 1.35 mm. Experimental conditions varied as shown (O7). a = rotational speed corresponding to critical state; b = rotational speed corresponding to equilibrium state; c = rotational speed corresponding to the state when all particles tend to rotate with the inner wall of mixer; D = diameter of mixer, m.; ○: $D = 0.25$; ●: $D = 0.30$; ×: $D = 0.35$. N = rotational speed of mixer, rpm; P = power consumption of the horizontal cylindrical mixer, kg.-m./sec.; X = per cent of mixer volume occupied by batch.

obtained as mentioned above. By changing one variable at a time, the effects of varying two quantities simultaneously cannot be determined. Thus the optimum angle of inclination which was determined at a certain speed may not be the same at another speed. (The statistical term for these effects is "interaction.") Also, extreme care is necessary when generalizing from these results to larger or smaller equipment. Furthermore, as was previously mentioned, the k's are based on the latter part of the rate plot, since there is a curved initial portion.

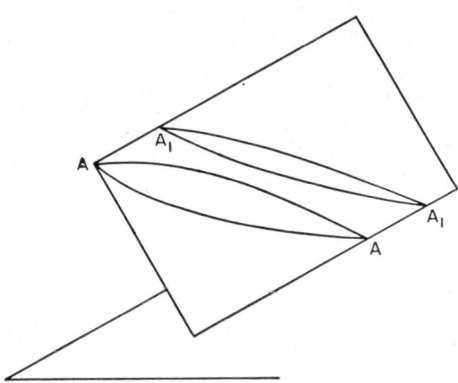

Fig. 40. Diagram showing maximum quantity of material which Coulson and Maitra felt could be effectively mixed (C3). This has surface AA. With surface A_1A_1, the authors state, "virtually no mixing will take place."

Yet, despite these limitations, there is some worthwhile information contained in a study of the rate coefficients. Graphs of these for different conditions are shown in Fig. 41. Figure 41(a) points to an optimum angle of inclination of approximately 14°, which the authors state can be shown to correspond with the maximum free surface. Furthermore, they state, that for angles less than 8° and greater than 30°, the mixture never achieves a high degree of dispersion. Figure 41(b) shows the effect of speed of rotation on k for an angle of 23°. (The authors do not explain why they chose this rather than the optimum of 14°.) Approximately 55 rpm was the optimum, although no runs were made at rpm's between this value and 80 rpm. Poor mixing was obtained at both very low speeds and very high speeds. Here a parallel can be drawn with Oyama's work, where the "cascading" motion at low speeds did not give good dispersion, and at high speeds the whole mass rotated, thus not promoting mixing.

Figure 41(c) indicates that faster dispersion was obtained with smaller particle sizes, both components having the same size. Where particles of different sizes were mixed, if the fines were put on the bottom no mixing was reported. With coarse material loaded on the bottom of the mixer, mixing reached an optimum degree of dispersion, followed by

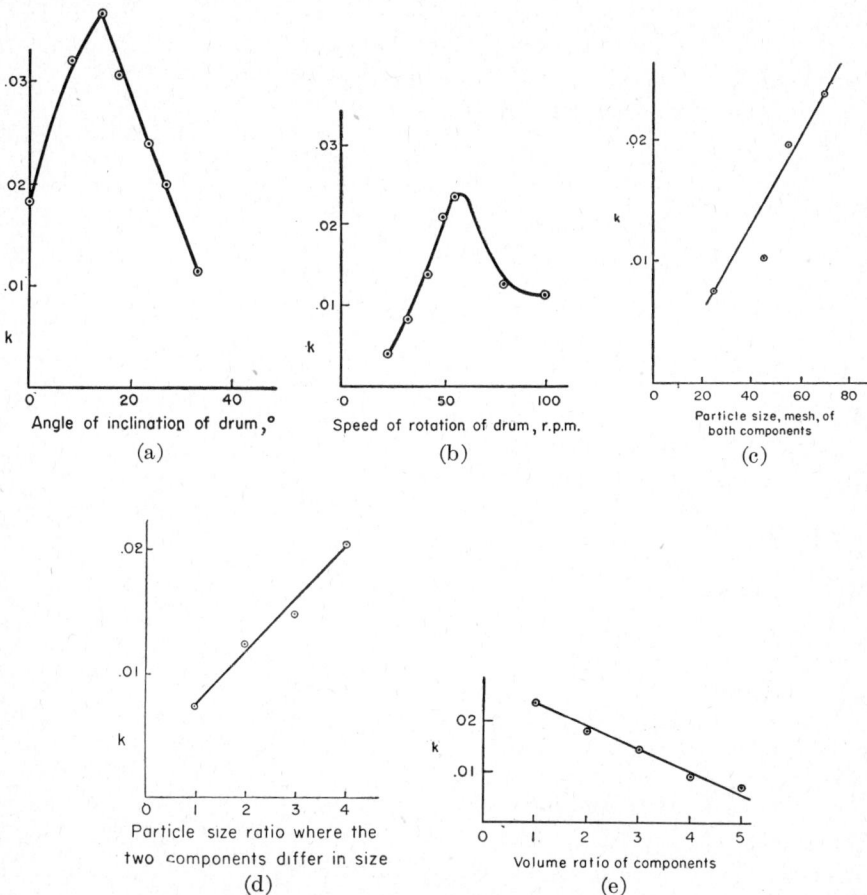

Fig. 41. Manner in which rate constant k, of Coulson and Maitra (C3), changes with different equipment or material variables.

unmixing, the rates of mixing and unmixing being approximately numerically equal. Thus, while a greater size ratio between the two materials being mixed would speed up mixing (see Fig. 41(d)), it would also speed up unmixing. These tests were made with 25-mesh coal, and salt ranging from 25-mesh in some tests to 95-mesh in others. The effects of density differences on mixing were determined with lead nitrate-coal and barium nitrate-coal systems, particle sizes being 70-mesh. The latter system gave better dispersion than could be obtained with the lead nitrate and coal. Also, with the lighter material (coal) loaded on the bottom, the lead nitrate-coal system went through a maximum degree of dispersion, and segregated upon continued mixing. A well mixed coal-lead nitrate system which was put into the mixer segregated upon mixing.

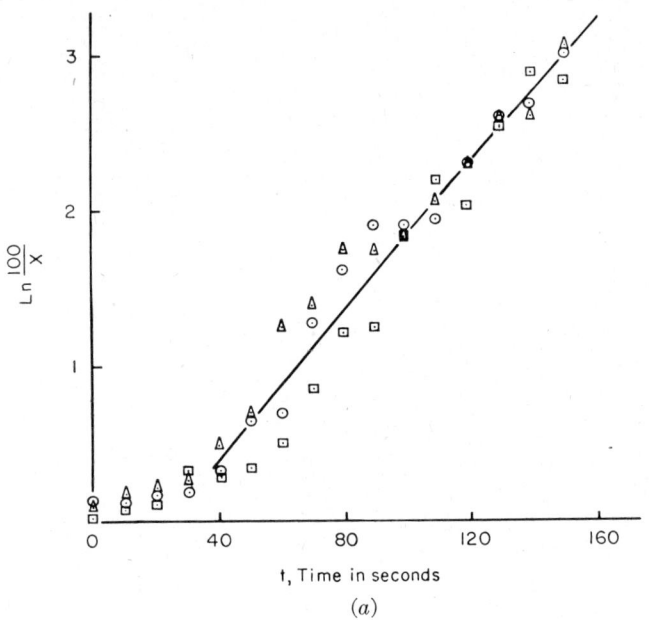

Fig. 42. Data from Coulson and Maitra (C3) plotted in several different ways. System: 70 mesh coal and 70 mesh salt mixed in a cylinder at a 23° angle with the horizontal. Speed 55 rpm. (a) Original graph given by Coulson and Maitra showing replications. X = per cent unmixed (C3). (b) Replot by Visman and Van Krevelen using Coulson and Maitra's method but with horizontal and vertical axes switched (V1). (c) Visman and Van Krevelen's probability paper replot of Coulson and Maitra's data (V1).

Figure 41(e) shows that increase in volume ratio of the two components causes k to decrease.

Although primarily concerned with a certain type of rate plot, Visman and Van Krevelen, in replotting the data of Coulson and Maitra, have postulated a reason why a break in their rate plot occurs. Since this has a bearing on equipment performance, it will be discussed here even though Visman and Van Krevelen did not actually perform any experiments themselves. Figure 42 shows the probability paper plot of Coulson and

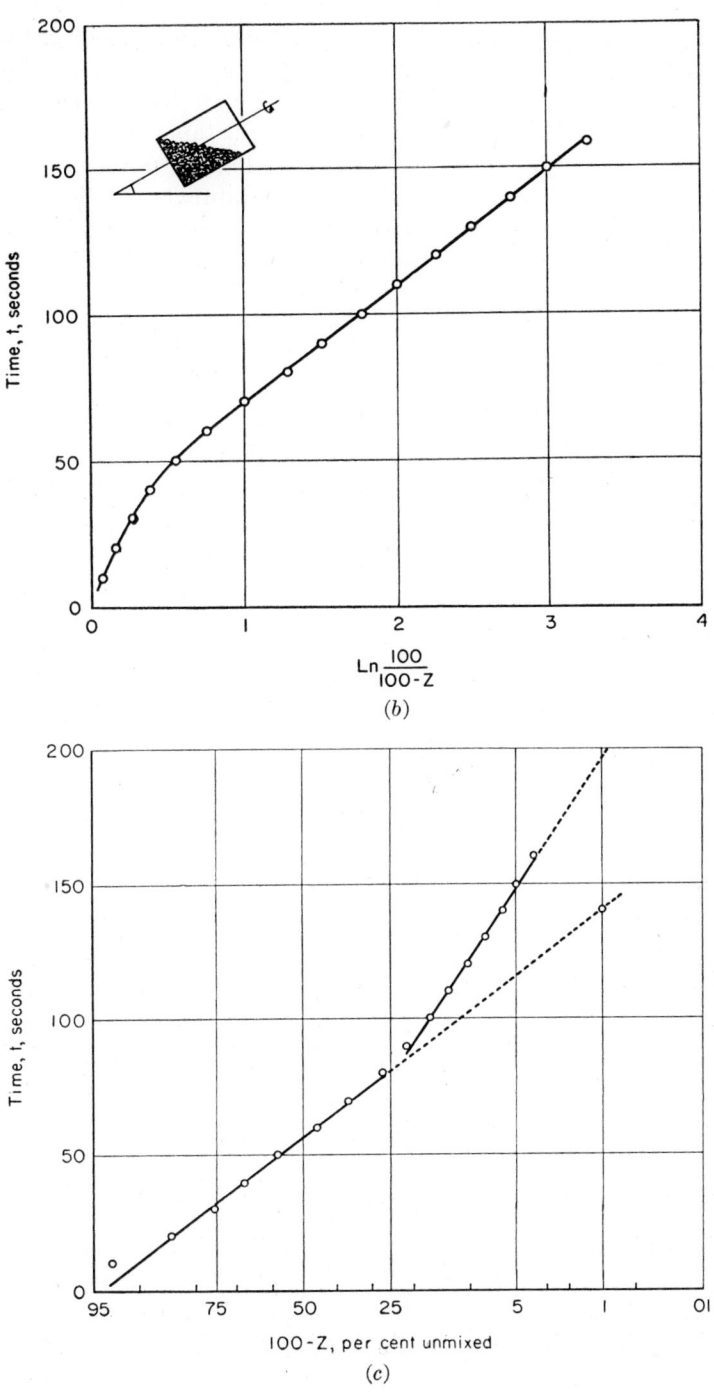

Fig. 42 (Continued)

Maitra's data, together with the original method of plotting, for a 70-mesh coal-salt system in a rotating drum at an angle of 23° with the horizontal and a speed of 55 rpm. Visman and Van Krevelen explain the break as follows: at the start of mixing one substance was loaded on top of the other so that the top surface of the sloping portion was almost entirely made up of one substance. When however, sufficient mixing had been obtained to give a mixture of two kinds of particles on this surface, an unmixing effect became appreciable due to differences in the manner in which the two substances roll down the surface. This segregating tendency slowed down the subsequent rate of mixing.

4. *Horizontal Rotating Cylinder* (W2)

Although this paper has been mentioned in several other sections, the segregation phenomenon reported warrants a separate discussion here under Equipment Performance. Using a horizontal rotating cylinder of $5^{13}/_{16}$ in. in diameter and $8^{5}/_{16}$ in. long, Weidenbaum and Bonilla mixed 40 to 50 mesh (U.S.) salt and sand, initially loaded on opposite sides of the cylinder. After a long period of mixing, it was found that the sand concentration was higher in the center of the mixer than at the ends. This segregation was attributed to differences in tumbling properties between salt and sand and to tumbling surface curvature near the end faces. The higher angle of repose of salt was believed to give a smaller horizontal component towards the center of the mixer to the salt particles than to the sand particles as the materials tumbled, thus giving more salt near the end faces and more sand in the center portion. Figure 43 illustrates the concentration gradient existing after mixing.

5. *Cylinder Rotating at Angle with Horizontal* (B4)

By using identical sands (except for color) in a cylinder inclined at an angle with the horizontal, Blumberg and Maritz (B4) showed that a random mixture of the two colored sands could be obtained. This is one further piece of evidence which, together with other similar studies of cylinders mentioned, points out that the problem in achieving a good mixture arises when there are dissimilar substances with segregating tendencies. When the substances are identical, except for color—as was the case here—a random mixture can be produced.

6. *Horizontal Rotating Cylinder* (O9)

Again, while phases of this work have been mentioned elsewhere, it is worth emphasizing one aspect related to equipment performance. This is the fact, that when identical but differently colored sands were mixed (Sōma sand—half colored with Rhodamine B and the other half colored with methylene blue), the rate equation mentioned in Section V, B was followed throughout mixing. However, when two different sands were mixed (Toyoura sand and Chigasaki sand), an irregularly shaped graph of $1 - M$ vs. t was obtained, rather than a straight line. Although large spot samples were taken in the case where two different sands were mixed, and small ones where sands differing only in color were used, these differences should not cause the change in the rate curve which occurred.

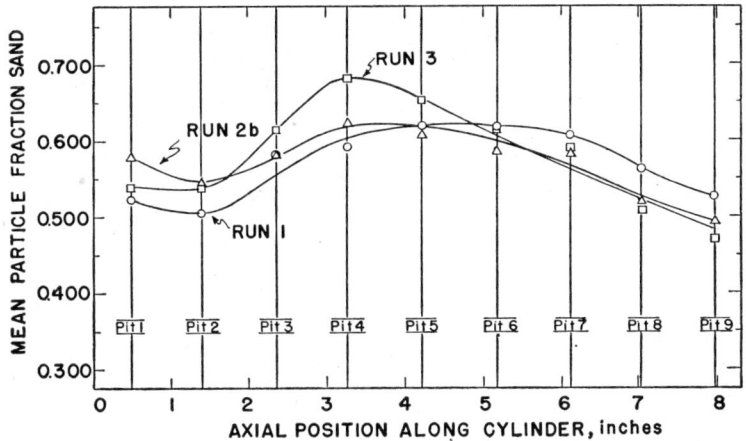

Fig. 43. Axial concentration gradient after a long period of mixing (W2). System: −40 to +50 (U.S.) mesh sand and salt in a horizontal cylinder rotating at 45 rpm; approximately one-third full.

Rather this is indicative of a segregating tendency between the two sands in this particular mixing system, and appears to be similar to the effects reported by Weidenbaum and Bonilla (W2).

7. *Comparison of Several Mixers* (A1)

Adams and Baker used the methods previously described in Section III, D, 21 and Section IV, F to test four common types of blenders. They blended natural polythene granules with black granules which contained additives. The particles were cubes of ⅛-in. side, or cylinders or spheres about ⅛ in. in diameter.

Their results, using the three test criteria, are summarized in Table V. The large rotating cubes are seen to have given the best results. However, since several sizes were not tried for the other types of equipment, no positive conclusions can be made on the basis of the data presented.

8. *Sigma Mixer* (S5)

In testing an equation which he gives for calculating the theoretical variance of samples taken from a randomly mixed batch of two materials, P and Q, Stange uses data for a laboratory mixer which he refers to as "an older model of the Werner & Pfleiderer firm, Stuttgart." This is a sigma-blade mixer, which ordinarily is used for sticky materials rather

TABLE V
Data on the Performance of Dry Solids Mixing Equipment[a]

System: Natural polythene granules (sp. g. = 0.92)—black polythene granules (sp. g. = 1.2) containing additives.
Size: Cubes of $\frac{1}{8}''$ side, and cylinders or spheres about $\frac{1}{8}''$ in diameter.

Type of blender	No. of turns	Fraction of samples outside of 90% "control" limits	If trend was indicated, No. of consecutive samples either above or below mean	Probability that the distribution of black particles in the samples came from a random mixture
Double cone	500	0.55	12 above mean 15 below mean	<0.01
Double cone (control)[b]	500	0.02	no trend	0.7
Ribbon blender, type A	not given	0.37	9 above mean 9 below mean	<0.01
Ribbon blender, type B	not given	0.22	7 below mean	<0.01
Twin-shell blender	100	0.18	7 above mean	0.025
Twin-shell blender	300	0.31	no trend	0.005
12" cube	320	0.25	no trend	<0.01
18" cube	250	0.14	no trend	0.2
48" cube	—	—	—	0.2 to 0.75 (range over a dozen results)

[a] Tabulated from Adams and Baker (A1).
[b] Materials mixed were almost identical except for color.

than dry solids. The systems mixed are shown in Table II. Using a simplified version of the equation for the standard deviation for a random mixture, Stange calculated this quantity for the three systems. He then showed graphically, using 90% and 98% confidence limits, that in all three cases a random mixture of the systems shown was achieved. The fact that this could be done even in the system for 3 F, in which there was a particle size ratio of approximately 4 to 1, is worthy of note. This was achieved, according to Stange, by the addition of 2 wt. % distilled water to those systems in which there was a difference in particle sizes between the two components being mixed.

9. *Comparison of Several Mixers with Different Materials* (G2)

Gray presented performance data for the mixing of three types of particulate solids in several of the commonly used kinds of dry solids mixing equipment. A probe containing a light and photocell was used for obtaining measurements related to composition of the material adjacent to the photocell, and the standard deviation of the reflectivity probe readings was used as a measure of the uniformity of composition of the solids mixtures. Three solid-solid systems were used as follows: sand-ilmenite, barium sulfate-ilmenite, and aluminum oxide–ilmenite. The principal conclusion reached was that the rate of mixing obtained depended on the properties of the solids mixed, as well as the type of equipment. When segregating tendencies were diminished, there were fewer differences among the various pieces of equipment. Detailed graphical and tabular comparisons were given by Gray for the various machines, and a summary of some of the highlights of Gray's data is shown in Table VI. The method of loading the materials was shown to affect the rate of mixing in the ribbon mixer.

An analysis of this work indicates the complexity and large number of variables involved in mixing. For example, Table VI shows that a V mixer rapidly mixed sand and ilmenite to a degree of mixing which did not change appreciably with further mixing. However, in the case of aluminum oxide and ilmenite, after 92 minutes the mixing still seemed incomplete. This apparent enigma might be explained by the possible segregation tendencies of sand and ilmenite balancing out the mixing effect after a certain level of mixedness was reached. However, where segregating tendencies were probably not as great (as with aluminum oxide and ilmenite), mixing would continually improve, although much more slowly as the driving force decreased due to the batch becoming better mixed. This one item, of which there are probably several others that careful scrutiny of the original paper will reveal, illustrates the complexity of analyzing the solids mixing operation.

10. *Finger-Prong Mixer with Materials Having Varying Moisture Content* (M4)

Using a small finger-prong mixer (7.2 cm. wide × 10 cm. long × 10 cm. deep), Michaels and Puzinauskas studied the effect of water content and volume of solids on mixer performance. Their work involved mixing a water soluble powder (dextrose), with a clay (kaolinite), and water. The material consistency varied from a dry powder to a liquid.

TABLE VI
Data on Performance of Dry Solids Mixing Equipment[d]

Type of mixer	Volume (cu. ft.)	Speed (rpm)	Total lbs. of material mixed	Range of % ilmenite in spot samples obtained, % ilmenite[a,b] after 10 min.[a]	Minimum range ilmenite[a,b]	Time to obtain this minimum, minutes[b]	Did unmixing occur after this minimum?[b,c]
System 1:[e] sand (65–200 mesh)—ilmenite (150 mesh–10 μ) (Weight ratio sand/ilmenite = 10, except where otherwise noted)							
Ribbon—ilmenite loaded at one end	3.2	90	174.9	2.3	2.2	22.3	Not continued much past minimum.
Ribbon—ilmenite loaded on top	3.2	90	108.3 to 174.9	2.3	2.2	22.3	Not continued much past minimum.
Muller	2.75	33.3	108.9	2.8	2.1	108	No, range still decreasing.
V mixer	0.5	32.6	18.15	3.2	3.0	1.22 to 1.84	Slightly, to about 3.6%.
Concrete mixer (sand/ilmenite wt. ratio = 9.1)	15	16.2	269.9	3.6	3.0	3.65	Yes, to about 6.0%.
Planetary paddle	0.42	60 (planetary speed)	17.71	4.2	3.8	38.4	No, range still decreasing.
Conical	3.2	25.8	145.2	5.0	4.7	12.2 to 18.4	Yes, to about 6.0%.
Cylinder	0.67	40	39.6	5.8	5.8	8.7	No
Conical	0.45	32.6	18.15	7.1	7.0	1.84 to 2.45	No

MIXING OF SOLIDS

System 2:[c] aluminum oxide (100–270 mesh)—ilmenite ore (65–270 mesh)

(Weight ratio $\frac{Al_2O_3}{ilmenite} = 4.0$)

V mixer	0.5	32.6	18.1	3.6	3.1	92	No, range still decreasing.
Conical mixer	0.45	32.6	18.1	4.1	3.8	2.75 to 3.08	Yes, to 6.3.

System 3:[c] barium sulfate (20 μ to finer than 0.5 μ)—ilmenite (80 μ to finer than 1.8 μ)

(Weight ratio $\frac{BaSO_4}{ilmenite} = 5.0$, except where otherwise noted)

V mixer (BaSO₄/ ilmenite wt. ratio = 4)	0.5	32.6	8.01	2.1	1.8	21.5	No, range still decreasing.
Muller	2.75	33.3	48.0	2.1	2.1	6.1	Yes, to 3.0.
Ribbon mixer	3.2	90	67.2 and 94.1	1.0	1.0	8.9 and 10	Very slightly, to 1.1%.
Hammer mill	—	single rapid pass	27.2 (mixture from ribbon mixer)	—	about 0.9	—	—

[a] For 90% of readings; based on interpolated standard deviation of meter readings and assumption that they are normally distributed. Because of the limitations of the photoelectric probe, small agglomerates may still be present even when a very low range is obtained.
[b] Read from graphs and converted by chart given in Gray's paper.
[c] The percentages given here are the ranges in % ilmenite after unmixing occurred.
[d] Tabulated from Gray (G2).
[e] Detailed size distribution of materials are available in (G2). Predominant size ranges are listed here.

going through the intermediate states (in the order of increasing water content) of pellet and powder, pellet, plastic state, and sticky state. Table VII summarizes the general trends in mixing characteristics. The method of measuring degree of mixing is shown in Section III, D, 20. The authors concluded that there was an optimum mixture volume at which *most rapid* mixing occurs for each clay system at a given water content, although the maximum efficiency of utilization of *mixing energy* was achieved at mixture volumes far below this optimum. As indicated in Table VII, the dry solids mixing operation gave the most rapid mixing with the most efficient utilization of energy.

TABLE VII

RELATION OF MIXING CHARACTERISTICS TO WATER CONTENT OF THE MIXTURE[a]

[System: dextrose-kaolinite-water.
Powder A to be blended with a mixture of B + liquid.]

Increasing liquid content of $B \rightarrow$

Physical appearance	powder	pellet + powder	pellet	plastic	sticky	liquid
Rate of homogenization	rapid	slow	very slow	rapid	slow	rapid
Power input	low	low to moderate	moderate	high	moderate	low
Efficiency of utilization of mixing energy	very high	fair	fair to poor	very poor	poor to fair	high

[a] From Michaels and Puzinauskas (M4).

The high liquid content system, if run at the proper mixture volume, also gave rapid mixing with efficient utilization of energy. However, the graphs show that at lower mixture volumes the high-liquid content system gave the poorest uniformity after mixing. Since the mixture volume was not varied for the dry solids condition, it was not possible to determine whether this state would have been as sensitive to changes in mixture volume. A variety of graphs is given to enable determination of the optimum conditions, when neither dry solids nor high-liquid content is used. For these intermediate cases, the most rapid mixing occurred when the water-clay ratio was in the neighborhood of 0.3, although this gave low utilization of energy efficiency. More efficient use of mixing energy and mixer capacity could be made at higher water contents without great sacrifice in rapidity of mixing.

This paper gives useful methods for aiding in the selection of optimum conditions for mixing. Also, the observations concerning the physical nature of solid systems of different water contents are of general use. Concerning the specific data given for this mixer, a quote from the paper is worth repeating (M4): "It should be kept in mind, however, that extrapolation of such data to mixing operations carried out in different-size or different-type apparatus, or with different solid-liquid systems, is open to serious objection." In particular, it is well to bear in mind the *small size of the mixer* used in this work.

11. *Machines to Mix Additives with Soil* (S3)

This paper gives performance data for machines used to mix additives with soil. Both batch and continuous mixers are discussed. Details of

TABLE VIII
Mixing Coefficients[a] for Various Natural Soils in Several Mixers[b]

Soil	Plasticity index	Batch mixers, 3 min. mix		Continuous mixers		
		2-arm kneader	Muller	Centrifugal disk	Reducing helix	Reciprocating conveyor
Sand (91% below 5 mesh, 12% below 200 mesh)	0	100	125[c]	15.9	—	38.5
Silt	3	67	29.4	22.2	20.4	45.5
Dunkirk clay	15	28.5	26.3	18.2	22.7	50
Marshall clay	33	20.8	16.7	—	27.8	83

[a] Crudely hand-mixed soil, with the additive still plainly visible, gave mixing coefficients of 7 to 12.
[b] Compiled from data of Smith (S3).
[c] 1.5 min. mix.

various mixer characteristics are given and their mixing action discussed. Using the measure of mixing defined in Section III, D, 22, Smith has tabulated mixing coefficients for the various mixers with natural soils.

Table VIII, which is a combination of two of Smith's tables, shows the necessity for choosing the proper mixer for the particular soil to be mixed; for example, although the two-arm kneader and muller mixed very well with sand, they did not do well with the Marshall clay. The reciprocating conveyor, however, while it did not mix sand too well, did the best job on Marshall clay.

Smith also gave mixing coefficient *vs.* time graphs for the soils in a laboratory kneader. He stated that whereas small quantities of soil for laboratory or small-scale field tests are mixed in stationary batch mixers such as tumbling drums, mullers, finger-prong mixers, or two-arm kneaders, the large quantities involved in field mixing require continuous machines such as continuous tumbling drums and pugmills, either single-shaft or twin-shaft.

Along with his own experimental results, Smith summarizes other papers related to soil mixing. He gives graphs from papers which, although evidently unpublished, had been presented at various meetings dealing with road construction problems. These graphs, from papers by Baker (B1) and Clare (C2), give quantitative evidence that the strength and

TABLE IX
Dimensions of Machines Used for Mixing Additives with Soil[a]

Mixer type	Dimensions
Centrifugal disk	14-inch disk rotating at 1750 to 3500 rpm.
Reducing helix	Chamber: 14 in. maximum diameter
	4 in. diameter at discharge.
	Agitator driven at 228 rpm by a 5 hp motor.
Reciprocating conveyor	Cylinder: 4 in. diameter and about 4 ft. long.
	Motor size: 10 hp.
	Shaft speed: 39 to 69 rpm.
2-arm kneader	No dimensions given, but it was called a laboratory kneader.
Muller	No dimensions given.

[a] Compiled from data of Smith (S3).

performance of soil-cement improves with increasing thoroughness of mixing. Again referring to Clare, Smith discusses when it is economically justifiable to incur added expense of thorough mixing, and when it is not.

He states that excessive amounts of power are not required to achieve high uniformity, even in plastic soils, in the appropriate mixer, although the mixing of dry, free-flowing soil requires less power than that of wet plastic soil. The maximum energy used for mixing by the *Ko-Kneader*, at various conditions of plasticity and at throughput rates between 200 and 750 pounds per hour, was 1 HP-hr. per ton of soil mixed (power load on the empty machine was 2 HP). Traveling in-place road mixers use less energy for mixing, Smith reported, the *Madsen Road Pug* requiring about 0.5 HP-hr. per ton and the *Wood Roadmixer* 0.75 HP-hr. per ton, for both mixing and propelling the vehicle. The *P and H Stabilizer* was reported as using 1 HP-hr. per ton for propulsion, excavation of the soil, and mixing.

Since mixer size was not varied for any machine, the effect of this variable on performance cannot be determined, although it probably would affect the mixing performance. The dimensions of the mixers tested are therefore listed in Table IX.

In his conclusions, Smith also covers some general information, besides the specific tests on mixers and soils of Table VIII. He states that a stationary mixer is preferable to a traveling mixer for treating heavy soils, especially with trace additives; and also that variation in consistency and moisture content of soils, plus the presence of occasional large stones, must be handled by a commercial mixer. He also states that precrushing of soil fed to a mixer with close clearance is needed, and that good control of the feed rates is important.

12. *Comparison of Muller and Ribbon Mixer* (L3)

In a study prepared for a muller manufacturer, Lofton *et al.* compared the ability of a muller and ribbon mixer to mix minerals, and also dairy and poultry feeds. They concluded that for the particular systems used, the muller produced greater uniformity in a shorter time, and that feeds mixed in the muller had less of a tendency to segregate upon further handling than those mixed in the ribbon mixer. Six of the 25 runs tabulated in the report (L3) are summarized in Table X, to give some idea of the type of data taken and the method of expressing results. There were several instances where the average of the samples was higher than the per cent tracer added, indicating that more representative sampling would have been desirable.

The segregation tendencies of the mixed feed materials were studied, utilizing the dry dairy feed for the test, and by employing a *Ro-Tap Soil Analyzer* and nested screens upon which the mixture was placed. On the basis of analysis for NaCl in one sample for each mixer, before and after shaking, it was concluded that there is considerably greater tendency for the feed mixed in the ribbon mixer to separate on handling than for that mixed in the muller. More samples would have been desirable in order to get an idea of the variation in NaCl content within the mixtures from each mixer, both before and after "tapping." This would have permitted a better evaluation of the significance of the difference between the mixtures after tapping.

Lofton *et al.*, mention some theoretical background for the case of mixing two materials, each with a known content of a common tracer material. In their introductory portion, they show diagrams to illustrate the effect of mixing on the distribution of tracer fraction in spot samples from a batch in which each of the two materials being mixed originally had a different fraction of a common tracer.

TABLE X
Data on Performance of Dry Solids Mixing Equipment[a,b]

Type of mixer	Dimensions	Material mixed	Rated capacity, lb.	Number of lb. mixed	Speed (rpm)	% Tracer added	Number of samples taken	After 1 minute mixing time, % tracer in samples	
								Average	Range
Muller	mixing chamber, 48" diam.	limestone (mineral feed)	500	500	42	0.90 KI	4	0.93	0.04
	depth 32"	dairy feed	200	190	25	1% NaCl, 5% molasses	4	0.97 (NaCl)	0.31
	2 rollers with 25" diam. and 5" face	poultry feed	200	200	39	1% NaCl (dry run)	4	1.05	0.04
Ribbon	36" long 20" wide at the top	limestone (mineral feed)	cover axle by approx. one inch	400	35	0.90 KI	3	1.18	0.48
	21¾" deep-dished bottom	dairy feed		100	35	1% NaCl, 5% molasses	3	0.90 (NaCl)	0.43
		poultry feed		100	35	1% NaCl (dry run)	3	1.39	0.27

[a] Only a few of the 25 runs are summarized here. Further details are given in the report (L3).
[b] Tabulated from Lofton et al. (L3).

13. *Helical Flight Mixer* (G3)

Data concerning helical flight mixers with a magnesium sulfate–sodium bicarbonate system were given by Greathead and Simmonds, together with a description of the types of particle movements which they felt occurred in this type of mixer. They mention coarse mixing, which takes place by movement of slugs of powder through the material, and fine mixing, which occurs at the boundaries between moving and stationary material. They postulate that local bulk density changes in the wake of the moving blades may cause segregation as the powder settles back behind the blade. This they compare to the ore-dressing operation known as "jigging."

The effects of several variables are depicted by plots of all the individual sample compositions *vs.* time. Although this procedure does have certain advantages in following changes in composition at individual sample locations, it is difficult to use in making overall comparisons. Thus, for example, it is not clear how the conclusions concerning the effects of particle size and density were arrived at. By computing standard deviations and using statistical methods to determine the significance of differences, a clearer picture of the effects of the several variables could be obtained. Where this was done, e.g., following the change in standard deviation during the further handling of the batch after mixing, the manner in which the conclusion was arrived at that there was segregation due to subsequent dumping and conveying, could be understood.

14. *Twin Shell Blender* (Y2)

In a paper dealing with the V-type mixer, Yano, *et al.* studied the mixing of $-100/+200$ mesh (Tyler) anhydrous sodium carbonate (Na_2CO_3) and polyvinylchloride. Two very small mixers were used, with working capacities of $\frac{1}{4}$ liter and 2 liters respectively. Several graphs were plotted to illustrate the influence of the per cent of the total volume occupied by the mixture, relative amounts of the two ingredients, method of loading, and rpm of the mixer. The optimum rpm for obtaining the smallest standard deviation among spot samples was reported at about 50 rpm for the larger mixer and about 60 rpm for the smaller one. Details on the other variables are given in graphs for which there are English titles, although the article is written in Japanese.

The graph of speed *vs.* degree of mixing, which, for the small mixer, extended from 7.8 to 158 rpm, showed that rather poor mixing was obtained at either very low or very high speeds. Although both of these mixers are on a laboratory scale, it would be of interest to see whether the actual operating speeds of the commercial units of comparable sizes are close to the optima reported by Yano *et al.* Of course, their optima apply to the specific system tested.

To help in evaluating this paper, the writer checked the speed of an available commercial unit which handles five laboratory sizes of the above type of blender. It ran at 34 rpm, which is a constant speed for all five sizes, since the same motor and gear drive are used in all cases; only the method of mounting the different size blenders varies. The commercially available blenders corresponding to the sizes mentioned by Yano *et al.* were about one-third full when loaded to the stated working capacity. A subsequent paper by Yano, Kanise, Hatano, and Kurahasi (Y1), indicated that an F/V percentage of 13.5, which was used throughout most of this study, would correspond to loading the mixer about one-third full. (F = real volume of the charged powder and V = inner volume of the mixer.) It might be worth investigating whether drives for several sizes of small blenders should have adjustable speeds to enable choice of the optimum speed for any size.

15. *Comparison of Several Mixers* (Y1)

A study of the comparative performance of several types of mixers was made by Yano *et al.* (Y1). They worked with two small sizes of each of the following: cube, double cone, horizontal cylinder, and twin-shell types. A short summary, as well as the titles of the figures and tables, were in English, but the remainder of the paper was in Japanese. Study of the graphs, tables, and English summary revealed a certain discrepancy between Conclusion 3 in their English summary and that indicated from their Table 2. Conclusion 3 stated that the cube and V-type (twin-shell) blender gave lower values of standard deviation than the double cone or horizontal cylinder mixers, but took a larger number of revolutions to reach their minimum standard deviations: Table 2 of the same paper indicated that a typographical error may have been made in that conclusion, since the opposite was the case.

From Table 2 of the above paper (Y1), the following conclusions were drawn:

1. In the very small sizes (inner volume = 741 cc.), the horizontal cylinder mixer gave the lowest standard deviation (1.1%), and the twin-shell blender the highest (2.6%). In the larger sizes (inner volume = 5,630 cc.), the lowest standard deviation attainable for the various blenders ranged from 0.9 (double cone) to 1.4 (twin-shell). It is not clear from visual observation alone which differences are significant. Confidence limits would have been helpful here for the interpretation of results. Thus, an approximately eight-fold increase in volume of all the blenders greatly reduced optimum standard deviation differences among them. For all

but the horizontal cylinder, which was not affected, this increase in volume was associated with a cut in the standard deviation of roughly half. This emphasizes the dangers of attempting to apply directly the results of experiments on small blenders to large ones, without adequate scale-up information.

2. The range of optimum volume per cent charged for the four types of mixers includes 31% in all cases.

Graphs were given to show the effect of rotational speed and volume per cent loaded, although they do not reveal how the values in Table 2 of (Y1) were arrived at, in certain cases.

C. Critical Evaluation of Published Performance Data

Because of the dearth of objective published data on the comparisons of various blenders, one may be tempted to place undue emphasis on the few papers that have been published. This may be harmful, because scrupulous attention to one or two particular tests, without considering the broader view of the capabilities of certain types of mixers, can lead to just as erroneous a conclusion as blindly choosing whatever particular mixer is handy when a certain job must be done. Just as certain questions should be asked when choosing a degree of mixing, as was previously pointed out, so should the following questions be asked when attempting to use published data as a guide to choosing a particular mixer for a specific application:

1. What are the major conclusions of the study? For example, although Gray (G2) did rate several mixers for several systems, his principal conclusions were that the rate of mixing depended on the properties of the solids mixed, as well as the type of equipment, and that when segregating tendencies were diminished, there was less difference in performance among the various pieces of equipment. Therefore, to attempt to use the graphs and tables given by Gray as a general guide to which mixer is "best," is to go far beyond the limitations of Gray's exploratory study.

2. Is the type of mixer used in a particular test best suited for the materials? For example, Bullock's discussion of mullers (B7), cautions against their use for materials which are too fluid or too sticky, although he states that if used in the proper ranges, mullers are well suited to the breaking down of aggregates. Obviously, therefore, a study of the mixing action of a muller, made on materials outside of the ranges advised for them, would not prove that the muller was not a good mixer just because the results of such a test indicated poor mixing.

3. Has the effect of size been considered in the comparisons of mixers? For example, in Adams and Baker's paper (A1), the larger cube machines

gave better results than the smallest one, but the size of the other blenders was not varied, and in fact, was not even stated.

4. Is the mixer being run at optimum conditions? For example, in the case of a commercial mixer which is not run at the recommended speed and volume fraction loaded, poor blending does not necessarily mean that the mixer is no good.

5. What method was used to measure the degree of mixing and how does this relate to the needs of the particular mixing operation for which the published data are being studied? For example, Bullock (B8) pointed out the limitations of the reflectivity probe used by Gray (G2) in differentiating between uniform mixes of aggregates and mixes of ultimate particles.

Despite the above precautions that are necessary in evaluating the published data, there is considerable value in the latter in providing objective bases for the choice of mixers. Because of the large variety of machines and materials, much more data on the performance of various solids mixing machines with different types of materials is desirable. A standard testing procedure for dry solids mixing equipment would help in the most efficient accumulation of meaningful data of this type. Such a procedure should pool the knowledge of equipment building and process points of view.

D. Summary and Conclusions and Related Thoughts—Equipment Performance

1. Several criteria have been used in various studies to judge the performance of dry solids mixing equipment: *a.* the uniformity of the batch with regard to either composition or some other property, *b.* the time required to achieve a certain uniformity, and *c.* the power required to mix the batch to a certain degree of uniformity. Most studies have dealt with either *a.* or *b.* Complete quantitative performance criteria should cover all three of these aspects.

2. There are several other factors besides these, however, which will influence the suitability of a particular mixer for a specific application. Such things as ease of cleaning in a case where contamination can be ruinous, or dust proofing, are typical practical considerations which must be taken into account. Although these were not discussed for the various performance tests, it should be borne in mind that they must automatically be considered before choosing a mixer.

3. Ambiguities in certain previously reported test results stress the need for using clear cut objective methods for determining mixing performance. (See Section III.)

4. Several tests on the performance of rotating mixing cylinders, both horizontal and inclined, have been made. A correlation, involving speed, diameter, and volume per cent loaded, has been given for horizontal cylinders. Conditions under which segregation during mixing has occurred are reported for several cases. This segregation is attributable to differences in properties between the materials being mixed. Such things as size, shape, and density differences can cause segregation. Where differently colored but otherwise identical sands were mixed in an inclined rotating cylinder, statistical tests on the data proved that a random mixture was achieved (B4). Graphical evidence has been published (O9) which indicates that, where differently colored but otherwise identical sands were mixed in a horizontal rotating cylinder, a randomly mixed batch was achieved.

5. A few equipment comparison studies have been made. In evaluating these, it is important to bear in mind certain check points which bring out limitations to the studies as well as to their use as guides. These studies do not all clearly point to any particular piece of equipment as being consistently superior. Rather, their careful scrutiny indicates that relative performance varies with the materials mixed as well as equipment size. Therefore, while the comparative studies may serve as rough guides for similar systems, the present limited published knowledge along these lines requires that an individual test be made when considering a particular application. It is very dangerous to generalize from the few reported results to other situations, since it is clear that the material to be mixed, as well as the kind of machine to be used, affects the flow patterns and effectiveness of mixing.

6. The effects of many different variables on the performance of different pieces of dry solids mixing equipment have been shown. Some of these are particle size, shape, density, and mixer speed and volume fraction loaded. Also the relative proportions of the materials being mixed has been studied for certain cases.

7. Published performance data on various pieces of equipment and different materials is still relatively scarce. More of this work would help in the selection of equipment and in the understanding of the effects of equipment and particle variables. Examples which clearly illustrate sound methods of measurement, calculation of product uniformity, and control of product quality; and photographs of mixing and particle movements, would also be helpful in developing this area.

8. A standard testing procedure for dry solids mixing equipment which was available as a means of evaluating equipment performance would help in the most efficient accumulation of meaningful data of this type. Points of view of the men who build and sell the equipment and of

those who must use it in processing should be taken into account by such a procedure. It would, of necessity, have to be flexible.

9. Scale-up relationships should be developed and experimentally investigated. Performance studies, in which various sizes and types of blenders have been used without provision for determining the effect of size, leave a big question mark as to the true relative merits of the different types of blenders.

10. Much can be gained in industrial practice by simply checking operational procedures where solids mixers are concerned. For example, a mixer rated at a certain capacity may, due to production expansion, be overloaded—with detrimental effects on the blend it produces. Or somehow, it may be underloaded, which, besides giving a less than optimum operating condition, can also be costly in that it does not make the fullest use of available plant capacity. These conditions may sometimes also affect related labor costs. Other "obvious" factors, such as making sure that the correct weights of materials are going into the mixer, are worth checking on. The possible loss of material as dust is another good point to check, particularly if there is strong suction in the dust collection apparatus for the mixer. And, as previously stated, the solids mixing operation must be checked from the point of view of delivering a well mixed batch to a certain point. Although this work has mainly dealt with the mixer itself, the subsequent travel of the mixed batch must be scrutinized for possible segregating points, such as long drops or flow through silos.

VII. Overall Concluding Comments

The published literature concerning the field of dry solids mixing is presently emerging from a primarily descriptive explanation of the equipment used, to methods for quantitatively dealing with this unit operation. Several points of view are involved in this transition. The process control point of view looks for statistical ways to treat the solids mixing operation, so as to produce an output within certain specifications. The equipment building point of view has tended to rely heavily on individual tests, with non-statistical judgment as to whether the final product is well mixed. A number of publications have dealt with the development of methods for evaluating extent of mixedness. Some of the concepts most fundamental to chemical engineers in other unit operations have barely been touched upon. Rate constants are still largely academic. Mechanisms of mixing are only sketchy hypotheses. However as more of the pieces of the puzzle are put together, usable methods are emerging which will enable a systematic attack on improving solids mixing performance. Elimination of inefficient mixers with dead spots may yet come out of rate studies. Easy check methods for quantitatively determining optimum operating conditions are breaking through the complicated and sometimes not too easily understood statistical procedures. Theoretical frequency distributions, which permit an understanding of the equilibrium state of the randomly mixed batch, have been investigated and checked against experimental data. They are no longer an oddity, but are on the threshold of broad application, and their proper use in certain problems, for example, the nut problem mentioned in Section III, C, 2, c, iv, can avert unnecessary experimental work and, perhaps, striving for impossibilities. Performance data, after years of neglect, are beginning to appear in increasing amounts in the published literature.

This all adds up to the following picture. Useful methods are now available to carry out fruitful investigation of both the experimental and theoretical aspects of the unit operation of solids mixing. They are still not widely used, but rather appear to be considered a novelty by many people with great practical experience who might make the best use of them. This is not surprising, because their presentation has not been in a streamlined, easy-to-use form, but rather it has involved a groping attempt to apply ideas without too much concern for making them easily understood and used.

What is now needed is much more performance data for many different types of mixers and combinations of materials, properly collected and interpreted, so as to put to effective use for industry the considerable amount of exploratory knowledge now available.

Acknowledgments

The author wishes to express his appreciation to the Corning Glass Works for the encouragement and assistance which helped to bring this work to fruition.

He also thanks Dr. Joseph B. Gray of E. I. du Pont de Nemours and Company for the many stimulating conversations in which he gave freely of advice and time, and Dr. Yoshitoshi Oyama of the Tokyo Institute of Technology, who has been very cooperative in supplying illustrations, English summaries, and additional helpful information.

And to his wife he gratefully acknowledges that her unmixing of material did much to aid the cause of mixing.

Nomenclature

Symbols are defined at the point of their introduction in the text.

References

A1. Adams, J. F. E., and Baker, A. G., *Trans. Inst. Chem. Engrs.* (*London*) **34**, 91 (1956).
A2. Anonymous, *Anal. Chem.* **29**, 19A (1957).

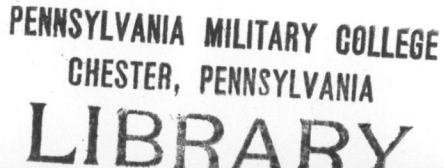

A3. Aspin, A. A., *Biometrika* **36**, 291 (Dec., 1949).
A4. ASTM Manual on Quality Control of Materials. Special Tech. Publ. 15-C (1951).
B1. Baker, C. N., Jr., "Strength of soil cement as a function of degree of mixing," presented at 33rd Annual Meeting, Highway Research Board, Washington, D.C., January 1954.
B2. Barrer, R. M., "Diffusion in and through Solids." Cambridge Univ. Press, London and New York, 1951.
B3. Beaudry, J. P., *Chem. Eng.* **55**, 112 (1948).
B4. Blumberg, R., and Maritz, J. S., *Chem. Eng. Sci.* **2**, 240 (1953).
B5. Brothman, A., Wollan, G. N., and Feldman, S. M., *Chem. & Met. Eng.* **52**, 102 (1945).
B6. Brown, C. O., *Ind. Eng. Chem.* **42**, 57A (1950).
B7. Bullock, H. L., *Chem. Eng. Progr.* **51**, 243 (1955).
B8. Bullock, H. L., *Chem. Eng. Progr.* **53**, 36 (1957).
B9. Burton, L. V., *Package Eng.* pp. 24–27 (Jan., 1956).
B10. Buslik, D., *ASTM Bull. No.* **165**, 66 (1950).
C1. Chilton, T. H., *Trans. Am. Inst. Chem. Engrs.* **31**, 128 (1935).
C2. Clare, K. E., "Some Problems in Mixing Granular Materials Used in Road Construction" presented at Public Works and Municipal Services Congress, London (1954).
C3. Coulson, J. M., and Maitra, N. K., *Ind. Chemist* **26**, 55 (1950).
D1. Dallavalle, J. M., "Micromeritics," 2nd ed. Pitman, New York, 1948.
D2. Danckwerts, P. V., *Appl. Sci. Research*, **A3**, 279 (1952).
D3. Danckwerts, P. V., *Research (London)* **6**, 355 (1953).
D4. Drew, T. B., "Notes on Diffusion and Mass Transfer." Columbia Univ., New York 1950.
D5. Drew, T. B., "Diffusion." Third Institute Lecture, Annual Meeting of AIChE, Atlantic City, New Jersey, 1951.
F1. Forscher, F., *J. Franklin Inst.* **259**, 107 (1955).
F2. Furnas, C., *U.S. Bur. Mines, Bull.* **307**, 47 (1931).
F3. Furnas, C., *Ind. Eng. Chem.* **23**, 1052 (1931).
G1. Grant, E. L., "Statistical Quality Control." McGraw-Hill, New York, 1946.
G2. Gray, J. B., *Chem. Eng. Progr.* **53**, 25J (1957).
G3. Greathead, J. A. A., and Simmonds, W. H. C., *Chem. Eng. Progr.* **53**, 194 (1957).
H1. Hald, A., "Statistical Theory with Engineering Applications." Wiley, New York, 1952.
H2. Herdan, G., "Small Particle Statistics." Elsevier, Amsterdam, 1953.
H3. Hillebrand, W. F., Lundell, G. E. F., Bright, H. A., and Hoffman, J. I., "Applied Inorganic Analysis," 2nd ed. Wiley, New York, 1953.
H4. Hixson, A. W., and Tenney, A. H., *Trans. Am. Inst. Chem. Engrs.* **31**, 115 (1935).
H5. Hoel, P. G., "Introduction to Mathematical Statistics." Wiley, New York, 1947.
J1. Johnson, J. D., *Trans. Inst. Chem. Engrs. (London)* **34**, 105 (1956).
K1. Kasai, K., *Bull. Inst. Phys. Chem. Research (Tokyo)* **11**, 793 (1932).
K2. Kirk, R. E., and Othmer, D. F., eds., "Encyclopedia of Chemical Technology," Vol. 9. Reinhold, New York, 1952.
L1. Lacey, P. M. C., *Trans. Inst. Chem. Engrs. (London)* **21**, 52 (1943).
L2. Lacey, P. M. C., *J. Appl. Chem. (London)* **4**, 257 (1954).
L3. Lofton, W. M., Jr., Moore, W. B., Jr., Goldsmith, S., and Cooperman, N.,

"Summary report to National Engineering Company: Mixing studies on the Simpson Intensive Mixer," University of Louisville Institute of Industrial Research (1948).

L4. Lowry, T. M., "The Incorporation of Amatol." Ministry of Munitions, London, 1917.
L5. Lundell, G. E. F., and Hoffman, J. I., "Outlines of Methods of Chemical Analysis." Wiley, New York, 1938.
M1. McAdams, W. H., "Heat Transmission," 2nd ed., p. 30. McGraw-Hill, New York, 1942.
M2. Maitra, N. K., and Coulson, J. M., *J. Imp. Coll. Chem. Eng. Soc.* **4,** 142 (1948).
M3. Mandelson, J., *Ind. Quality Control* **XIII,** 31 (1957).
M4. Michaels, A. S., and Puzinauskas, V., *Chem. Eng. Progr.* **50,** 604 (1954).
O1. Oyama, Y., *Bull. Inst. Phys. Chem. Research (Tokyo)*, Rept. 1, **12** (12), 953 (1933). In Japanese.
O2. Oyama, Y., *Bull. Inst. Phys. Chem. Research (Tokyo)*, Rept. 2, **14** (7), 570 (1935). In Japanese.
O3. Oyama, Y., *Bull. Inst. Phys. Chem. Research (Tokyo)*, Rept. 3, **14** (9), 770 (1935). In Japanese.
O4. Oyama, Y., *Bull. Inst. Phys. Chem. Research (Tokyo)*, Rept. 4, **15** (6), 320 (1936). In Japanese.
O5. Oyama, Y., *Bull. Inst. Phys. Chem. Research (Tokyo)*, Rept. 5, **18** (8), 600 (1939). In Japanese.
O6. Oyama, Y., *Bull. Inst. Phys. Chem. Research (Tokyo)*, Rept. 6, **19** (8), 1070 (1940). In Japanese.
O7. Oyama, Y., *Bull. Inst. Phys. Chem. Research (Tokyo)*, Rept. 7, **19** (8), 1088 (1940). In Japanese.
O8. Oyama, Y., *Sci. Papers Inst. Phys. Chem. Research (Tokyo)* **37** (951), 17 (1940). In English.
O9. Oyama, Y., and Ayaki, K., *Kagaku Kikai* **20** (4), pp. 6–13 (alternate pagination 148–155) (1956).
P1. Perry, J. H., "Chemical Engineers' Handbook," 3rd ed., p. 1195. McGraw-Hill, New York, 1950.
P2. Pierce, D. E., *Ind. Eng. Chem.* **48,** 65A (1956).
Q1. Quillen, C. S., *Chem. Eng.* **61,** 178 (1954).
S1. Sakaino, T., *J. Ceram. Assoc. Japan* **65,** 171 (1956).
S2. Scott, R. A., *in* "Chemical Engineering Practice" (H. W. Cremer and T. Davies, eds.), Vol. 3, p. 362. Butterworths, London, 1957.
S3. Smith, J. C., *Ind. Eng. Chem.* **47,** 2240 (1955).
S4. Stange, K., *Chem.-Ingr. Tech.* **26** (3), 150 (1954).
S5. Stange, K., *Chem.-Ingr. Tech.* **26** (6), 331 (1954).
T1. Tooley, F. V., and Tiede, R. L., *J. Am. Ceram. Soc.* **27,** 42 (1944).
V1. Visman, J., and Van Krevelen, D. W., *Ingenieur (Utrecht)* **63,** 49 (1951).
W1. Weidenbaum, S. S., "A Fundamental Study of the Mixing of Particulate Solids," PhD Thesis in Chemical Engineering, Columbia University, New York, June, 1953.
W2. Weidenbaum, S. S., and Bonilla, C. F., *Chem. Eng. Progr.* **51,** 27J (1955).
W3. Westman, A. E. R., and Hugill, H. R., *J. Am. Ceram. Soc.* **13,** 767 (1930).
W4. Westman, A. E. R., and Hugill, H. R., *J. Am. Ceram. Soc.* **19,** 127 (1936).
W5. Work, L. T., *Chem. Eng. Progr.* **50,** 476 (1954).

W6. Wornick, R. C., "Premixing Micro-ingredients." Charles Pfizer Lecture Series No. 4, Terre Haute, Indiana, 1956.
Y1. Yano, T., Kanise, I., Hatano, Y., and Kurahasi, S., *Kagaku Kikai* **21** (7), pp. 14–19 (alternate pagination 420–425) (1957).
Y2. Yano, T., Kanise, I., and Tanaka, K., *Kagaku Kikai* **20** (4), pp. 14–20 (alternate pagination 156–162) (1956).
Y3. Young, R. S., and Snaddon, R., *Chem. Eng.* **58**, 160 (1951).

ERRATA

"Theory of Diffusion," Chapter in Vol. 1 of *Advances in Chemical Engineering*

R. B. Bird

p. 166 In Eq. (27) change $+k_i$ to $-k_i$

p. 166 In the fifth line after Eq. (27) change "fifth and sixth terms" to "terms containing $p - T(\partial p/\partial T)$"

p. 168 In Eq. (35) replace $c^2/\rho RT$ by $-c^2/\rho RT$

p. 169 An important relation among the D_{ij} should be added:

$$\sum_{i=1}^{\nu} \{M_i M_h D_{ih} - M_i M_k D_{ik}\} = 0 \qquad (36a)$$

p. 173 In Table III change $\mathbf{j}_A + \mathbf{j}_A = 0$ to $\mathbf{j}_A + \mathbf{j}_B = 0$.

p. 173 In Table III change $\omega_A(\mathbf{n}_A + \mathbf{n}_A)$ to $\omega_A(\mathbf{n}_A + \mathbf{n}_B)$.

p. 184 The location of the minimum in the potential energy curve should be labeled $r = r_m$ and $\varphi = -\epsilon$.

p. 185 In line 5 change (6-12) to (6-exp)

p. 201 In the 4th line after Eq. (121) change $= 1$ to $= p$.

p. 202 Eq. (127) should read:

$$Q_A = 4\pi r_1^{-n/2}[r_1^{-1-\frac{n}{2}} - r_2^{-1-\frac{n}{2}}]^{-1}\left(1 + \frac{n}{2}\right)\cdot\left(\frac{p\mathfrak{D}_{AB,1}}{RT_1}\right)\ln\frac{p_{B2}}{p_{B1}}$$

p. 204 In the line just after Eq. (134b) change $\bar{V}_B = \bar{V}_B$ to $\bar{V}_B = \tilde{V}_B$.

p. 205 On the left side of Eq. (134e) replace (x_{AL}/x_{A0}) by (x_{A0}/x_{AL}).

p. 212 In Eq. (161) change c_{A0} in numerator of left hand side to c_A

p. 214 Eq. (169) should read

$$N_A = k_L(c_{A0} - c_A)_{l.m.}$$
$$= k_L\left[\frac{\bar{c}_{A2} - \bar{c}_{A1}}{\ln\left(\dfrac{c_{A0} - \bar{c}_{A2}}{c_{A0} - \bar{c}_{A1}}\right)}\right]$$

p. 222 In Eq. (203) change v_{A0} and v_{B0} to u_{A0} and u_{B0}

AUTHOR INDEX

Numbers in parentheses are reference numbers and are included to assist in locating references in which the author's names are not mentioned in the text. Numbers in italics indicate the page on which the reference is listed.

A

Acrivos, Andreas, 177
Adams, E. Q., 174, 178, *206*
Adams, J. F. E., 235, 237, 238, 256, 268, (A1), 269(A1), 270(A1), 272, 273 (A1), 288(A1), 289, 305(A1), 306, 317, *321*
Adamson, A. W., 153(B8), 160(B9), 163 (B8), 166(B8), 170(B9), 176(B8), 188, 198(B8), *205, 206, 207*
Addoms, J. N., 13(M3), 17(M3), *31*
Aikman, A. R., 42, *79*
Alberda, G., 179(K6), 183(K6), *207*
Amundson, N. R., 69, *79*, 153, 182(K1), 183(A4, L3), 185, 186, *205, 206, 207*
Anderson, C. R., 92(A1), *115*
Anzelius, A., 153(A3), 180, *206*
Aris, R., 183(A4), *206*
Aspin, A. A., 228(A3), *322*
Ayaki, K., 254, 271(O9), 272, 274, 275, 277, 278(O9), 279(O9), 283, 285, 290, 292, 294(O9), 304(O9), 319(O9), *323*

B

Baddour, B. F., 166(G2), 186, 197, *206*
Baker, A. G., 235, 237, 238, 256, 268(A1), 269(A1), 270(A1), 272, 273(A1), 288 (A1), 289, 305(A1), 306, 307, *321*
Baker, C. N., Jr., 312, *322*
Ballantine, D. S., 108(B1), *115*
Banchero, J. T., 16(B1), 17, *30*
Barker, G. E., 16(B1), 17(B1), *30*
Barker, K. H., 199(G4), *207*
Barrer, R. M., 164, *206*, 285, *322*
Barrow, R. F., 182, *206*
Bauman, W. C., 151, 160, *206*
Bay, T., 75, *80*
Beaton, R. H., 153(B5), 181, *206*

Beaudry, J. P., 246, *322*
Beers, N. R., 89(B2), *115*
Benedict, M., 121(B3), *145*
Bernath, L., 26, *30*
Bilous, O., 69, *79*
Blasewitz, A. G., 103, *115*
Bliss, H., 157, 175, *206, 208*
Blodgett, K., 94, *116*
Blumberg, R., 255, 264(B4), 272, 277, 285, 304, 319(B4), *322*
Bohart, G. S., 174, 178, *206*
Boll, R. H., 16(B1), 17(B1), *30*
Bonilla, C. F., 213, 238, 247, 255, 263 (W2), 270(W2), 272, 274, 275, 276, 285, 304(W2), 305, *323*
Bosanquet, C. H., 101, *115*
Boyd, G. E., 153(B8), 160, 163, 166(B8), 167, 170, 176, 198, *206*
Bradner, M., 64, *79*
Bretton, R. H., 93(M2), *116*
Breyer, F. G., 120(B1), *145*
Bright, H. A., 215(H3), *322*
Brinkley, S. R., Jr., 180(B10), *206*
Brodkey, R. S., 8(B4), 11(B4), *30*
Bromley, L. A., 8, 10, 11(B4), 12(B3), 13(B3), 19(B3), 20(B5), *30*
Brooks, F. R., 152, *206*
Brothman, A., 240, 248, 253, 274, 275, 276, 286, *322*
Brown, C. O., 288, *322*
Brown, G. G., 169(B11), *206*
Brown, G. P., 125, 127, *145*
Brown, G. S., 42, 45(B4), 64(B4), *79*
Brown, R. E., 99, 100(B5), *115*
Brownell, L. E., 169, *206*
Brunauer, S., 150(B12), 155, *206*
Bryant, L. W., 101, *115*
Buchberg, H., 27, *30*
Bullock, H. L., 289, 317, 318, *322*
Burton, L. V., 290, *322*
Buslik, D., 247, 260(B10), *322*

C

Caddell, J. R., 179(C1), *206*
Campbell, D. P., 38, 42, 45(B4), 64(B4), *79*
Carey, W. F., 101(B4), *115*
Carl, R., 25(M4), 26(M4), *31*
Carslaw, H. S., 43, 45(C2), *79*
Cassidy, H. G., 149, 150, *206*
Castles, J. T., 10(C1), *30*
Catheron, A. R., 72, *79*
Ceaglske, N., 42, *79*
Cheng, G. K., 125, 127(B2), *145*
Chestnut, H., 42, *79*
Chilton, T. H., 170, *206*, 240, *322*
Christiansen, C., 113, *115*
Church, P. E., 89(C2), *115*
Churchill, R. V., 43, 45(C6), *79*
Clare, K. E., 312, *322*
Clark, J. A., 26, *31*
Close, C. D., 63, *79*
Coates, J. I., 165(G5), 170(G5), 176, *207*
Cohen, G. H., 74, *79*
Cohen, R. K., 148(H3), 168(H2, H3), 195 (H2), *207*
Cohen, W. C., 52, *79*
Colburn, A. P., 10, *30*, 170, *206*
Cooke, N. E., 133, *145*
Coon, G. A., 74, *79*
Cooperman, N., 313(L3), 314(L3), *322*
Corrigan, T. E., 156, *207*
Coulson, J. M., 213, 238, 239(M2), 248, 274, 275, 276, 285, 298(C3, M2), 300, 301, 302, *322*, *323*

D

Dallavalle, J. M., 273, *322*
Danby, C. J., 182(B3), *206*
Danckwerts, P. V., 249, 250(D2), 251 (D2), *322*
Davidson, W. F., 101, *115*
Davis, B. I., 13, *31*
Davoud, J. G., 182(B3), *206*
Day, R. S., 13(M3), 17(M3), *31*
Deming, L. S., 155(B14), *206*
Deming, W. S., 155(B14), *206*
DeVaney, F. D., 150(D1), *206*
DeVault, D., 154, 173, 193(D2), *206*
Dew, J. E., 22, 25(M4), 26(M4), *30*, *31*
DiNardo, A., 125(B2), 127(B2), *145*

Dobbins, W. E., 114, *115*
Dodge, F. W., 165, *206*
Dougherty, E. L., 7(D2), *30*
Douglas, J., 175, *206*
Draper, C. S., 42, *79*
Drew, T. B., 2, 4(D3), *30*, 165, 169, 175, *206*, 211, 281(D4), *322*
Dryden, C. E., 165, 175, 177, *206*
Dwyer, O. E., 84(D3), *115*

E

Eagleton, L., 157, 175, *206*
Eckman, D. P., 56, *79*
Edwards, H. E., 180(B10), *206*
Eichhorn, J., 151, 160, *206*
Einstein, H. A., 183(E2), *206*
Ekedahl, E., 166(S4), 174(S4), 178(S4), *208*
Emmett, P. H., 155(B13), *206*
Epstein, P., 197(B1), *206*
Ergun, S., 169, *206*
Evans, W. R., 64, *79*

F

Farmer, W. S., 14, *30*
Farrington, G. H., 42, 48, 61, 64, *79*
Feldman, S. M., 240(B5), 248, 253, 274 (B5), 275(B5), 276(B5), 285, *322*
Ferguson, D. E., 84(F1), *115*
Fishman, N., 8(B4), 11(B4), *30*
Fleming, E. P., 100, *116*
Flügge-Lotz, I., 58, *79*
Forscher, F., 270(F1), *322*
Foust, A. S., 15(L2), 18(L2), 19(L2), 27, *30*, *31*
Franks, R. G., 75, *79*
Fredericks, E. M., 152, *206*
Fujita, H., 177, *206*
Furnas, C. C., 153(B5), 181, 182, *206*, 244, *322*

G

Gaffney, B. J., 165, *206*
Generaux, R. P., 169, *206*
Gilliland, E. R., 131, 186, *145*, 166(G2), *206*
Gloyna, E. F., 114, *115*
Glueckauf, E., 82, 83, 99, *116*, 165(G5), 170, 176, 193, 198, 199, *206*, *207*

Goldsmith, S., 313(L3), 314(L3), *322*
Goldstein, D. J., 197(B1), *206*
Goldstein, S., 182, 186, 197, *207*
Goodman, T. P., 51, *79*
Gosline, C. A., 89(C2), *115*
Grant, E. L., 235(G1), *322*
Gray, J. B., 257, 307(G2), 309, 317, 318, *322*
Greathead, J. A. A., 315(G3), *322*
Greenberg, A. E., 113(K1), *116*
Greenfield, M., 27(B6), *30*
Gregor, H. P., 160, 161, 162, 163(G7), *207*
Grossman, J. J., 189, *205*, *207*
Gunther, F. C., 21, 22, 23(G1, G2), 26, *30*

H

Hainsworth, B. D., 72, *79*
Hald, A., 229(H1), *322*
Halton, E. M., 101(B4), *115*
Harnett, R. T., 69(W1), *80*
Hatano, Y., 316, 317(Y1), *324*
Hatch, L. P., 82, 113, 115(H2), *116*
Hawthorn, R. D., 197(B1), *206*
Healy, T. V., 99, *116*
Herdan, G., 232, 235, 236, 237, 253, 276 (H2), 282, *322*
Hickman, K. C. D., 28, *30*, 132, *145*
Hiester, N. K., 148(H3), 159(V4, V5), 160(V5), 161(H6), 162, 166, 167 (H4), 168, 171, 181(H6), 185, 186, 188(V4), 189(V4), 190(H6), 191(H4, H6), 192(H4), 195(H2, V4), 196, 197, 198, 199(V4), 200(V4), 201(V4), *207*, *208*
Higgins, I., 97, *116*
Hillebrand, W. F., 215(H3), *322*
Hinshelwood, C. N., 182(B3), *206*
Hittman, F., 113(H2), 115(H2), *116*
Hixson, A. W., 238, 239, *322*
Hoel, P. G., 220(H5), 222(H5), 223(H5), 225(H5), 226(H5), 262(H5), *322*
Hoffman, J. I., 215(H3, L5), *322*, *323*
Holly, C. E., 94(R3), *116*
Horrigan, R. V., 93(M2), *116*
Hougen, J. O., 46, *80*
Hougen, O. A., 153(H7), 165, 171, 179, 182, *206*, *207*, *208*
Huffman, E. H., 162, 168(V6), 177, *208*
Hugill, H. R., 245, *323*
Hurt, D. M., 171(H8), *207*

J

Jacques, G., 183(J1), *207*
Jaeger, J. C., 43, 45(C2), *79*
Janssen, J. M. L., 70(J1), *80*
Jens, W. H., 25, 26, 27, *30*
Johnson, E. F., 42, 52, 75, *79*, *80*
Johnson, J. D., 296, *322*
Judson, B. F., 103, *115*

K

Kanise, I., 258, 315(Y2), 316, 317(Y1), *324*
Kasai, K., 245, *322*
Kasten, P. R., 182, *207*
Katz, D. L., 15(L2), 18(L2), 19(L2), *31*, 169(B11), *206*
Kaufman, W. J., 113, *116*
Kennel, W. E., 25(M4), 26(M4), *31*
Keulemans, A. I. M., 152, *207*
Kirk, R. E., 288(K2), *322*
Kitt, G. P., 199(G4), *207*
Klein, G., 113(K1), *116*
Klinkenburg, A., 182, *207*, *208*
Klotz, I. M., 163, 193(K4), *207*
Knudsen, M., 125(K1), 132, *145*
Koble, R. A., 156, *207*
Kochenburger, Ralph, 68, *80*
Kramers, H., 179(K6), 183(K6), *207*
Kreith, F., 21, 22(G2), 23(G2), 24(K2), 27, *30*
Kunin, R., 151, 162, *207*
Kurahasi, S., 316, 317(Y1), *324*
Kyte, J. R., 130(K2), *145*

L

Lacey, P. M. C., 246, 247(L1), 253, 254, 259, 260(L1), 261(L1), 274, 275, 276 (L2), 281(L2), 282, 283(L2), 284(L2), 285(L2), 289, 292, *322*
Langmuir, I., 94, *116*, 132, *145*, 155(L1), *207*
Lapidus, L., 162, 182(K1), 183(L3), 185, *207*, *208*
Lawrance, R. B., 128, 129, 144, *145*
Ledoux, E., 182, *207*
Lees, S., 42(D1), 46, *79*, *80*
LeRoy, N. R., 8(B5), 20(B5), *30*
Lipkis, R., 27(B6), *30*

Lloyd, B. A., 94(R3), *116*
Lofton, W. M., Jr., 313(L3), 314, *322*
Lottes, P. A., 25, 27, *30*
Lowery, A. J., Jr., 6(L1), *31*
Lowry, T. M., 232, 233(L4), 236, 253, 282, *323*
Lundell, G. E. F., 215(H3, L5), *322, 323*
Lyon, R. E., 15, 18(L2), 19(L2), *31*

M

McAdams, W. H., 2, 5(M2), 6(M2), 13(M3), 17(M3), 25, 26, *31*, 280(M1), *323*
MacColl, L. A., 68, *80*
McCune, L. K., 165, *207*
McGregor, W. K., 69(S2), *80*
McKay, W., 42(D1), *79*
Madden, A. J., 130(K2), *145*
Mair, B. J., 193, *207*
Maitra, N. K., 213, 238, 239(M2), 248, 274, 275, 276, 285, 298(C3, M2), 300, 301, 302, *322, 323*
Mandelson, J., 238(M3), *323*
Manowitz, B., 93, 108(M1), 113(H2), 115(H2), *116*
Mantell, C. L., 149, *207*
Mareck, L. J., 97(W1), *116*
Maritz, J. S., 255, 264(B4), 272, 277, 285, 304, 319(B4), *322*
Marshall, W. R., 153(H7), 171, 179, 182, *207*
Marshall, W. R., Jr., 133, *145*
Martin, A. J. P., 152(M4), 171, *207*
Marx, J. W., 13, *31*
Matheson, L. A., 200, *207*
Mawson, C. A., 100(M3), *116*
Mayer, R. W., 42, *79*
Mayer, S. W., 170, 171, 200, *207*
Michaels, A. S., 166(M7), 175(M7), *207*, 256, 278, 286, 307(M4), 310, 311(M4), *323*
Minden, C. S., 25(M4), 26(M4), *31*
Miyauchi, T., 183(M8), *207*
Moison, R. L., 179(C1), *206*
Moore, W. B., Jr., 313(L3), 314(L3), *322*
Morton, R. J., 100(S2), *116*
Mueller, A. C., 2, 4(D3), *30*
Myers, L. S., Jr., 153(B8), 163(B8), 166(B8), 176(B8), 198(B8), *206*
Myers, R. J., 151, 162, *207*

N

Nachod, F. C., 149, 150, 163(N1), *207*
Nelson, R. L., Jr., 166(N2), 166(H4), 167(H4), 168(H4), 191(H4), 192(H4), *207*
Nichols, N. B., 74, *80*
Normand, C. E., 125, *145*
Nukiyama, S., 13, *31*
Nyquist, H., 41, *80*

O

O'Connell, H. E., 93, *116*
Offord, A. C., 173(W3), 193(W3), *208*
O'Gara, P. J., 100, *116*
Oldenbourg, R. C., 52, 70, *80*
Oldenburger, R., 42, 71, *80*
Opler, A., 186, *207*
Othmer, D. F., 288(K2), *322*
Oyama, Y., 241, 242, 244(O2), 245, 254, 271(O9), 272, 274, 275, 277, 278(O9), 279(O9), 283, 285, 290, 292, 293(O2), 294(O9), 295, 297(O5, O7), 298(O4), 299(O7), 304(O9), 319(O9), *323*

P

Parker, H. M., 99(B5), 100(B5), *115*
Perry, J. H., 56, *80*, 260(P1), 288(P1), *323*
Pettyjohn, E. S., 93, *116*
Phillips, R. C., 148(H3), 168(H2, H3), 195(H2), *207*
Picornell, P. M., 25(M4), 26(M4), *31*
Pierce, D. E., 289, *323*
Piret, E. L., 130(K2), *145*
Poiseuille, J. L. M., 125(P1), *145*
Powell, J. E., 193(S6), *208*
Pramuk, F. S., 7(P1), *31*
Puzinauskas, V., 256, 278, 286, 307(M4), 310, 311(M4), *323*

Q

Quillen, C. S., 289, *323*

R

Radding, S. B., 166(H4), 167(H4), 168(H4), 191(H4), 192(H4), *207*

AUTHOR INDEX

Ragazzini, J. R., 70, *80*
Ranz, W. E., 133, *145*
Regan, W. H., 113(H2), 115(H2), *116*
Reid, D. G., 86(R1), *116*
Reswick, J. B., 51, *79*
Rinaldo, P. M., 13, 17(M3), *31*
Robbers, J. A., 8(B5), 20(B5), *30*
Rodebush, W. H., 94, *116*
Rodger, W. A., 82, 83, *116*
Rohrmann, C. A., 92(A1), *115*
Rohsenow, W. M., 26, *31*
Romie, F., 27(B6), *30*
Rose, A., 69(W1), *80*
Rosen, J. B., 182, 184(R1), 198, *207*, *208*
Rossini, F. D., 193(M2), *207*
Rupp, A. F., 98, *116*
Russell, D. W., 69(S2), *80*
Rutherford, C. I., 42, *80*

S

Sakaino, T., 258, 275, 279, *323*
Santangelo, J. G., 2(W2), 3(W2), 9(W2), 15(W2), *31*
Sartorius, H., 52, 70, *80*
Schwent, G. V., 69, *80*
Schubert, J., 160(B9), 170(B9), *206*
Schumann, T. E. W., 180, *208*
Scott, R. A., 287, 289, *323*
Selke, W. A., 175, *208*
Shedlovsky, L., 150(S3), *208*
Sherwood, T. K., 125(B2), 127(B2), 131, *145*
Sillén, L G., 166(S4), 174, 178, *208*
Silverman, L., 103, *116*
Simmonds, W. H. C., 315(G3), *322*
Sips, R., 155, *208*
Smith, J. C., 257, 311, 312, *323*
Smith, J. M., 99(B5), 100(B5), *115*
Smith, R. W., Jr., 180(B10), *206*
Snaddon, R., 289, *324*
Soldano, B. A., 167, *206*
Spedding, F. H., 193, *208*
Spooner, F. M., 175, *206*
Stange, K., 230, 231, 232, 233(S4), 234, 249, 265(S5), 266(S5), 267(S5), 282, 306(S5), *323*
Stauffer, R. A., 136(S1), *145*
Staveley, L. A. K., 182(B3), *206*
Steel, E. W., 114(G1), *115*
Stene, S., 198, *208*

Straub, C. P., 100(S2), *116*
Struckness, E. G., 100(S2), *116*
Summerfield, M., 24(K2), *30*
Synge, R. L. M., 152(M4), 171, *207*

T

Takahashi, Y., 52, *80*
Tanaka, K., 258, 315(Y2), *324*
Teller, E., 155(B13, B14), *206*
Tenney, A. H., 238, 239, 240, *322*
Thomas, H. C., 153(T2), 179, 180(T1), 185, 186, 188, *208*
Tiede, R. L., 279, *323*
Tiselius, A., 194, *208*
Tompkins, E. R., 170, 171, 200, *207*
Tooley, F. V., 279, *323*
Torpey, W. A., 28(H3), *30*
Treybal, R. E., 149, 175(T4), *208*
Trimmer, J. D., 43, *80*
Truxal, J. G., 42, 68, *80*
Tsien, H. S., 42, *80*

V

Van Arsdel, W. B., 152(V1), *208*
Van Deemter, J. J., 182, *208*
Van Krevelen, D. W., 248, 274, 275, 276, 285, 298(V1), 302, *323*
Vermuelen, T., 159(V5), 160(V5), 161(H6), 162, 166(H4), 167(H4, V3), 168, 171, 176, 177(V3), 178(V2), 181(H6), 183(J1), 186, 188(V4), 189(V4), 190(H6), 191(H4, H6), 192(H4), 195(V4), 197, 198, 199(V4), 200(V4), 201(V4), *207*, *208*
Visman, J., 248, 274, 275, 276, 285, 298(V1), 302, *323*
Vorkauf, H., 93, *116*

W

Walter, J. E., 173, 178, 193(W1), *208*
Walters, W. R., 97, *116*
Walton, H. F., 162, *208*
Weidenbaum, S. S., 213, 214(W1), 238, 241, 247, 255, 262(W1), 263(W2), 270(W2), 272, 274, 275, 276, 281(W1), 282, 285, 304(W2), 305, *323*
Weiser, D. W., 97(W1), *116*
Weiss, J., 173, 193(W3), *208*
Westhaver, I. W., 193(M2), *207*

Westman, A. E. R., 244, *323*
Westwater, J. W., 2(W2), 3(W2), 7(P1), 9(W2), 15(W2), *31*
Wheeler, A., 150, 165, *208*
Wicke, E., 174, 176, 179, 182, 183(W5), *208*
Wilhelm, R. H., 165, *207*
Wilke, C. R., 165, *208*
Williams, C., 121(B3), *145*
Williams, T. J., 69, *80*
Wilson, J. N., 173, 193(W7), *208*
Wilson, S., 162, *208*
Winsche, W. E., 198, *208*
Woolan, G. N., 240(B5), 248, 253, 274 (B5), 275(B5), 276(B5), 285, *322*

Work, L. T., 289, *323*
Worley, C. W., 75, *79*
Wornick, R. C., 289, *324*

Y

Yano, T., 258, 315(Y2), 316, 317(Y1), *324*
Young, A. J., 42, 61, 70(Y1), 72(Y1), *80*
Young, R. S., 289, *324*

Z

Zadeh, L. A., 70, *80*
Ziegler, J. G., 74, *80*
Zuiderweg, F. J., 182(V2), *208*

SUBJECT INDEX

A

Adsorption, 149–208
  equilibrium in, see Equilibrium, in adsorption processes operations, 149
  isotherms, 155–158
  in radioactive waste treatment, 96
  rates, 165, 166
  separation methods by, 147–208
Aerosols, in radioactive waste disposal, 102
Agitation, effect on heat transfer, 7
Air
  cleaning equipment, 104–105
  as coolant, 85
Atomic energy plants
  sites, air pollution aspects, 88–90
Automatic process control, see Process control

B

Beds
  fixed, 148, 152, 167–194
  fluid, 149
  radial, 185
  thickness modulus, 170–171
  variables of, 167–172
Binomial distribution, see Statistics
Biological hazards of radioactive wastes, see Radioisotopes, biological hazards of
Boiling, 1–31
  of benzene, 14
  Bromleys equations for, 6, 10, 11
  bumping in, 27–29
  film, see Film boiling
  heat transfer in, 1–31
    see also Heat transfer coefficients, 11, 12, 19
  measurement of, 29
  nucleate, 4
    effect of pressure on, 4
    of surface active agents on, 4, 6

    nucleation theory of, 28
  of subcooled liquids, 21–27
    data and correlations of, 22–27
    effect of pressure on, 24
      of velocity on, 24–25
    with forced convection, 22
    photographic studies of, 20
    vapor bursts, slugs and rods, 318
Breakthrough behavior, of columns, 184–185

C

Carbon tetrachloride, in film boiling, 14
Centrifugal collectors, see Collectors, centrifugal
Cesium, in radiochemical wastes, 83, 108–113
Chi-square test, 224–225, see also Statistics
Chromatograms, 195–200
  non trace case, 196–203
  trace case, 198–203
Coating in vacuum, see Vacuum
Collectors, centrifugal, 104
Colloids, in radioactive waste disposal, 96
Columns, see also Adsorption
  capacity parameter, 170–171
  dynamics of, 179–185
  in mass transfer operations, 147–208
  in radioactive waste treatment, 95
Concentration ratio, 169
Confidence limits, 229–235, see also Statistics
Contamination, radioactive, 81–116
Continuity, equation of, in material balance, 172
Control, automatic, see Process control, automatic
Cycling, in separations, 201–203
Cyclone separators, 95, 102, 104
Cylinder mixers, see Mixing equipment

D

Deionization, in radioactive waste treatment, 97
Diffusion, see also Mixing
　external, 175, 178, 189–193
　in fixed beds, 175–179
　internal, 176, 178, 190–193
　pore, 177, 189–190
Dispersion, longitudal, in bed, 182–183
Disposal, of radioactive wastes, see Wastes, radioactive
Distribution, statistical, 216–274, see also Statistics
Distribution ratio, in fixed beds, 169–170
Drying of solids, 152
　freeze, in high vacuum, 120

E

Elution, column, 188–189
Entrainment
　problems in radioactive waste evaporation, 92–96
Equilibrium
　in adsorption processes
　　behavior, limiting cases, 173–179
　　constant pattern case, 174–179
　in ion exchange, 160–162
　parameter in fixed bed, 169
　proportionate-pattern case, 172–173
　in separation performance, 153–162
　in mixing, 228
　reactions in high vacuum, see Vacuum, high, reaction equilibrium
Evaporators, in radioactive waste treatment, 93
Exchange systems, binary, 186–188
Extraction, by fixed beds, 152

F

Film boiling, 8–21
　active centers in, 8
　on carbon, 16
　of diphenyl ether, 14
　effect, of agitation on, 20
　　of geometric arrangement on, 16–17
　　of impurities on, 21
　　of pressure on, 17–18
　　of surface tension on, 18
　　ethanol, 14
　experimental values of, 13–21
　and forced convection, 20
　on horizontal plates, 14
　on horizontal tubes, 8–10
　of liquid metals, 14
　of mercury, 15
　of methanol, 8, 14
　of n-pentane, 14
　photographic studies of, 10
　on stainless steel, 16
　theoretical treatment of, 10
Filters
　fiberglass, 95
　in radioactive waste treatment, 95, 105, 106, 114
Fission products, 81–116
　permissible concentration of, 82
　recovery from wastes, 198–213
Fixed beds, see Beds, fixed
Fluid beds, see Beds, fluid
Fluid flow, in high vacuum technology, 125–129
Foaming, problems in radioactive waste evaporation, 95
Frequency
　distribution, statistical, 259–274
　response, in process control, 42

G

Gages, high vacuum, see Vacuum, high, gages
Gases, with radioactive contamination, 85–86
Gaussian shaped zones, in chromatography, 198–200
Glass filters, see Filters

H

Hazards
　biological, from radioactive wastes, 82, 83, 87, see also Radioisotopes and Radiotoxic isotopes
Heat transfer
　in boiling liquids, 1–31
　　data and correlation for subcooled liquids, 22–27
　　data in nucleate and transition boiling, 4–8

SUBJECT INDEX

experimental values, 13–21
  theoretical treatment, in film boiling, 10–13
  of heat regenerator, 184
  in high vacuum technology, 129–131
  regenerative, 152–153
High vacuum, see Vacuum, high
Hydrofluoric acid, in radioactive waste treatment, 86

I

Iodine, as radioactive contaminant, 86
Ion-exchange, 150–151
  in radioactive waste treatment, 86, 96, 97
  rates, 166–167
Irradiation, neutron, of reactors' coolant, 85
Isotopes
  radioactive, 81–116
  separation by high vacuum, 121

J

J function, 180–187

K

Krypton, as radioactive contaminant, 86

L

Leaching, 152
Lead, as hazardous isotope, 87
Leaks, in nuclear equipment, 86
Legislation
  concerning permissible exposure to radioisotopes, 87

M

Mass balance, in high vacuum fluid flow, 125–129
Mass transfer, 147–208
  in high vacuum technology, 131–134
Material balance, in fixed beds, 172
Mean, statistical, 220, 222
Melting, under high vacuum, 121
Metallurgy, high vacuum, 120, 122–123
Mixers, see Mixing equipment

Mixing
  of solids, 209–324
  coefficients, 311
  degree of, 212, 238–259
  diffusion analogy in, 282
  equipment, 287–320
    finger-prong mixer, 307–311
    helical flight mixer, 315
    horizontal cylinders, 274–285, 290–305
    Muller mixer, 313, 314
    ribbon mixer, 313–314
    Sigma mixer, 306–307
    twin shell blender, 315
  of granular solids, 241
  literature references, 321–323
  phases in, 270–271
  of powder, 288, 316–324
  rate equations, 271–287
  reflectivity probe in, 257–258
  in rotating cylinder, 254
  samplers in, 213, 214
  sampling in, 209–324
  sampling considerations in, 213–216
  segregation intensity, 249–253
  statistics, 216–274
Modulus, in adsorption, 170–171
Multicomponent saturation, in beds, 193–194

N

Neutron, flux in reactors' coolants, 85
Normal distribution, see Statistics, distribution
Nuclear, radiation hazards, control of, 81–116
Nuclear reactors, 84
Nuclear separation plants, 86
Nuclides, see Isotopes
Nusselt's equation, 10, 11

O

Oxygen, in reactors' coolants, 85

P

Parameter
  of column capacity, 170
  of solution capacity, 171

Particles
  in mixing, 209–324
  removal from radioactive gases, 102
Particulate matter, as contaminant, 86, 94, 101–103
Permissible amounts of radioisotopes in human body, 87
Poisson, distribution, see Statistics, distribution
Pollution
  of air, by radioactivity, 81–116
  by particulate matter, see Particulate matter
Polonium
  as hazardous isotope, 87
Precipitators
  electrostatic, in radioactive waste treatment, 165
Probability, see Statistics
Process control, automatic, 33–80
  analysis of problems in, 40, 42
  Bode diagram, 48
  cascading systems in, 50
  characteristics of, 36–38
  complex systems, 70
  components of, 38
  control, criteria of, 70
  control loop behavior, 64–70
    linear systems in, 64–67
    non-linear systems in, 67–68
    sampled data in, 69
  control systems, 68–69
    derivative control, 60
    integral control, 59–61
    proportional control, 59–60
    three mode control, 61
  controller specification, 73
  controllers, 57–62
  definition of, 36
  elements in, 43–55
  feed back in, 39
  frequency response, 42, 71
  magnitude ratio, 46
  measuring elements, 55–57
    composition of, 57
    liquid flow, 56
    pressure, 56
    temperature, 56
  Nyquist diagram, 48
  optimum controller setting, 74–75
  in process industries, 35

  pulse forcing in, 46
  ramp forcing in, 46
  random forcing in, 51
  recommended characteristics, 71–73
  regulating units in, 62–63
  series of first order lag in, 50
  sinusoidal forcing in, 46
  step forcing in, 45
  time constant in, 44
  time delays in, 54
  transmission lines in, 64
Pumps, high vacuum, 136–142

Q

Quality control charts, 235–237

R

Radiation
  dosage, 81–116
  effect of, on condensation, 11
  hazards, control of, 81–116
Radioactive wastes, see Wastes, radioactive
Radioactivity in reactors' coolants, 85
Radiochemical rare earths, 96
  in waste treatment, 81–116
Radioisotopes, biological hazards from, 81–116
Radiotoxic isotopes, isolation of, 83
Radiotoxicity
  tolerances, 82–84
  wastes, sources of, 84–86
Randomness, statistical, 267, see also Statistics
Reaction equilibrium, in high vacuum, 134–136
Reaction units, in fixed beds, 170–171
Reduction
  under vacuum, of calcium, 121
  of magnesium, 120–121
Ripples, during film boiling, 8

S

Samplers, 213–214
Samples, see also Statistics
  analysis of, 215
  spot, 215, 218, 227
  statistics, 216–274

SUBJECT INDEX

Sampling, see Mixing, sampling in
Saturation, see Multicomponent saturation
Scrubbers
 in radioactive waste treatment, 105, 106, 107
Segregation, in mixing, 212
Separation
 by adsorption, 147–208
 chromatographic, 194–203
 factor, 158–160, 185–193
  in fixed beds, 169
  "trace," 198–203
Separators, 104–105
Site selection, concerning waste disposal, 87–90
Solids, mixing of, see Mixing of solids
"Speed," in high vacuum technology, 124
Spray washers, 104
Stacks, for radiation waste disposal, 88–90, 100–101
Standard deviation, 220, see also Statistics
Statistical test of significance, 221
Statistics, for mixing of solids, 216–274
Stoichiometric capacity, of columns, 162
Strontium, as hazardous isotope, 83, 87, 108–113

T

Tanks, for holding liquid radioactive wastes, 90
Text, significance, statistical, 271, 272
Throughout ratio, in fixed beds, 171
Tolerances, for safe isotope dilution, 82–83
Towers, packed in radioactive waste treatment, 103–104
Transfer units, in fixed beds, 170–171
Transuranium elements, as hazardous isotopes, 87

U

Uranium, as a reactor fuel, 84–85

V

Vacuum
 high, 117–145

applications of, 120–124
chemical engineering of, 124–136
coating in, 122
fluid flow in, 125–129
freeze drying, 120
gages, 136, 142–145
 Alphatron, 143
 hot filament ionization, 143
 McLeod, 143
 Philips, 144
 thermocouple, 144
heat transfer in, 129–131
historical development, 119
isotope separation by, 121
mass balance in, 124
melting under, 121
metallurgy, 120, 122–123
nomenclature and standards, 119
pumps
 diffusion, 146
 mechanical, 137–139
 mechanical booster, 139
 multistage steam ejectors, 139
 oil ejectors, 140
reaction equilibrium in, 134–136
reduction of calcium under, 21
 of magnesium under, 120
ultrahigh, 119
Vacuum, drying, 123
Vapor, in liquid boiling
 bursts and slugs, 3
 rods, 8
Variance, statistical, 220, 222

W

Wastes, radioactive, 81–116
 gaseous, 100–103
  disposal by stock, 88–90, 100–101
  geological considerations in, 88
 liquid, 90–100
  collection and pretreatment of, 90–91
  concentration of, 92–99
  cooling of, 91
  corrosion of containers by, 91–92
  dispersion, 99–100
  evaporation of, 92–96
  high level, 90
  low level, 90

storage of, 83–84, 91–92
meteorological aspects of, 88–90
nomenclature of, 115–116
recovery of fission products from, 108–113
solid, 103–108
  disposal of by burial, 107
    by incineration, 106
    by stacks, 88–90, 100–101
strength of streams of, 86
treatment and disposal practices, 90–108

Water, as nuclear reactors' coolant, 85

X

Xenon, as radioactive contaminant, 86

Z

Zirconium
  uranium alloys, as nuclear fuel, 86

Date Due

DOES NOT CIRCULATE